Geomorphological Landscapes of the World

2. Granite Landscapes of the World
by Piotr Migoń

Granite Landscapes of the World

Piotr Migoń

OXFORD
UNIVERSITY PRESS

Library
Quest University Canada
3200 University Boulevard
Squamish, BC V8B 0N8

OXFORD
UNIVERSITY PRESS

Great Clarendon Street, Oxford OX2 6DP
Oxford University Press is a department of the University of Oxford.
It furthers the University's objective of excellence in research, scholarship,
and education by publishing worldwide in

Oxford New York

Auckland Cape Town Dar es Salaam Hong Kong Karachi
Kuala Lumpur Madrid Melbourne Mexico City Nairobi
New Delhi Shanghai Taipei Toronto

With offices in

Argentina Austria Brazil Chile Czech Republic France Greece
Guatemala Hungary Italy Japan Poland Portugal Singapore
South Korea Switzerland Thailand Turkey Ukraine Vietnam

Oxford is a registered trade mark of Oxford University Press
in the UK and in certain other countries

Published in the United States
by Oxford University Press Inc., New York

© Piotr Migoń 2006

The moral rights of the author have been asserted
Database right Oxford University Press (maker)

First published 2006

All rights reserved. No part of this publication may be reproduced,
stored in a retrieval system, or transmitted, in any form or by any means,
without the prior permission in writing of Oxford University Press,
or as expressly permitted by law, or under terms agreed with the appropriate
reprographics rights organization. Enquiries concerning reproduction
outside the scope of the above should be sent to the Rights Department,
Oxford University Press, at the address above

You must not circulate this book in any other binding or cover
and you must impose the same condition on any acquirer

British Library Cataloguing in Publication Data
Data available

Library of Congress Cataloging in Publication Data

Migoń, Piotr.
Granite landscapes of the world/Piotr Migoń.
   p. cm.
   Includes bibliographical references and index.
   ISBN 0–19–927368–5 (alk. paper)
   1. Granite. 2. Granite outcrops. 3. Geomorphology. I. Title.
QE462.G7M49 2006
552'.3–dc22   2005025704

Typeset by SPI Publisher Services, Pondicherry, India
Printed in Great Britain on acid-free paper by
Antony Rowe Ltd, Chippenham

ISBN 0–19–927368–5   978–0–19–927368–3

1 3 5 7 9 10 8 6 4 2

This book is dedicated to Edyta

Preface

Some of the most spectacular scenery on Earth is supported by granite. Granite geomorphology is typically associated with inselbergs rising above vast savanna plains and with castellated tors which give a special flavour to treeless European uplands. But granite landscapes do not have one face, but many. No less distinctive, and for a geomorphologist no less intriguing, are granite mountains, such as the Sierra Nevada of California, extensive tracts of closely spaced convex hills present in many subtropical regions, and the perfect rock-cut plains of Australia. Superimposed on these grand sceneries are minor landforms, which occur in an amazing variety of shapes and settings.

In this book I primarily intend to show the diversity of granite landscapes, at the scale both of individual landforms, as well as of regional landform assemblages. More challenging is to take a step further and provide an explanation of this variety. Despite a long history of research, our understanding of granite geomorphology is still incomplete and some of the most typical granite landforms, such as for example weathering pits, are as mysterious as they were 100 years ago. Granite is a tough rock which changes slowly over time. Therefore the efficacy of individual geomorphic processes shaping granite outcrops is more often hypothesized rather than measured and quantitatively assessed. Nevertheless, I will at least attempt to explain how and why different granite landscapes may have formed, even if some of the proposed conceptual models may fail the test of time. In particular, I will try to emphasize the role of the rock itself, which I believe is crucial in the evolution of rock-cut landscapes.

This book is not the first to focus on granite geomorphology, but it has a few important predecessors. Herbert Wilhelmy was the first who dared to offer a comprehensive treatment of granite landforms, publishing his *Klimamorphologie des Massengesteine* in 1958. As the title shows, his approach was climatic and he was seeking an explanation of the geomorphological variety primarily in past and present environmental conditions. Alain Godard's *Pays et paysages du granite* appeared in 1977 and remains distinctive through its balanced coverage of individual landforms and larger landform assemblages, although its limited size affected the depth of discussion. Rowland Twidale's *Granite Landforms* published in 1982 remains a standard reference book for all granite geomorphologists and contains an extensive collection of photographs. In particular, the coverage of small-scale landforms is very comprehensive. The last one to mention is *Basement Regions* edited by Alain Godard and his colleagues, published in 2001 as an

English, updated version of *Régions de Socle* from 1993. Although the chief idea behind this book was to present the achievements of French geomorphologists, so that references are almost exclusively French, and granites are considered alongside other crystalline rocks, the volume deserves high acclaim. One of its strengths is that various instances of rock control are exposed to an extent unparalleled before.

So why write yet another book? One reason is that significant advances have been made in geomorphology during the last 20 years or so, many of them highly relevant to granite areas. For example, the evolution of rock slopes is now much better understood, the patterns of mass movement in weathered terrains are better known, the issue of landform inheritance is better constrained, and the efficacy of many weathering processes has been re-assessed. Another reason is that most of the previous volumes focused on individual landforms, whereas the distinctiveness of granite terrains is as much a matter of impressive minor and medium-sized forms, as it is the result of repetitive patterns of these landforms. Chapter 9 of this book is an attempt to look at granite geomorphology from this perspective. Furthermore, a few topics have been strikingly absent from the existing publications, except Godard, such as processes in deeply weathered low-latitude terrains, geomorphic characteristics of granite coasts, or the role of humans in transforming many granite landscapes.

Despite what the title might suggest, the structure of this book does not follow a regional approach. This is for the following reasons. First, many granite regions are poorly covered by geomorphological literature and any regional survey would be unbalanced in coverage and highly biased to a few areas which happened to be better researched than the others. Second, notwithstanding the variety, similar granite landforms occur in different parts of the world. Third, and most importantly, a regional approach would make it rather difficult to integrate the explanatory part with the rest of the book. Instead, this book adopts a systematic approach and, after briefly introducing aspects of geology of granite relevant to the origin of landforms, discusses weathering processes and products, residual landforms due to selective weathering, slope form and evolution, coasts, and cold-climate landform assemblages. After that, geological controls at different spatial scales are presented, followed by the typology of granite landscapes, which hopefully adequately addresses the geomorphic variability of granite terrains. The book concludes with a consideration of selected aspects of human presence within, and shaping of, granite landscapes.

This approach has implications for the selection of examples and references. It is not possible to quote every single publication on geomorphology of granite areas that has appeared, nor to present each interesting granite area in sufficient detail. The geographical coverage is evidently biased towards my own experience from granite regions, and to areas well served by the existing literature. Nevertheless, I hope that no characteristic granite landforms and their assemblages have been ignored and that this volume will serve as a valuable guidebook to the variety of granite geomorphology of the world.

Acknowledgements

This volume is the outcome of my fascination with granite landforms, which has a history of 20 years or so. Over this period, many people were kind enough to spend time guiding me through granite areas and discussing their geomorphology, to send me their publications, to invite me to join their research projects, and, most recently, to read various parts of the draft of this book. My special thanks go to fellow researchers, with whom I have had the privilege to carry out research in granite areas across the world: Irasema Alcántara-Ayala, Kerstin Ericson, Andrew Goudie, Magnus Johansson, Karna Lidmar-Bergström, Mats Olvmo, Philip Ringrose, Pavel Roštinský, Michael Thomas, and Gonçalo Vieira. I would also like to acknowledge various individuals who, in different ways, helped me to see and understand granite landscapes. These are, in alphabetical order: Peter Allen, Marie-Françoise André, Janusz Czerwiński, Chris Green, Karel Kirchner, Cliff Ollier, Andreas Peterek, Vlastmil Pilous, Bernt Schröder, and Sławek Tulaczyk. My own research in granite regions has been supported, through various grants and research fellowships, by the University of Wrocław, the State Committee of Scientific Research in Poland, the Royal Society, the Fulbright Commission, the University of Göteborg, Instituto de Geografía and Programa de Intercambio Académico, CTIC, UNAM, Mexico. Andrew Goudie, Michael Thomas, Marie-Françoise André, Yanni Gunnell, and Irasema Alcántara-Ayala read different parts of the book and offered valuable suggestions for improvement. Needless to say, they are in no way responsible for any of this book's weaknesses. Małgorzata Wieczorek and Wojtek Zalewski drew most of the line diagrams. Last but not least, I want to thank my wife Edyta and my parents for constant support without which the book would not have been completed. After all, they had no choice but to see so many granite regions with me!

The following publishers agreed to allow their copyrighted material to be used in this book: Blackwells Publishing—Figs. 3.3, 4.5, 7.6; Elsevier—Figs. 2.2, 3.6, 5.8, 5.9, 8.7; Fondo de Cultura Economico, Mexico—Fig. 9.8; Gebrüder Borntraeger—Figs. 3.1, 3.4, 3.5, 3.9, 3.11, 3.12, 4.3, 4.8, 5.5, 7.4, 9.2, 9.3, 9.7; Geographical Society of Finland—Fig. 9.4; Geological Society of America—Figs. 2.4, 9.9; Geological Society of London—Fig. 2.11; Geologists' Association—Figs. 2.12, 8.2, 8.3; Institute of British

Geographers—Fig. 2.10; Institute of Geonics—Fig. 8.9; International Association of Sedimentologists—Figs. 2.9; International Glaciological Society—Fig. 5.6; Macmillan Press—Fig. 8.6; Quaternary Research Association—Fig. 5.11; V. H. Winston and Son—Fig. 7.2; John Wiley and Sons—Fig. 5.2.

Contents

List of Colour Plates   xvi
List of Plates   xvii
List of Figures   xxiii
List of Tables   xxvii

1. Geology of Granite   1

Granite Rocks among the Igneous Rock Family   2
Petrological Classifications of Granitoid Rocks   4
    Granite-Forming Minerals   4
    Chemical Composition and Geochemical Classification of Granite   5
    Inferred Parameters Classification   6
Granite Textures   7
Origin of Granite   8
Magma Emplacement and Shapes of Granite Bodies   11
Tectonic Settings of Granite   14
Fractures   16
    Primary Fractures   17
    Master Fractures   19
    Secondary Fractures   20
    Microcracks   22
Vein Rocks   22

2. Granite Weathering   24

Weathering Processes   25
    Patterns and Products of Granite Breakdown   25
       Surface Roughness   26
       Granular Disintegration   27
       Block-by-Block Breakdown   28

Exfoliation and Related Phenomena	29
Block Splitting	33
Chemical Weathering and Denudation	35
Weathering Indices	40
Lithological and Structural Controls on Granite Breakdown	42
Weathering Susceptibility of Granite	44
Frost (Freeze-Thaw) Weathering	44
Salt Weathering	45
Thermal Weathering and Fire Effects	46
Biological and Biochemical Weathering	48
Granite Weathering in Comparison to other Common Rock Types	50
Deep Weathering	52
Deep Weathering Profiles on Granite	53
The Weathering Front	56
Corestones	57
Lithology and Mineralogy of Granite-Derived Weathering Mantles	61
Grus	63
Weathering versus Hydrothermal Origin of Deeply Weathered Granite	65
Thickness and Rate of Deepening of Weathering Profiles on Granite	67
Spatial Patterns of Deep Weathering	73
Regional Scale	73
Slope Scale	78
Evolutionary Pathways of Weathering Mantle Development	80
Economic Significance of Deeply Weathered Granite	81

3. Boulders, Tors, and Inselbergs   83

Problems of Terminology	84
Boulders	88
Distribution and Structural Control	88
Origin and Significance	90
Tors	96
Distribution and Appearance	96
Origin of Tors	102
Tor Evolution at the Surface	105

Tors and Changing Environments	108
Tors in Glaciated Terrains	110
Inselbergs and Bornhardts	**111**
Diversity of Form	111
Origin of Inselbergs	121
Geological Controls	126
Further Development of Inselbergs	129

4. Minor Landforms — 132

Weathering Pits	132
Tafoni and Alveoles	139
Polygonal Cracking	145
Karren	149
Flared Slopes	153
Microforms as Climate-Related Features	156

5. Slope Development in Granite Terrains — 161

Rock Slopes	**161**
Granite Rock Slopes—Form and Geological Control	161
Rock Slope Failures	169
Particle Fall	169
Rock Mass Failures	171
Gravitational Spreading	178
Caves Associated with Rock Slope Development	179
Pediments	182
Slope Development in Weathered Granite Terrain	**184**
Types of Movement and Association with Weathering Zones	186
Corestone Movement	188
Landslides	189
Debris Flows	197
Gully Erosion	199
Slope Sediments	**203**
Mixed Slope Deposits in Temperate and Cold Uplands	203
Block Fields and Block Streams	208
Granite Colluvium in Humid and Seasonally Humid Low Latitudes	212
Pediment Mantles	216

6. Granite Coasts — **218**

- Granite Coasts—Are They Specific? — 218
- Cliffs and Platforms — 220
- Inlets, Sea Caves and Arches — 226
- Granite Weathering in Coastal Settings — 227
- Non-littoral Inheritance — 230

7. Cold-Climate Granite Landscapes — **236**

- Granite Periglacial Landscapes — 237
- Glaciated Granite Terrains — 245
  - *Mountain Glaciation* — 246
  - *Ice Caps* — 250

8. Geological Controls in the Evolution of Granite Areas — **254**

- Tectonic Setting — 254
  - *Orogenic Settings* — 256
  - *Eroded and Rejuvenated Ancient Orogenic Belts* — 257
  - *Passive Continental Margins* — 262
  - *Cratonic Continental Interiors* — 264
- Lithological Variation — 268
  - *Texture* — 268
  - *Mineralogy and Geochemistry* — 272
  - *Veins, Enclaves, Borders, and Roof Pendants* — 277
  - *Ring Dykes* — 278
- Discontinuities — 278
  - *Master Fractures* — 278
  - *Jointing* — 280
  - *Microfractures* — 283
- Hierarchy of Geological Controls — 284
  - *Krumlovský Les, Czech Republic* — 285
  - *Serra da Estrela, Portugal* — 286

9. Typology of Natural Granite Landscapes — **290**

- Approaches to Granite Landscapes — 290
- Etching Concept and its Application — 294
  - *Etching and Etchplains* — 295
  - *Regional Examples* — 297

The Concept—Relevance and Limits	304
Characteristic Granite Landscapes	306
Plains	307
Plains with Residual Hills	308
Multi-Convex Topography	310
Multi-Concave Topography	311
Plateaux	313
Dissected Plateaux	314
Joint-Valley Topography	315
All-Slopes Topography	317
Stepped Topography	318
Rock versus Climatic Control	322

10. Granite Landscapes Transformed   **327**

Megalithic Granite Landscapes	328
Origin of Moorlands	331
Rural Landscapes	333
Defensive Aspect of Granite Geomorphology	336
Impact of Mining	338
Large-Scale Granite Carvings: From Sri Lanka to Mt Rushmore	340
References	344
Index	379

List of Colour Plates

(between pages 196 and 197)

    I Coarse Revsund granite from Sweden, with large potassium feldspar phenocrysts
   II Deeply weathered granite, with abundant joint-bound corestones, at Lake Tahoe, Sierra Nevada, USA
  III Boulders within a hilly relief of the Erongo Massif, Namibia
  IV Haytor in Dartmoor, south-west England
   V Spitzkoppe—one of the highest inselbergs in the world, Namib Desert, Namibia
  VI Granite landscape of the Serra da Estrela, Portugal
 VII Plain with residual hills in the Mojave Desert (Joshua Tree National Park), California
VIII High-mountain granite landscape, Yosemite National Park, Sierra Nevada

List of Plates

1.1 Coarse-grained porphyritic granite with large potassium feldspars, Margeride, Massif Central, France — 5
1.2 Fracture control on granite landforms: tors at Königshainer Berge, eastern Germany — 17
1.3 Surface-parallel sheeting joints, Yosemite Valley, Sierra Nevada, California — 20
1.4 Aplite vein cut through medium-grained massive granite, Mojave Desert, California — 23
2.1 Rough surface of granite in a natural outcrop, Karkonosze, south-west Poland — 26
2.2 Grus apron below scalloped surfaces subject to granular disintegration, Mojave Desert, California — 28
2.3 Block field built of predominantly angular blocks of medium-grained granite — 29
2.4 Multiple flakes on granite boulders, near Aswan, Egypt — 30
2.5 Pseudobedding on a granite outcrop, Karkonosze, south-west Poland — 31
2.6 Widespread sheeting on granite slopes of the Tenaya Canyon, Yosemite National Park, Sierra Nevada, California, accounting for frequent slab failures — 33
2.7 Gigantic split granite boulder, marginal desert environment, Erongo, Namibia — 34
2.8 Boulder cluster affected by various breakdown processes: granular disintegration, flaking, and splitting, near Mono Lake, California — 35
2.9 Granite sculptures in Mahabalipuram, southern India, severely affected by salt weathering — 46
2.10 The upland of the Serra da Estrela, Portugal, is frequently swept by intentionally set fires, which contribute to the widespread flaking of granite boulders — 48
2.11 Lichen mats on granite boulders, Pierre de Jumaitres, Massif Central, France — 49

2.12 Deeply weathered granite, disintegrated into a sandy residuum, southern Sierra Nevada, California — 53

2.13 Very sharp weathering front in granite, Valley of the Thousand Hills, South Africa — 58

2.14 Rounded corestones within a deep weathering profile on coarse granite, Miłków, Karkonosze, south-west Poland — 59

2.15 Less weathered compartments concentrated in the upper part of the weathering profile, Kings Canyon National Park, Sierra Nevada, California — 60

2.16 Lateral contact of solid and weathered granite, Kings Canyon National Park, Sierra Nevada, California — 61

2.17 Deeply weathered granite in a mountainous subtropical environment, south-east Brazil — 71

2.18 Uneven weathering front exposed in a former kaolin mine, Ivön, southern Sweden — 74

3.1 Granite terrain dominated by boulders, Aubrac, Massif Central, France — 88

3.2 One of many enormous boulders on the high plateau of the Serra da Estrela, Portugal — 89

3.3 Corestones of various sizes and shapes co-existing in one outcrop, Harz, Germany — 91

3.4 Partly exposed boulders on hillslopes, Sierra Madre del Sur, Mexico — 92

3.5 Granite boulder fields near Gobabeb, Central Namib Desert — 94

3.6 Castellated tor, defined by an orthogonal fracture set, Karkonosze, south-west Poland — 100

3.7 Lamellar tor, dominated by closely spaced horizontal partings (pseudobedding), Bellever Tor, Dartmoor, England — 101

3.8 Massive dome-like tor, Serra da Estrela, Portugal — 101

3.9 Pedestal rock, Jizerské hory, Czech Republic — 106

3.10 Vogelfederberg—one of bornhardt-type inselbergs in the Central Namib Desert — 112

3.11 Domed inselberg in Kerala, India — 113

3.12 Group of closely spaced domes, which during degradation gradually acquire the morphological features of a block-strewn inselberg, or nubbin, Mojave Desert, California — 118

3.13 Castellated inselberg defined by an orthogonal fracture pattern, Mojave Desert, California — 119

3.14 Dilatation operating on the slope of the domed inselberg of Mirabib, Central Namib Desert — 128

4.1	Twinned weathering pit with overhanging rims, Serra da Estrela, Portugal	133
4.2	Giant weathering pit, Erongo, Namibia	138
4.3	Tafoni on a boulder inselberg, Mojave Desert, California	140
4.4	Big tafone cut into the summit part of a boulder inselberg, Spitzkoppe group, Namib Desert	141
4.5	Alveolar weathering in granite, Žulova granite massif, Czech Republic	145
4.6	Irregular polygonal cracking on a boulder, Karkonosze Mountains, south-west Poland	146
4.7	Extensive polygonal cracking associated with surface crusting, Erongo, Namibia	147
4.8	Impressive karren networks on granite boulders, Pierre de Jumaitres, Massif Central, France	151
4.9	Flared side of an outcrop, Serra da Estrela, Portugal	154
4.10	Large weathering pits in the Spitzkoppe group, Namib desert	158
5.1	Granite rock slopes in the Tenaya Valley, Yosemite National Park, Sierra Nevada, California	163
5.2	Different geometries of granite domes, Rio de Janeiro, Brazil	164
5.3	Rock slopes cut by ravines and chutes, through which the material to scree cones is supplied, Tatra Mountains, Slovakia	165
5.4	The south-facing escarpment of the Erongo Massif, Namibia	167
5.5	Recent collapse of a basal cave in massive granite, Erongo, Namibia	169
5.6	Dome-like residual hill (Hen Mt.) in the Mourne Mountains, Northern Ireland	171
5.7	Dome form and talus resultant from large-scale rock slides, Spitzkoppe group, Namibia	175
5.8	Opened fractures superimposed on the dome form, Witosza hill, south-west Poland	177
5.9	Many granite pediments in the Mojave Desert, California, have very rough relief, with numerous boulders and low rock platforms	184
5.10	Subtropical landscape near Rio de Janeiro, Brazil	185
5.11	Shallow landslides and subsequent gullying in the weathered terrain near Bananal, south-east Brazil	186
5.12	Lag deposit of massive boulders, the legacy of catastrophic corestone movement down the channels during Hurricane Pauline, 1997	189
5.13	Deforested hills built of weathered granite subject to shallow slides and gullying, Petropolis, Serra do Mar, south-east Brazil	195

5.14	Landscape around Bananal, south-east Brazil, with abundant signs of past and recent slope instability in weathered rock	196
5.15	Deep gullies (donga) in colluvium derived from weathered granite, Swaziland	199
5.16	Expanding gully network within a valley floor, near Bananal, south-east Brazil	200
5.17	'Drowned landscape' near Bananal, south-east Brazil	202
5.18	Mixed slope deposits in Dartmoor, south-west England	205
5.19	Long planar slopes underlain by a mantle of mixed slope deposits, with occasional tors, typical of granite uplands in western Europe (Dartmoor, south-west England)	208
5.20	Summit block field in fine-grained granite, Les Monts de la Marche, north-west Massif Central, France	209
5.21	Block slope on a fine-grained granite, Serra da Estrela, Portugal	211
5.22	Hilltop boulder pile in coarse porphyritic granite, Les Monts de la Marche, north-west Massif Central, France	213
5.23	Colluvium deposit near Bananal, south-east Brazil	214
5.24	Rampa complex consisting of an amphitheatre-like hollow cut into the slope and a planar depositional surface below, near Bananal, south-east Brazil	215
5.25	Bouldery colluvial deposit below a steep slope, Acapulco, Mexico	216
6.1	Granite cliffs, stacks, and arches, Land's End, south-west England	219
6.2	Headlands and bays, Land's End coast, south-west England	221
6.3	Steep cliffs in densely jointed granite at La Quebrada, Acapulco, Mexico	222
6.4	Stacks off the cliff line at Point Lobos, California	223
6.5	Marine caves and tunnels cut along fracture lines, Point Lobos, California	225
6.6	Diversified microrelief on coastal outcrops, Trégastel, Brittany, north-west France	229
6.7	Fluted boulders on the coast of the Seychelles	230
6.8	Granite skerries along the coast of Bohuslän, south-west Sweden	231
6.9	Coastal landscape of northern Brittany	234
7.1	Block streams and angular tors, Dartmoor, south-west England	240
7.2	The northern mountains, Isle of Arran	246
7.3	Glacial trough of the Vale do Zêzere, Serra da Estrela, Portugal	249
7.4	Royal Arches, Yosemite Valley, California—an example of large-scale sheeting, likely to have formed in response to erosional unloading	250

7.5	Intermontane basin of Rannoch Moor, Scotland	252
7.6	Plucked lee side of a granite dome, Lembert Dome, Sierra Nevada, California	253
8.1	High-mountain topography developed in granodioritic rock, Karakoram	257
8.2	Upland topography of Bodmin Moor upland, south-west England, dominated by saucer-like basins and broad ridges with occasional tors	259
8.3	Dissected plateau of the Cairngorms, Scotland	259
8.4	All-slopes topography of Yanshan Mountains, north of Beijing, China	261
8.5	Intermontane basin of the Tuolumne Meadows, Sierra Nevada, California	262
8.6	Co-existence of deeply weathered hills (foreground) and monolithic domes (background) near Petropolis, Serra do Mar, south-east Brazil	263
8.7	Domes around the Guanabara Bay, Rio de Janeiro, Brazil	264
8.8	Precambrian planation surface preserved as an unconformity beneath the Cambrian sedimentary succession, south-west Jordan	265
8.9	Mountainous topography of the Dead Sea Transform shoulder, developed through dissection of an ancient exhumed plain, north-east of Aqaba, south-west Jordan	266
8.10	Inselberg landscape, Kora area, Kenya	267
8.11	Residual hills associated with stock-like intrusions of fine-grained granite, Jelenia Góra Basin, south-west Poland	269
8.12	Rock shelter at Haytor, Dartmoor, signifying differential weathering of fine (below) and coarse (above) variants of granite	270
8.13	Planar granite landscape developed upon medium-grained Curral do Vento granite, Serra da Estrela, Portugal	271
8.14	Massive granite domes, Nigeria	276
8.15	Raised, iron-enriched rims left after weathering of a granite boulder, Gobabeb, Namib Desert	284
8.16	Massive domes in the coarse Seia granite, Serra da Estrela, Portugal	288
9.1	Etchsurface topography, near Sortelha, eastern Portugal	300
9.2	Plains cut across the Salem Granite, Central Namib Desert, Namibia	309
9.3	Multi-convex topography in weathered granite, Cameroon	311
9.4	Multi-concave topography, Serra da Estrela, Portugal	312
9.5	Steep slopes of Dalhejaberg, Blekinge, southern Sweden, one of the features of etched joint-valley topography	316
10.1	Neolithic stone alignments in Carnac, Brittany, north-west France	329
10.2	Standing stone (menhir) in St Uzec, Brittany, north-west France	330

10.3	Bare granite upland of Dartmoor	332
10.4	Granite village of Monsanto, Portugal	333
10.5	Boulders collected from the field have been piled up to form an artificial tor. A natural granite outcrop in the background, Aubrac Plateau, Massif Central, France	334
10.6	Great Zimbabwe	335
10.7	The Great Wall of China, built on granite outcrops, Badaling, China	337
10.8	Granite quarries, Dartmoor, south-west England.	339
10.9	Granite rock carvings in Mahabalipuram, southern India	342
10.10	Large-scale rock carvings at Mount Rushmore, South Dakota, USA	342

List of Figures

1.1	Classification of granitic rocks according to IUGS recommendation	3
1.2	Bowen crystallization series for granite-forming minerals	9
1.3	Types of granite intrusions	12
1.4	Distribution of granite massifs across the world	15
1.5	Typical fracture patterns in granitic rocks	18
2.1	Mineral-stability series	37
2.2	Relationships between the size of principal granite-forming minerals and their compressive strength	43
2.3	Geomorphology of contact zone between the Karkonosze granite and the country rock	52
2.4	Partition of a typical weathering profile on granite	54
2.5	Diagram to explain the terminology of weathering mantles and profiles	57
2.6	Corestone development through progressive weathering of joint-bounded compartments within saprolite	59
2.7	Geographical distribution of deep grus weathering	64
2.8	Diversity of evolutionary pathways and interpretations of grus weathering profiles	66
2.9	Distribution of deep weathering sites in Scandinavia	72
2.10	Variations in the thickness of weathering mantle in northern Nigeria	75
2.11	Deep weathering patterns in Hong Kong	77
2.12	Patterns of grus distribution and its relation to topography in the Karkonosze granite massif, south-west Poland	79
3.1	Origin of boulder cluster through the decay of a tor	95
3.2	Tor distribution around the world	98
3.3	Conceptual model of tor origin	102
3.4	Hypothetical pathways of subaerial evolution of tors	105
3.5	Model of tor destruction through the development of weathering pits	107
3.6	Tors in glaciated terrains, using the example of Aurivaara plateau	111
3.7	Global distribution of inselbergs and domed hills	114
3.8	Types of slope/plain junctions around inselbergs	116
3.9	Detailed geomorphology of a bornhardt	117

3.10	Expected relationships between the conical form of an inselberg and fracture density	120
3.11	Model of dome evolution	122
3.12	Origin of inselbergs and escarpments during long-term planation lowering on a tilted block	124
4.1	Plan of a weathering pit cluster	134
4.2	Global distribution of weathering pits	136
4.3	Tentative pathways of weathering pit development, based on observations from the Namib Desert	138
4.4	Global distribution of tafoni	142
4.5	Rock properties within and outside tafoni in Korea	144
4.6	Global distribution of polygonal cracking	148
4.7	Global distribution of karren	150
4.8	Development of flared slopes	155
5.1	Diversity of rock slope profiles encountered in granite areas and their relationship to fracture density	162
5.2	Relationships between slope relief and rock mass strength, on the example of a composite bornhardt profile from the Namib Desert	168
5.3	Contrasting development of talus slopes in massive granodiorite and jointed diorite, Yosemite Valley, Sierra Nevada, California	172
5.4	Categories of toppling relevant to granite rock slopes	173
5.5	Model of dome growth and disintegration	176
5.6	Pattern of deep clefts within a granite dome, Hunnebo, Bohuslän, south-west Sweden	180
5.7	Geomorphic setting of granite caves on the slopes of Witosza hill, south-west Poland	182
5.8	Landslide distribution in weathered granite in the Hiroshima area, Japan	191
5.9	Distribution of major landslides in Hong Kong in relation to bedrock geology	192
5.10	Inverted weathering profile and resultant profile of slope deposits	206
5.11	Stone stripes around Great Staple Tor, Dartmoor, south-west England	210
6.1	Evolution of stacks from pre-weathered granite	224
6.2	Sequential development of slope-over-wall cliffs	224
6.3	Origin of a cauldron at Bullers of Buchan, north-east Scotland	228
6.4	Isles of Scilly—an example of inherited coastal geomorphology	233
6.5	Lithological control on the development of sheltered bays, Connemara, Ireland	234

7.1	Granite terrains in relation to recent and Pleistocene periglacial environments	238
7.2	Composite profile of a Dartmoor slope	241
7.3	Periglacial and glacial phenomena in granite uplands in western and central Europe, including localities from Balkan and Sinai	242
7.4	Distribution of glacial landforms in the Karkonosze Mountains, Poland/Czech Republic	247
7.5	Distribution of glacial landforms in the Serra da Estrela, Portugal	248
7.6	Glacial modification of a pre-Pleistocene etched granite relief	252
8.1	Relationships between tectonic settings of origin of granite masses and the setting of granite landscapes in relation to present-day global tectonics	255
8.2	Contrasting surface roughness in different parts of the Karkonosze Mountains, south-west Poland, depending on structural and textural characteristics of granite	272
8.3	Relationships between relief and mineralogical composition of granite, Jelenia Góra Basin, south-west Poland	274
8.4	Bedrock–topography relationships in the eastern Drakensberg, South Africa	277
8.5	Geomorphological expression of a granite ring dyke, Kudaru Hills, Nigeria	279
8.6	Master fractures and granite relief, Nigeria	281
8.7	Fracture-controlled pattern of topographic basins, Bohuslän, south-west Sweden	282
8.8	Hierarchy of geological controls in relation to spatial scales and relevant components of granite landscapes	285
8.9	Hierarchy of geological controls in the granite upland of Krumlovský les, Czech Republic	287
9.1	Etchplains and etchsurfaces—a global view	298
9.2	Weathering front exposed in a quarry at Ivön, south Sweden	300
9.3	Etchplain development in the Laramie Mountains	301
9.4	Spatial distribution of different types of etchplains in Finland	305
9.5	Model relating weathering types and products to the rate of uplift in different tectonic settings	306
9.6	Tentative pathways of granite landscapes evolution	307
9.7	Development of etchsurfaces according to Büdel	308
9.8	Granite amphitheatre of the Veladero Massif, Acapulco, Mexico	319

9.9 Stepped topography of the Sierra Nevada, California 321
10.1 Ground plan of the Bolczów Castle, south-west Poland 336
10.2 Impact of china clay mining on the natural landscape of
     St Austell Massif, south-west England 341

List of Tables

1.1	Representative chemical compositions of granitoid rocks	6
1.2	Types of granitoids using the inferred parameters classification approach	7
1.3	Selected characteristics of granite plutons related to the depth of emplacement	10
1.4	Tectonic settings of granitoid intrusions and selected examples	16
2.1	Suggested terminology for surface-parallel rock detachment	32
2.2	Relationships between primary rock-forming minerals, secondary clays, and leaching intensity	36
2.3	Granite chemical weathering and denudation rates in selected catchments located in different climatic zones	39
2.4	Selected weathering indices used to characterize the degree of weathering of granitoid rocks	41
2.5	A weathering profile on granite according to Ruxton and Berry (1957)	54
2.6	The scale of weathering grades of rock mass by Dearman *et al.* (1978)	56
2.7	Criteria to distinguish between weathering and hydrothermal kaolinitic mantles	67
2.8	Alteration processes in the St Austell granite massif, south-west England	68
2.9	Thickness of weathering mantles recorded on granitoid rocks	69
3.1	Suggested terminology of tors and inselbergs	87
4.1	Selected morphometric parameters of weathering pits in the Spitzkoppe area, Namib Desert	134
5.1	Characteristics of selected large-scale rock slope failures in the Yosemite Valley	174
5.2	Types of mass movement in different types of granite-derived weathering mantle	186
5.3	Strength of weathered granite approximated by *in situ* Schmidt hammer values and uniaxial compressive strength	187
5.4	Death toll associated with landslides and associated debris flows in weathered granite terrains in the mountains of Japan	192

5.5	Gully erosion studies in granite areas	200
5.6	Reference list for studies of granite-derived slope deposits in temperate zone uplands	204
5.7	Facies of slope deposits in south-west England	204
8.1	Rock–landform relationships in Serra da Estrela	288
9.1	Climate–landforms relationships according to Wilhelmy (1958)	291
9.2	Characteristic granite landforms according to Godard (1977)	292
9.3	Characteristic types of granite landscapes according to Godard (1977)	293
9.4	Relief classes in Sweden according to Lidmar-Bergström (1995)	294

1

Geology of Granite

The unifying theme for granite landscapes of the world is the granite itself, hence it is logical to start with a brief account of granite geology. For obvious reasons of space and relevance, this chapter cannot provide a comprehensive and extensive treatment of granite as a rock. Rather, its aim is to provide background information on those aspects of granite geology which are relevant to geomorphology and may help to explain the variety of landforms and landscapes supported by granite.

The survey of literature about the geomorphology of granite areas reveals that in too many studies the lithology of granite and the structure of their intrusive bodies have not received adequate attention, especially if a ruling paradigm was one of climatic, or climato-genetic geomorphology. Granites were usually described in terms of their average grain size, but much less often of their geochemistry, fabric, or physical properties. Even the usage of the very term 'granite' may have lacked accuracy, and many landforms described as supported by granite may in fact have developed in granodiorite. On the other hand, it is true that granite may give way to granodiorites without an accompanying change in scenery. In the Yosemite National Park, Sierra Nevada, California, these two variants occur side by side and both support deeply incised valleys, precipitous slopes and the famous Sierran domes. Likewise, wider structural relationships within plutons and batholiths, and with respect to the country rock, have been considered in detail rather seldom. In analyses of discontinuities, long demonstrated to be highly significant for geomorphology, terms such as 'joints', 'faults', and 'fractures' have not been used with sufficient rigour. But it has to be noted in defence of many such geologically poorly based studies that adequate geological data were either hardly available or restricted to a few specific localities within extensive areas, therefore of limited use for any spatial analysis of granite landforms.

Notwithstanding the above, there exist a number of studies in which landforms have been carefully analysed in their relationships to various aspects of the lithology, structure, and tectonics of granite intrusions. For example and at the regional level, Lageat (1978) has shown how differences in altitude and general landscape within the Archaean basement complex of Barberton in South Africa relate to the differences in lithology within the basement suite, which themselves have influenced rates of weathering and erosion. Similarly, Pye *et al.* (1986) were able to show that inselberg landscapes in north-eastern Kenya do not occur randomly but are tied to specific lithological units within the crystalline basement. At the more local level, Ehlen (1992) has examined the petrological and fabric properties of the Dartmoor granite in south-west England and showed how they control the distribution and shape of tors. Many further examples from French geomorphology can be found in Godard *et al.* (2001).

Granite Rocks among the Igneous Rock Family

There is a fair degree of confusion around the use of the word 'granite' in the scientific literature, not to speak about popular science and layman's language. Apparently, it has its roots in the early history of geology when any light-coloured igneous rock was readily described as 'granite', as followed from the definition offered by Playfair (1802: 82): 'Granite is an aggregate stone, in which quartz, feldspar and mica are found distinct from one another and not disposed in layers'. Moreover, continental crust was spoken of as 'granitic' and contrasted with 'basaltic' oceanic crust, although many different types of rock are involved in its build-up.

Nowadays, apart from examples of evidently erroneous use, two meanings of the term persist. Strictly speaking and adhering to the classification approved by the International Union of Geological Sciences (IUGS), granite is a plutonic rock defined by the following proportions among the main rock-forming minerals. Quartz constitutes 20–60 per cent of the sum quartz + alkali feldspar + plagioclase, whereas plagioclase accounts for 10–65 per cent of the total feldspar (Streckeisen, 1976). The position of granite defined in this way on the diagram showing the IUGS classification is indicated in Figure 1.1. In addition, granites have been subdivided into two classes, defined by the percentage of plagioclase among total feldspar, namely syenogranite (10–35%) and monzogranite (35–65%). However, these more specific terms are rarely

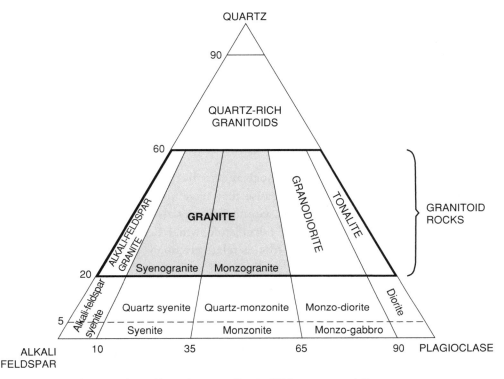

Fig. 1.1 Classification of granitic rocks according to IUGS recommendation

used by geomorphologists. Granodiorite differs from granite through the higher percentage of plagioclase in respect to potassium feldspar (Fig. 1.1).

The IUGS classification was not consistent with the schemes previously adopted in various parts of the world, which also differed from one another. In particular, it was at significant variance with traditional North American schemes, according to which granite s.s. had a narrower range of composition and was defined by the feldspar ratio (percentage of alkali feldspar divided by percentage of alkali feldspar plus percentage of plagioclase) within the range 67–100 per cent. For rocks with the feldspar ratio between 33 and 67 per cent the name 'quartz monzonite' was used. Hence the majority of rocks formerly classified as quartz monzonites are granite according to the IUGS classification. 'Adamellite' was a name synonymous with 'quartz monzonite' but it has in practice dropped out of usage since.

Granite as defined by IUGS belongs to the group of plutonic rocks called granitoids. All of these have a quartz content within the range 20–60 per cent but the percentage of plagioclase in the total feldspar may vary from 0 to 100 per cent and accounts for different specific rock names. However, it has become customary to refer to large bodies of acid plutonic rocks as being 'granite'. How much this can be misleading has been shown by Pitcher (1978) for the 'granite' batholith of Peru. In reality granite *s.s.* comprises there less than 5 per cent of the total outcrop.

To increase the potential for confusion and misuse, the diagram (Fig. 1.1) is occasionally referred to as showing the classification of 'granitic and allied rocks'. It includes rocks such as diorite, syenite, and monzonite which are relatively common and occasionally considered as members of the granite family by individual authors, although they are very much distinguished by their low content of quartz. For example, ancient Egyptian statues carved from syenite are often labelled to be from 'black granite'.

At the more general level of mineralogical classification, granites and other granitoids are described as felsic rocks because they have abundant feldspar and quartz, which are classified as felsic minerals. Granitoids may also contain iron- and magnesium-rich minerals such as biotite and hornblende, which belong to the group of mafic minerals, but their percentage in the total is relatively small.

Petrological Classifications of Granitoid Rocks

Granitoid rocks can be classified in many ways using different criteria, mineralogical, chosen by IUGS as a basis of classification, geochemical, as well as textural. In fact, Barbarin (1999) noted that about 20 different classification schemes exist in the literature. Although the geochemical approach emphasizing isotopic composition appears to dominate in modern granite petrology because it usually bears on the environment of origin of a given rock, it appears less useful for geomorphology. Names emphasizing mineralogy and textural features are preferred by geomorphologists since it is mineral size and proportions of more or less weatherable minerals which are often decisive for rock resistance against exogenic processes and hence, for landforms.

Granite-Forming Minerals

Granitoids are composed of many kinds of minerals, from major to accessory, but it is invariably plagioclase feldspar, potassium feldspar, and quartz which are dominant (Plate I). Plagioclase is usually represented most and may occur as sodic plagioclase, usually albite, and calcic plagioclase, most often oligoclase, although proportions between them may vary widely. Potassium feldspars are orthoclase or microcline and they often attain a very large size, forming phenocrysts as long as 10 cm (Plate 1.1). Both potassium and sodium feldspar are alkali feldspars. In addition, minerals of the mica group, pale muscovite and dark biotite, may occur in significant proportions. Hornblende and augite are representatives of the amphibole and pyroxene group, respectively, but their content is normally limited to a few per cent of the total rock volume. Various secondary and accessory minerals may be present in granite and

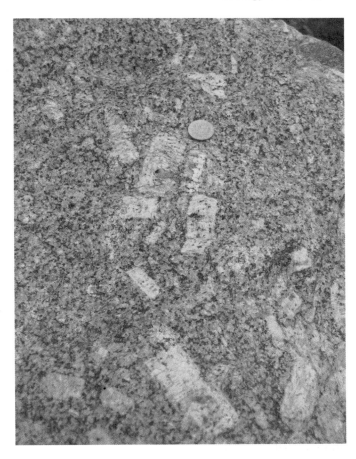

Plate 1.1 Coarse-grained porphyritic granite with large potassium feldspars, Margeride, Massif Central, France

although they rarely account for more than 1 per cent of the total, they may be significant enough to give the rock a specific name, such as tourmaline granite.

The presence or absence of certain minerals, particularly of mica, is occasionally reflected in the names of granite types in particular areas. For example, denominators such as biotite granite, two-mica granite, hornblende granite, and so on are not uncommon in regional studies, where they have been devised mainly for the purpose of field mapping. They have some usefulness for geomorphology as they emphasize the occurrence of minerals such as biotite, which is crucial for many weathering processes.

Chemical Composition and Geochemical Classification of Granite

The chemical composition of granite reflects its mineralogy, and is dominated by silica (usually more than 60% by weight) and alumina (more than 10% by weight). Feldspar

Table 1.1 Representative chemical compositions of granitoid rocks, on the example of the Lusatian Massif, Germany (in percentage by weight)

	Lusatian granodiorite	Two-mica granite	Rumburk granite	Königshain granite
SiO_2	65.60	69.81	75.65	75.08
Al_2O_3	15.09	14.94	13.03	12.36
Fe_2O_3	0.73	0.48	0.70	1.30
FeO	4.37	2.19	0.51	0.83
MgO	2.14	1.10	0.30	0.26
CaO	3.12	2.24	0.54	0.81
Na_2O	3.38	3.30	4.09	3.87
K_2O	3.56	4.18	3.88	4.68
TiO_2	1.18	0.35	0.08	0.15

Source: Möbus (1956). Some minor chemical compounds have been omitted.

constituents such as potassium, sodium, and calcium oxides account for a few per cent each, with more K_2O in granite *s.s.* and more CaO and Na_2O in granodiorite. In addition, magnesium and iron oxides may occur in significant quantities. Table 1.1 shows typical chemical compositions of representative granitoid rocks.

The geochemical classification of granite is based on the concept of alumina saturation and uses a ratio that relates molar quantities of alumina to those of calcium, sodium, and potassium combined according to the formula (molar $[Al_2O_3/CaO + Na_2O + K_2O]$) (Clarke, 1992). This ratio, written as the A/CNK ratio, is unity in all varieties of feldspar, therefore any additional minerals in granite move the ratio either above or below 1. Consequently, three basic types of granite are distinguished: peraluminous (A/CNK > 1), metaluminous (A/CNK < 1), and peralkaline, in which A < NK.

It has been further argued that the above-mentioned types of granite magma are generated in specific tectonic environments if seen in the plate tectonics context. Peraluminous granites form in continent–continent collision settings where thickening of continental crust is involved. Metaluminous varieties typify subduction-related settings, continental and island arcs, whereas peralkaline granites are related to post-tectonic or anorogenic extension within the continents.

Inferred Parameters Classification

An 'inferred parameters' approach relies on a number of indexes, including the A/CNK ratio used in the geochemical classification. The other two indexes considered are isotopic signatures of strontium ($^{87}Sr/^{86}Sr_i$) and oxygen ($\delta^{18}O$). All these taken together may indicate the source of magma, which explains why the inferred parameters classification is being constantly debated by petrologists. It was originally proposed to distinguish between granites derived from igneous (I-type) and sedimentary sources

Table 1.2 Types of granitoids using the inferred parameters classification approach

Type of granitoid	Index values	Implied source and tectonic setting
I-type	A/CNK < 1.1 $^{87}Sr/^{86}Sr_i$ < 0.705 $\delta^{18}O$ < 9%	rocks of mafic to intermediate **i**gneous composition or infractrustal derivation continental margins
S-type	A/CNK > 1.1 $^{87}Sr/^{86}Sr_i$ > 0.707 $\delta^{18}O$ > 9%	**s**edimentary rocks or supracrustal protoliths
M-type	A/CNK < 1.0 $^{87}Sr/^{86}Sr_i$ < 0.705 $\delta^{18}O$ < 9%	**m**antle, either indirectly through partial melting of subducted oceanic crust or directly by extended fractional crystallization of basalt oceanic island arcs
A-type	A/CNK > 1.0 $^{87}Sr/^{86}Sr_i$ and $\delta^{18}O$ comparable to other types	**a**norogenic settings (stable cratons and rift zones)

Source: Adapted from Clarke (1992) and Cobbing (2000).

(S-type) (Chappel and White, 1974) and later extended through the addition of two other types. Consequently, four petrogenetic types of granite are now distinguished (Clarke, 1992; Table 1.2). Because each type of granite has been given the letter code, this approach has become known as 'alphabet granitoids'. However, its usefulness is much debated among geologists (Clarke, 1992; Cobbing, 2000).

Granite Textures

Granitoid rocks are described in terms of their texture and structure but no universal agreement exists about the precise meaning of these two notions. Texture is usually defined as the term pertinent to the degree of crystallinity, grain size and shape, and the geometric relationships between constituent minerals. The latter feature is also called 'fabric'. Hence, textures are small-scale features which can be observed and described at a hand specimen scale, whereas 'structure' refers to larger features such as layering or zonation within igneous bodies.

Cobbing (2000) defines two main classes of granite textures and names these as equilibrium textures and disequilibrium textures. The former arise if crystallization from the melt proceeds uninterrupted whereas the latter indicate the opposite. For geomorphological purposes, however, texture classification which emphasizes the size of individual minerals seems to be more appropriate as it is the crystal size and strength of grain boundaries which directly bear on the resistance of rock, and hence on the resultant landforms. A standard distinction is between fine-grained, medium-grained, and coarse-grained granitoids. In igneous petrology fine-grained rock would be one with an average crystal diameter of less than 1 mm whereas average diameter in excess of 5 mm would classify the rock as coarse grained. If the average crystal size exceeds 3 cm, such rocks are spoken of as very coarse. Moreover, a mix of crystals of widely different sizes is often present in granitoid rocks. Textural variants with large potassium feldspars (megacrysts), more than 10 cm long in some plutons, amidst much smaller crystals are not uncommon and such granitoids are called 'porphyritic' (Plate 1.1). Another possible approach is to distinguish between granitoids which are more or less equigranular, that is, constituent minerals are of similar size, and those which are not. Granites subject to metamorphism generally retain their original mineralogical composition but respond to the stresses imposed by the development of foliation and deformation of individual crystals. In this way, they acquire a gneissic texture, that is, minerals are arranged in crudely defined parallel layers. The name granite-gneiss is used to account for granite that shows textural features indicative of metamorphism.

Spatial arrangement of constituent minerals within granite may often appear as non-ordered and chaotic. In reality, certain minerals, particularly K-feldspars, do show directional arrangement as disclosed by statistics and this feature has long been used as a magma flow indicator. A most characteristic linear textural feature of many granites, however, is the presence of streaks of dark minerals, mainly biotite and hornblende, called schlieren (Cloos, 1925; Cobbing, 2000). They are particularly common in marginal zones of plutons and may show a variety of geometries; contacts with the remaining rock mass are gradational. But there is no universal agreement about the usage of the term 'schlieren'. Didier and Barbarin (1991) use it to describe elongated and lenticular enclaves with no sharp margins and prefer the term 'layering' for linear concentrations of mafic minerals as referred to above. Enclaves, in turn, are any fragments of rocks enclosed in homogeneous igneous rock, ranging in composition from single crystals (xenocrysts) to large polymineralic aggregates. Accordingly, they differ widely in size from a few millimetres to hundreds of metres and may assume different shapes. As to their origin, they could be xenoliths of country rock, fragments of source rock that resisted melting in S-type granites, or relics of invading mafic magma from beneath. Cobbing (2000) provides a lengthy discussion about the origin of enclaves but the real wealth of information can be found in the volume edited by Didier and Barbarin (1991). Boundaries of enclaves are commonly very sharp. Their mineral composition may be significantly different from that of the host granite, which is the textural feature favouring selective weathering.

Origin of Granite

Granite is a plutonic rock, which means it solidifies from magma at a certain depth and at the time of origin does not have anything in common with conditions at the surface, although it is now widely recognized that granite magma emplacement within the crust may be associated with rhyolitic volcanism. The primary source of granitic magma is crustal material, which begins to melt if heated to temperatures in excess of 600°C. However, some granites are mantle-derived, whereas in other cases mixed sources are demonstrated (Leake, 1990). This variety of sources is accounted for in the 'alphabet' classification of granitoid rocks presented earlier. The exact nature and thoroughness of melting depends not only on temperature, but also on H_2O content and pressure, and is typically analysed using *P-T-t* (pressure-temperature-time) path diagrams. The source material needs to be heated above 760°C to provide enough melt to become mobile, whereas complete melting is only attained if the temperature rises significantly above 1,000°C. Anatexis is the term used to describe widespread melting of primary igneous and metasedimentary rocks. The required magnitude of heating occurs if significant thickening of the crust takes place as it does in orogenic subduction zones, or if heat production from the mantle increases as it does in rift zones. Consequently, granitoids do not occur at random, but in specific geotectonic settings as will be discussed below.

Granite-type melt is light, hence it is able to move buoyantly upward due to density contrasts. After it reaches a lower temperature zone, crystallization is initiated and the melt solidifies into a solid rock. This usually proceeds slowly, leaving ample time for crystallization of constituent minerals. The order of crystallization follows the well-known Bowen series (Fig. 1.2). Biotite, hornblende, and plagioclase form at higher temperatures and belong to the high-melting minerals, whereas quartz, muscovite, and alkali feldspar are low-melting minerals.

The depth of crystallization of granite magma may be highly variable. Some granites may have solidified a mere *c*.1–1.5 km below the topographic surface, whereas others

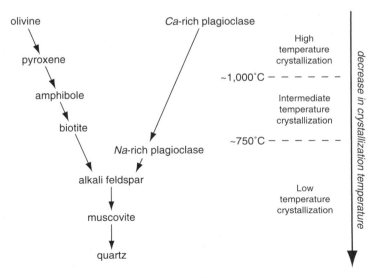

Fig. 1.2 Bowen crystallization series for granite-forming minerals

Table 1.3 Selected characteristics of granite plutons related to the depth of emplacement

	Epizonal	Mesozonal	Catazonal
Contact with country rock	typically discordant	discordant and concordant	typically concordant
Roof pendants	common	common	rare
Foliation	generally absent	common	common and parallel to regional trends
Local deformation at contact	common	may be present	absent
Contact metamorphism	very common	uncommon	absent
Chilled margins	common	absent	absent
Associated migmatites	none	few	common

Source: Buddington (1959).

were emplaced at depths in excess of 15 km. Such a vast range of emplacement depth is mirrored by the variability of structural features in individual intrusions, which have been divided into shallow epizonal (less than 6–7 km), intermediate mesozonal (7–15 km), and deep catazonal (more than 10–12 km) (Buddington, 1959). Each type has its own dominant structural characteristics (Table 1.3).

In some field situations rocks which conform to the definition of granites occur alongside metamorphic rocks, usually gneiss, forming concordant veins, lenses, and sheets, locally as thin as a few centimetres. Their occurrences are evidently too small to be considered as individual intrusions and cannot be separated cartographically. For these granite-metamorphic rock complexes the term migmatite, meaning 'mixed rock', was coined by a Swedish geologist, J. J. Sederholm, in 1907. Migmatites have been interpreted in various ways but the favoured explanation nowadays seems to be one holding that they are the result of partial melting of the metamorphic rocks at low temperatures ($< 700°C$), whereby granite veins are the melted portions (Hall, 1996). Migmatitic complexes tend to be hard and resistant and may support morphology similar to that on granitoids.

In the 1950s the concept of 'granitization' (Read, 1957) was discussed as an alternative to the magmatic origin of granite and similar rocks. In essence, granitization would have involved transformation of crustal rocks into granite by addition, or removal, of certain constituents without passing through the melting stage. Therefore it was regarded as a metamorphic process, specifically one of metasomatism. For example, a directional change sediment → slate → schist → gneiss → granite was envisaged by proponents of granitization as one of the ways granites originate. It was particularly applied to the origin of extensive bodies of granitic rocks in shield areas. However, Tuttle and Bowen (1958) have demonstrated experimentally that the lowest-temperature silicate melts co-existing in equilibrium with quartz and alkali feldspar have a composition such that subequal proportions of quartz, potassium feldspar, and albite form after crystallization. Such a composition of the product is identical with the range of composition observed in granitic plutons which provides a strong argument for the magmatic origin of granite, especially because solid-state granitization is incapable of producing a similar final product. The commonly held view nowadays is that the vast majority of granitoids are of magmatic origin and metasomatism is exceptional both in time and space, and recent general treatments of granite geology refer to granitization only as a subject of historical interest (Clarke, 1992; Pitcher, 1997; Cobbing, 2000).

Magma Emplacement and Shapes of Granite Bodies

Granites form from melts which buoyantly rise within the crust due to their lower density. The approximate density of granitic magma is only 2.2–2.3 g/cm^3 which is 10–20 per cent less than the density of other crustal igneous and metamorphic rocks. Therefore magma of granitic composition is gravitationally highly unstable.

In detail, five different mechanisms of granite magma emplacement are envisaged, each involving a buoyant rise into shallower crust and applicable in site-specific situations (Best, 2003; Fig. 1.3). Stoping is a mechanism favoured in shallower and cooler brittle crust and is an example of passive emplacement of magma. Ascending magma causes joint-bound blocks of overlying country rock to break off and gradually sink as the lighter magma moves upward. Xenoliths, locally as large as 1 km across, typify intrusions formed by stoping. The root of the term itself derives from mining practices where prising of masses of ore-bearing rock from the roof and walls of a gallery was called 'stoping'.

A variant of stoping is ring-fracture stoping when a large slab of crust sinks and magma intrudes into the void. Depending on the level of subsequent erosion, the intrusion may appear as a circular one at shallow depth or as a ring dyke if deeply eroded and the level of sunken slab is reached. Large-scale examples of apparent concurrent country rock subsidence and magma ascent are called cauldron-subsidence. This is the preferred mode of emplacement for plutons forming the Coastal Batholith in Peru (Pitcher, 1978). Ring dykes are known to support distinctive geomorphology, mainly because of significant lithological and structural contrast between adjacent rocks.

Brecciation denotes emplacement of sub-vertical columns of magma, called pipes, plugs, or stocks, depending on their size, into the brittle upper crust. They are small in diameter, usually less than 1 km. The term 'plug' is used for magmatic bodies less than 100 m across, many of which are deeply eroded conduits of former volcanic vents.

Forceful emplacement mechanisms include doming, which is accomplished by the invasion of overpressured magma which pushes country rock up and creates

12 Geology of Granite

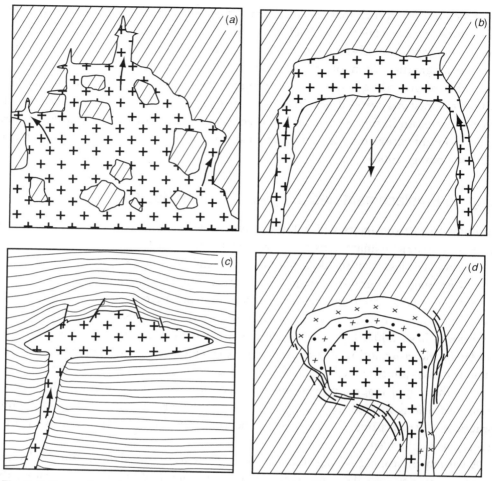

Fig. 1.3 Types of granite intrusions: (a) stoping, (b) ring-fracture stoping, (c) doming and formation of laccolithic bodies, (d) ballooning

room for itself. Intrusions due to doming are typically flat-floored whereas the upper surface is convex-upward and concordant with the rocks into which magma intruded. Many granitoid bodies due to doming are extensive in horizontal direction, but the connection with deeper crust is provided by a relatively narrow, funnel-shaped mass. They are called laccoliths and may be up to 2 km thick. Roofs of laccoliths are frequently fractured and faulted (Jackson and Pollard, 1988).

Multi-phase intrusions are accomplished by ballooning. Its essence is continuous, radial inflation of a magma chamber as additional magma is rising from below. The evidence for ballooning includes a circular to elliptical outline of the igneous body, more or less concentric and contact-concordant magmatic foliation within the pluton, increasing strain towards the pluton/host rock contact, and concentric compositional zonation within the pluton.

The fifth mechanism is magma invasion into tectonically favoured zones such as fault, zones, bends along active faults, or hinges of active folds. Large granite masses in orogenic settings, such as the Sierra Nevada Batholith of the USA and the Coastal Batholith of Peru, have probably formed along zones of local extension within the overall compressional/strike-slip regime. They may also assume a laccolithic-like shape and extend horizontally as lobate bodies (Hamilton and Myers, 1967). Petford and Clemens (2000) note that the evidence for tabular shapes of granite plutons is growing and argue that the model involving the occurrence of a narrow vertical conduit and a large horizontally extended body is most realistic. Intrusion thickness (T) and its length (L) seem to obey a power-law $T = cL^a$, being typically 0.6 ± 0.1. An adoption of the tabular pluton model provides an answer to the 'room problem', that is, of space available to accommodate the newcomer in the crust. As Petford and Clemens (2000) argue, a sufficient amount of space can be created by lifting the roof rocks and depressing the pluton floor.

The recognition of laccolithic shapes of granitic plutons also bears on the relationships between the size of outcrop and level of erosion. At intermediate levels of erosion spatially extensive plutons may be exposed, but if erosion reaches significantly deeper, then only small isolated bodies will be present, the conduits leading to long-eroded masses of granitic rocks. It was suspected that the scarcity of granitoids in the Appalachians in comparison to the Sierra Nevada reflects much deeper erosion of the former area (Hamilton and Myers, 1967).

The mechanisms of emplacement presented above refer to individual granite bodies called plutons. A pluton is a body of solidified magma which intruded in one discrete episode and rarely exceeds 30 km in diameter. Plutons can be classified in different ways. Simple plutons are those composed of a single lithological unit whereas composite plutons contain more than one lithology and reflect emplacement and/or differentiation sequence. A sub-category of composite plutons are zoned plutons in which parts of contrasting composition are arranged more or less concentrically. A typical arrangement is the presence of a dioritic or tonalitic outer part and granodioritic or monzogranitic core (Cobbing, 2000).

Plutons may occur in isolation but more commonly they cluster together to form much larger bodies of granitoid rocks called batholiths. The largest batholiths may extend over hundreds of kilometres and consist of many individual plutons. For example, the Coastal Batholith of Peru is $c.500$ km long and 20–30 km wide and is composed of about 1,000 plutons of variable composition and mode of emplacement.

Tectonic Settings of Granite

The association of intrusive processes with tectonics has long been recognized and emphasized in the terms such as pre-, syn-, and post-tectonic used in reference to plutons. Another approach has been to classify plutons into syn-, late-, and post-kinematic (Marmo, 1971). The advent of plate tectonics, however, brought the tectonic context of granitoids into a different light and assisted in the better understanding of the complex relationships between tectonics, source of magma, geochemistry, and textural characteristics of granites. However, in geomorphology it may still be useful to describe granites in the context of temporal relation to tectonic events. This is because of the implications for the history of deformation, mainly fracturing, which in turn profoundly affects the evolution of granite landforms.

From the plate tectonic point of view granitoid rocks may form and occur in a diverse variety of settings, from oceanic island arcs through orogenic belts to anorogenic interiors of continental cratons (Fig. 1.4). Extensive discussions of the subject can be found in a number of specialized papers and book chapters (cf. Pitcher, 1982, 1997; Maniar and Piccoli, 1989; Barbarin, 1999), whereas a more general approach has been offered by Hall (1996). In the most comprehensive classification Maniar and Piccoli (1989) discriminate between seven types of tectonic settings: island arcs, continental arcs, continental collision zones, post-orogenic zones, rifts, areas of continental epeirogenic uplift, and oceanic ridges and islands (Table 1.4). The present-day distribution of granitoids reflects these contrasting environments of origin but it has to be borne in mind that the growth of continental crust through time was accomplished by accretion of orogenic belts. Therefore many granite batholiths located far inland from the collision zones of today were in fact formed in long extinct collision belts such as the Mid-European Variscan belt of extensive magmatism.

However, the abundance of plutons within each setting is not identical. This is because these settings differ in the availability of magma sources capable of producing a melt of granitic composition. In island arcs at the junction of two oceanic plates there is a very limited supply of magma rich enough in quartz and alkali constituents to produce granite intrusions or their volcanic rhyolitic equivalents. Consequently, plutons are rare and never attain larger dimensions. Likewise, oceanic islands are very seldom built of granitoid rocks. However, if an island arc is a surface expression of subduction of oceanic plate under continental crust, melting of the latter may generate a sufficient amount of acid magma to crystallize as granite or erupt as rhyolites. The prime example of this relationship is the Aleutian Arc, which shows very little acid magmatism in the western part and contains granite bodies in the east, precisely because it is only the eastern one-third of the arc where continental crust is present. By contrast, orogenic continental margins abound in huge granite batholiths (Fig. 1.4).

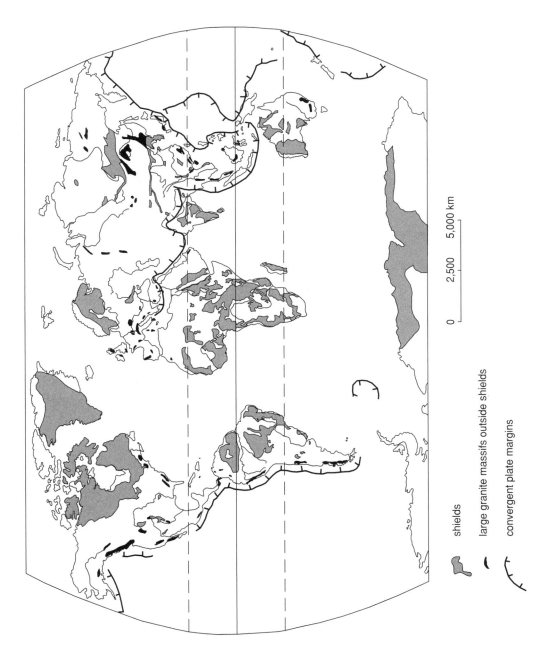

Fig. 1.4 Distribution of granite massifs across the world. Within shields granite typically occurs alongside other igneous and metamorphic rocks, but their cartographic separation would be difficult at the scale adopted in the diagram

Table 1.4 Tectonic settings of granitoid intrusions and selected examples

Tectonic setting	Examples
Orogenic	
Island arc	Eastern Pacific islands: Bougainville, Solomon Islands, Papua New Guinea
Continental arc	North America: Sierra Nevada, Coastal Batholith of British Columbia; South America: Coastal Batholith of Peru
Continental collision zone	High Himalaya
Transitional	
Post-orogenic zone	Late Caledonian plutons of Ireland and Scotland, Late Variscan plutons of western and central Europe, Basin and Range province in SW USA
Anorogenic	
Rift	British Tertiary Igneous Province, Oslo Rift
Continental epeirogenic uplift zone	Ring complexes of Nigeria
Oceanic ridge and islands	Ascension, Reunion

Source: After Pitcher (1982) and Maniar and Piccoli (1989).

Granites of the world vary considerably in age. They occur within both Precambrian and Phanerozoic mobile belts (Pitcher, 1982). The latter range in age from early Palaeozoic Caledonide granites in Scotland and the northern Appalachians of North America to Miocene granites in parts of the Andes and the Himalaya. In some oceanic arc settings such as in the Pacific Islands granite intrusions as young as late Neogene/early Quaternary are exposed at the surface (Pain and Ollier, 1981).

Fractures

Granites are cut by discontinuities. These may be planar or curved, of variable length, orientation, and distinctiveness. In any given rock there are usually many such discontinuities showing various geometrical relationships to each other. Even a cursory glance at a bare granite landscape will reveal that they exert a profound influence on the shape of individual landforms and the appearance of entire landform assemblages (Plate 1.2). Hence it may be inferred that discontinuities are key controls of geomorphic processes operating within granite landscapes.

Discontinuities in granite are usually referred to as 'joints' and geomorphologists tend to speak about 'joint-controlled' geomorphic features. However, this is not always correct because joints, strictly speaking, encompass discontinuities which are purely extensional (dilatational) and exhibit no shear along their strike. If a shear component occurs, however small, the term 'fault' would apply (e.g. Price and

Geology of Granite 17

Plate 1.2 Fracture control on granite landforms: tors at Königshainer Berge, eastern Germany

Cosgrove, 1990). In field situations it could be very difficult, if not often impossible, to identify this shearing component and the presence of joints could be overemphasized and that of shearing overlooked. Therefore, the use of the more neutral term 'fracture' is recommended to avoid confusion. A 'fracture' is any discontinuity breaking the coherence of a rock mass. Faults and joints of any origin all fall into the family of fractures.

Fractures form in response to stress imposed on a rock mass and indicate that the strength of the material was insufficient to withstand the stress. Granite masses are typically subjected to a multitude of stresses throughout their history, hence fractures of different origin and generations may be present in them. They may form in response to either compressive stresses associated with collision settings, or tensile stresses brought about by cooling of the magma, de-pressurizing of the magma chamber, or regional crustal extension.

Primary Fractures

Fractures are divided into two main groups (Price and Cosgrove, 1990): systematic and non-systematic. The former are planar and their traces are straight, whereas the latter have their traces curved. Parallel systematic fractures form sets and granite outcrops typically show extremely well developed sets of this kind (Plate 1.2). One of the first to seek explanation and order among fractures in granite was a German geologist, Hans Cloos (1925). He defined four main groups of primary discontinuities, mainly on the basis of observations from the northern part of the Karkonosze (Germ. *Riesengebirge*) massif in present-day Poland. These fracture classes are (Fig. 1.5):

(a) cross fractures, formed in the direction perpendicular to the flow of magma. He regarded them as mainly extensional but the presence of slickensides indicates that a shear component must have been present too. They were denoted as 'Q' type fractures;

(b) longitudinal fractures, perpendicular to cross fractures and striking parallel to the flow. They are younger than cross-fractures, hardly show shear displacement, and are probably generated by cooling of the magma. Cloos described them as 'S' type fractures;

(c) flat-lying fractures ('L' type fractures) typify higher parts of intrusions and are more or less parallel to their roofs. They are horizontal or dip at low angles ($< 20°$) and form probably due to a decrease in magmatic pressure;

(d) diagonal fractures strike at an angle of about $45°$ to the flow structures and indicate compression normal to the flow and extension parallel to the flow.

Q, S, and L type fractures cross each other at right angles, hence together they form an orthogonal fracture system which is repeatedly present in many granitoid bodies of the world and responsible for the common breakdown of rock mass into cubic blocks.

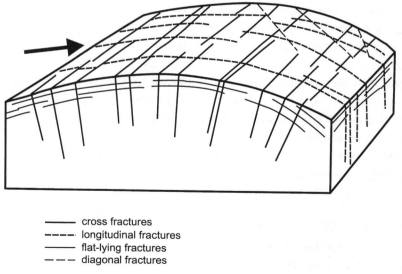

——— cross fractures
- - - - longitudinal fractures
——— flat-lying fractures
— — — diagonal fractures

Fig. 1.5 Typical fracture patterns in granitic rocks

However, fracture spacing in each set may be unlike in the other sets, and may significantly change within a single set if analysed in a wider spatial context. Segall and Pollard (1983) carried out detailed analysis of two granodiorite outcrops in the central part of the Sierra Nevada, USA, and demonstrated that the distance between primary joints belonging to a single N10–20°E set varies between 0.2 m and 25 m, hence by two orders of magnitude. Highly variable spacing of fractures is probably attributable to local variability in the mechanical properties of an intrusion and to the absence of mechanical layering, which controls more or less uniform joint spacing in sedimentary rocks. Furthermore, vertical fractures are not necessarily normal to each other, but criss-cross at acute angles. For example, in many parts of the Sierra Nevada Batholith two principal directions of vertical fractures are NNE-SSW and WSW-ENE, indicating fracture formation under significant compressive stress in an orogenic setting (Bateman and Wahrhaftig, 1966).

Master Fractures

Aerial photographs typically show fractures which are continuous over many kilometres, or even for tens of kilometres. They have been noted in both shield areas in different continents (Mabbutt, 1952; Thorp, 1967a; Jeje, 1974; Johansson *et al.*, 2001a), as well as in mountainous terrains (Ericson *et al.*, 2005). These long and prominent fractures usually control a variety of geomorphological features. Therefore they are of significant interest to geomorphologists, who tend to call them 'master joints'. Indeed, Blenkinsop (1993: 350) offers the following definition of a master joint: 'Master joint is a joint whose extent is appreciably greater than the average of the set in which it occurs, and against which less prominent joints terminate'.

However, the above definition suffers from ambiguity about the size of the structure, and in the light of the previously outlined reservations about the use of the term 'joint' it is not certain if these structures are indeed joints. As 'master joints' are identified from air photos, then given their scale it is unlikely that they represent single discontinuities. Rather, they are zones of closely spaced fractures, eroded to form negative topographical features. In the Sierra Nevada, California, 'master joints' are probably strike-slip fault zones as described by Segall *et al.* (1990), 0.5 m to 10 m wide, bounded by two boundary faults enclosing a zone of highly fractured and shattered rock. They originally developed through linking certain smaller faults and, while growing, they put the surrounding rock into a 'shear stress shadow'. Hence displacements became more and more localized on the longer faults (Martel *et al.*, 1988). An increase in shear strain rate led to the development of cataclastic zones and to significant alteration of primary minerals. Chlorite, epidote, and sericite largely replaced hornblende, biotite, and feldspars. These structural and lithological circumstances favour erosion, which becomes focused on the fault zones, exploiting them as 'master fractures'.

Secondary Fractures

All fracture types described so far are 'primary', in the sense that they have formed in the crust and were not influenced by conditions at the topographic surface, nor by the geometry of this surface. Granitoid rocks, however, also display an array of secondary fractures, which are usually considered to form at the surface or in shallow subsurface in response to pressure release (Gilbert, 1904; Jahns, 1943; Johnson, 1970; Holzhausen, 1989). These fractures are described using terms such as 'pressure release joints', 'unloading joints', 'exfoliation joints', or 'sheeting joints', and their common characteristic is a general parallelism to the topographic surface (Plate 1.3). For flat-lying fractures, the term 'pseudobedding' has been proposed (Waters, 1954). This multitude of names and their often loose application is unfortunate, because the effects of decompressing a rock mass are still insufficiently understood and doubts have been expressed as to whether large-scale sheet structures are really entirely related to pressure release (Twidale, 1973, 1982; Vidal-Romani and Twidale, 1999). Whereas the first two terms given above imply the cause of joint formation, the other two do not bear such genetic connotations and should be preferred until their origin due to unloading can be demonstrated. However, they face problems too as neither the relationship of sheeting to exfoliation, nor their definitions, are universally agreed. Exfoliation is a particularly ambiguous term, as will be shown in the next chapter. Vidal-Romani and Twidale (1999) suggest the value of 0.5 m as an arbitrarily defined minimum thickness of sheeting slabs.

Plate 1.3 Surface-parallel sheeting joints, Yosemite Valley, Sierra Nevada, California

Although sheeting had already been noted in the nineteenth century (see Holzhausen, 1989, for a list of early references), the unloading hypothesis was first elaborated by Gilbert (1904) in respect to extensive sheeting present on granite domes in the Sierra Nevada, California. He proposed that sheeting develops in rocks formed under high compressive stress deep in the crust, in response to the removal of overlying load as denudation brings the rock mass ever closer to the surface. The relief of primary compressive stress would have induced tensile strain, manifested in layer-like and surface-parallel separation of the rock mass. Increase of sheet thickness with depth and its eventual disappearance observed on truncated domes in the Sierra Nevada was cited as the key observational evidence supporting the unloading hypothesis. Later works in granite quarries in New England provided further field evidence of sheet thickening with increasing depth (Jahns, 1943; Holzhausen, 1989). The physics of the process has been subsequently considered by Johnson (1970), but fully developed by Holzhausen (1989). In his interpretation, sheet structures form under conditions of high differential stress in which two principal stresses are parallel to the surface and the third one is perpendicular to the surface and is close to zero. These conditions are satisfied if erosional unloading and decrease of compression normal to the surface takes place. He further argues that sheeting does not form at greater depth, under high triaxial compression, nor can it develop after complete three-dimensional unloading. The latter assertion is debatable, however, as many inselbergs do show well-developed sheeting planes dipping as steeply as 50–60° (Selby, 1982b).

The identification of secondary pressure release fractures is not always an easy task. First, they should generally be normal to the direction of pressure release and the relevant topographic boundary surface. But these surfaces could be from horizontal (a planation surface supported by the flat-lying roof of an intrusion) to vertical (sides of former glacial troughs), hence the resultant discontinuities may have various orientations. Indeed, in the Cairngorms Mountains of Scotland two sets of unloading joints have been recognized, one parallel to the gently rolling plateau surface and another one parallel to the steep sides of glacial valleys (Glasser, 1997). In the Sierra Nevada Huber (1987) provided examples of planar unloading joints disposed horizontally and vertically, as well as of curved joints in convex and concave variants. Second, pressure release joints may be superimposed on the existing pattern of primary fractures and therefore not easily discernible. However, they may be expected to be non-systematic and to terminate against primary structures. Third, there is a problem of scale. Unloading has been demonstrated to affect rocks since they are at least $c.100$ m from the surface (Holzhausen, 1989) and major unloading fractures may begin to form at this depth, being only roughly concordant with the surface. Further denudation would expose these rock masses and induce the development of a new set of pressure release fractures, subordinate to the previous one but with a better fit to the contemporaneous surface geometry. Fourth, as noted earlier, there is a problem of terminology, the result of which is that features of different origin may be described using the same term. The problem of pressure release joints has a direct bearing on the evolution of granite domes and rock slopes (Chapters 4 and 5), where a follow-up of the discussion can be found.

Finally, there are many non-systematic, local fractures cutting granitoid masses which have developed due to weathering processes. These include closely spaced discontinuities responsible for flaking of rock surfaces, fractures attributable to frost weathering and thermal shocks, cracks induced by the removal of lateral support, and others. They will be the subject of more detailed examination in subsequent chapters.

Microcracks

Apart from fractures observed at the outcrop and regional scales, granitoid rocks may show an extensive network of microcracks. Microcracks, or microfractures, are defined as planar discontinuities which are too small to be seen within a hand specimen (Engelder, 1987). Their length is typically of the order of 100 μm. Three types of microcracks may be present within a rock: grain boundary, intergranular and intragranular cracks (Simmons and Richter, 1976), the latter usually following cleavage planes if such are present. Cracks begin to form when local strength is exceeded by local stress. Therefore any pre-existing points or planes of mechanical weakness, such as crystal boundaries, cavities, or cleavage planes, where stress is concentrated, favour crack development (Kranz, 1983). Likewise, in granitoid rocks they can easily be induced by thermal changes because constituent minerals have different coefficients of thermal expansion. Other cracks are initiated at grain scale and reflect volumetric strain within crystals. Seo *et al.* (2002) have found that pre-existing cracks were abundant and multidirectional in quartz, whereas they were much fewer in feldspar and biotite. These differences were attributed to the different thermomechanical properties of minerals and greater volume change in quartz (4–5%) as opposed to feldspar and biotite (1–2%), during cooling below 600°C. Cracks, once initiated, may propagate through the rock. Under stress, microcracks in quartz open, which induces stress at the adjacent quartz/feldspar boundary and initiates microcrack growth within the feldspar. In biotite grains stress results in sliding along cleavage planes and again is transmitted to adjacent feldspars, where it results in low-angle microcracks.

Vein Rocks

Granite masses *s.l.* commonly host a variety of veins of different composition and texture. Some may be monomineralic, such as quartz veins, but the majority are polymineralic and these are typically divided into two groups, the dividing line being their texture. Aplites are fine- and generally even-grained. Individual crystals do not exceed a few millimetres across. Aplite veins are rather narrow, often a mere few

Plate 1.4 Aplite vein cut through medium-grained massive granite, Mojave Desert, California

centimetres wide and rarely more than 1 m (Plate 1.4). Quartz, alkali feldspar, and muscovite dominate in their composition, which is thus similar to that exhibited by typical granite. However, some aplites are sodic, with albite rather than potassium feldspar being present. By contrast, pegmatites are coarse-grained and form much larger bodies, up to a few hundred metres across. Individual feldspar crystals may be as much as 1 m long and examples of 10 m long minerals are known. They are also much more complex rocks than aplites, take a variety of shapes, and may display clear zonation from muscovite-rich outer edge to quartz-dominated core. Pegmatites may also form lenses and various irregular bodies within the host granitoid rock.

The enrichment of vein rocks in low-melting constituents of granite such as quartz and alkali feldspar and depletion in high-melting ferromagnesian and calcic plagioclase components indicates that they crystallize from final liquid fractions remaining after crystallization of the main body of magma. These late-stage melts typically use newly established fractures in the cooling mass, therefore strikes of veins may be similar to the strike of predominant fractures. Their lateral extension may reach more than 10 km. Many generations of veins may be present, with the younger ones cutting across the older veins, intruded at earlier stages of magmatic activity.

The occurrence of vein rocks, as well as of schlieren and enclaves, is important in the context of surface weathering of granite and the development of microforms.

2

Granite Weathering

Weathering is a necessary precursor for landform development. However, in the context of granite it acquires a particular importance for various reasons.

First, many granite terrains show an extensive development of deep weathering profiles, which can be extremely varied in terms of their depth, vertical zonation, degree of rock decomposition, and mineralogical and chemical change. Moreover, the transitional zone between the weathering mantle and the solid rock, for which the term 'weathering front' is used (Mabbutt, 1961b), may be very thin. There is now sufficient evidence that many geomorphic features of granite landscapes, including boulders, domes, and plains, have been sculpted at the solid rock/weathering mantle interface and they are essentially elements of an exposed weathering front. Therefore, the origin of granite landscapes cannot be satisfactorily explained and understood without a proper understanding of the phenomenon of deep weathering.

Second, granites break down via a range of weathering mechanisms, both physical and chemical, which interact to produce an extreme diversity of small-scale surface features and minor landforms. In this respect, it is only limestones and some sandstones which show a similar wealth of weathering-related surface phenomena.

Third, both superficial and deep weathering of granite act very selectively, exploiting a variety of structural and textural features, including fractures, microfractures, veins, enclaves, and textural inhomogeneities. In effect, the patterns of rock breakdown may differ very much between adjacent localities, and so the resultant landforms differ. In the context of deep weathering, selectivity is evident in significant changes of profile thickness and its properties over short distances, and in the presence of unweathered compartments (corestones) within an altered rock mass.

Fourth, it is emphasized that granites are particularly sensitive to the amount of moisture in the environment (Bremer, 1971; Twidale, 1982). They alter very fast in moist environments, whereas moisture deficit enhances rock resistance and makes it very durable. Hence, a bare rock slope shedding rainwater and drying up quickly after rain will be very much immune to weathering, whereas at its foot a surplus of moisture will accelerate decomposition. Such local differences explain a contrast typical of many granite areas, where massive outcrops of intact rock are surrounded by surfaces underlain by a thick mantle of decomposed, but still *in situ* material. Likewise, rock surfaces long kept moist by rain and fog are expected to weather more effectively.

Although systematic accounts of weathering in geomorphology usually discriminate between physical and chemical weathering processes and their effects, such an approach is followed in this book, which focuses on landforms and landscapes as a whole. These two groups of processes operate in synergy, reinforcing each other, whereas landforms are most fruitfully seen as a combined effect of various weathering (and other) processes acting upon rock surfaces. Therefore, this chapter will primarily explore the following issues:

- peculiarities of weathering acting on granite. Although there is perhaps no weathering mechanism endemic to granite, the results of weathering of granite surfaces are very distinctive. This suggests that granite lithology has a profound influence on mechanisms and effects of otherwise 'standard' weathering processes;
- environmental and topographical controls on the course of granite weathering, both at and beneath the ground surface;
- particular emphasis will be given to the phenomenon of deep weathering, because of its profound implications for landscape development.

Weathering Processes

Patterns and Products of Granite Breakdown

Even a cursory glance at a natural granite outcrop reveals that the rock has suffered from superficial breakdown which has released smaller particles in a variety of sizes, from silt to massive angular blocks. At the elementary level, rock disintegration is evident in the light of two parallel observations. First, the surface of the rock itself shows effects of weathering such as pitting, flaking, and cracking. Second, grains and larger rock pieces accumulated at the foot of an outcrop indicate detachment from the host rock surfaces. A worldwide literature survey shows that granite disintegration patterns appear similar across climatic zones. On the other hand, systematic studies of surface weathering patterns such as those by Swantesson (1989, 1992) in south-west Sweden provide clear evidence that a variety of disintegration effects can be found within a relatively limited area. Given the obvious climate-dependent constraints of many weathering processes, this situation suggests that different weathering mechanisms are able to produce similar patterns of rock disintegration. Hence the relationships between products and processes of breakdown are by no means straightforward.

Surface Roughness

An overwhelming majority of granite surfaces exposed to the air, except for those with recent glacial polish, are rough and show micro-relief of the order of a few millimetres to perhaps one centimetre (Plate 2.1). The surface roughness, also called pitting, evidently relates to the mineralogy of granite. Quartz crystals usually form asperities and stand proud, as do large potassium feldspars, whereas plagioclases and mica are visible in concavities between those protruding minerals. These mineralogy–surface relief relationships indicate that pitting is a consequence of differential weathering of the constituent minerals, among which quartz and potassium feldspar are more resist-

Plate 2.1 Rough surface of granite in a natural outcrop, Karkonosze, south-west Poland

ant. Crystal size also affects the degree of roughness. Rock surfaces of coarse granite usually show more relief than those of fine-grained granite, in which all minerals have small dimensions.

There is convincing evidence that an increase in surface roughness is a function of time. At the microscale (2 × 2 cm reference surfaces), Swantesson (1992) compared granite outcrops newly emergent from the sea and found that surfaces exposed to subaerial weathering for 350 years are significantly less rough than those subjected to weathering attack for 1,500 years. In the granite area around Narvik, Norway, subject to strong isostatic uplift in the Holocene, differential surface lowering of 8–12 mm has been measured at altitudes of 90 m a.s.l., whereas it is only 4–8 mm at 50 m a.s.l., and less than 1 mm at sea level (Dahl, 1967). Likewise, polished surfaces of granite *roches moutonnées* in the same area, dated to 10 ka, have largely disappeared in the meantime, except for localized aplite and pegmatite veins (André, 1995, 2002).

Another view on the origin and significance of pitting, based mainly on observations from the Eyre Peninsula in southern Australia, has been presented by Twidale and Bourne (1976). After examining rock surfaces recently cleared of soil grus cover, they concluded that pitting develops in a subsurface position due to preferential weathering in a moist micro-environment and only becomes exposed subsequently. They also noted that orthoclase is usually preferentially weathered out, attributing this phenomenon to selective feldspar weathering at the weathering front. It seems that rough granite surfaces can evolve in both surface and subsurface positions, and the behaviour of potassium feldspar may hold the clue to their origin. Upstanding orthoclases and quartz would indicate surface weathering, whereas a subsurface origin might be inferred if quartz is the sole mineral standing proud.

Granular Disintegration

Granular disintegration, that is, the breakdown into individual grains, which can be mono- or polymineralic, is extremely widespread on granite. Detached grains are usually found in the gravel and sand fractions, but significant silt production in excess of 20 per cent may occur too and has been documented both in the field and experimentally (Power *et al.*, 1990; Wright, 2002). Coarse granites in particular are very susceptible because of their less tight fabric. Following an initial loosening of the rock mass, grains may remain attached to underlying surfaces, or become detached and fall down to accumulate on the ground surface. If grain release is sufficiently fast and plant growth is suppressed because of low temperatures or scarce rainfall, unvegetated aprons of loose debris develop around rock outcrops and these can be rather easily eroded by water, wind, or sorted by frost heave (Vieira, 1999). Such superficial covers of loose crystal-size debris occur in all climatic zones, from the Arctic to the warm deserts of Africa (e.g. Schattner, 1961; Folk and Patton, 1982; Vieira, 1999; Bjornson and Lauriol, 2001) (Plate 2.2). This observation suggests that granular disintegration is

Plate 2.2 Grus apron below scalloped surfaces subject to granular disintegration, Mojave Desert, California

caused by a variety of mechanisms and the prime factor governing grain-by-grain disintegration appears to be the texture of the rock itself.

Microcracks appear to play a key role in granular disintegration of granite and similar rocks (Whalley *et al.*, 1982; Pye, 1986). Widening of grain boundary cracks initiates the breakdown, which is subsequently enhanced by the growth of intra-granular cracks and associated etching along crack margins. Further development of cracks within individual quartz grains may ultimately result in the production of silt-size particles (Power *et al.*, 1990).

The causes of microcrack opening, in turn, are variable and include compressive stress release, swelling pressure exerted by water in capillaries, chemical alteration of mineral boundaries, stresses induced by heating/cooling cycles, salt crystallization, and frost action (Pye, 1986). This multitude of reasons explains why granular disintegration is a universal, climate-independent mechanism of granite breakdown.

Block-by-Block Breakdown

Breakdown into angular blocks is less common in granite, as demonstrated by the virtual absence of such blocks in many granite landscapes. This is because granular disintegration and flaking operating along pre-existing joint surfaces usually precede the detachment. Hence a block at the moment of separation has already acquired smooth edges and is more or less rounded. In addition, coarse granites are typically cut by widely spaced fractures defining large surface areas, hence stresses generated within those fractures by freezing water or salt crystals are insufficient to disrupt the rock. This is not to say that superficial mantles composed of angular granite blocks do not exist, and densely fractured fine-grained granite seems particularly prone to this kind of breakdown (Plate 2.3). Angular block covers have been reported from many present-day and past cold environments, including the Old Crow Batholith in Yukon, Canada (Bjornson and Lauriol, 2001), Dartmoor, England (Ballantyne and Harris, 1994), the French Massif Central (Valadas and Veyret, 1974), Corsica (Klaer, 1956) and the Karkonosze Mountains, Poland (Dumanowski, 1961), and their origin is usually attributed to frost weathering. However, as will be shown later, the efficacy of frost weathering has recently come into question and alternative mechanisms have been proposed such as thermal shock fatigue (Hall, 1995; Hall and André, 2001; Hall *et al.*, 2002). In specific cases, block fields may be inherited and represent a residual deposit derived from a pre-Quaternary weathering mantle, left after finer material had been eroded away (Whalley *et al.*, 1997).

Exfoliation and Related Phenomena

Another common means of granite disintegration is the peeling off of platy fragments from outer rock surfaces, called spalling or flaking (Plate 2.4). Individual flakes are

Plate 2.3 Block field built of predominantly angular blocks of medium-grained granite. In the background (upper right) there is a rock outcrop which may have supplied angular material to the block field, Great Staple Tor, Dartmoor, England

Plate 2.4 Multiple flakes on granite boulders, near Aswan, Egypt

from a few millimetres to a few centimetres thick. Flaking is most common on the undersides of boulders and rock overhangs, as well as within concavities on rock surfaces. As with granular disintegration, various mechanisms such as hydration of expandable minerals (Dragovich, 1967), salt attack (Goudie and Viles, 1997), and temperature changes, including those generated by fire (Ollier and Ash, 1983), are probably responsible for flaking, which is particularly common in arid and semi-arid areas (Schattner, 1961; Ollier, 1978b; Migoń and Goudie, 2003).

Outer rock layers subject to detachment may attain considerable thickness, much in excess of a few centimetres. In such cases the term 'flaking' is no longer used, but tends to be replaced by 'exfoliation'. Exfoliation, however, is a poorly defined term in respect to the size and the cause of the phenomenon, hence it has generated much confusion. A simple statement that exfoliation is 'weathering of a rock by peeling off of the surface layers' (Goudie, 1994: 198) encompasses both a millimetre-scale flaking as described above, as well as separation of massive, many metres-thick sheets from rock slopes. In addition, the above definition does not specify whether the exfoliating layer is to be curved or not, but many authors apparently restrict the usage of the term to curved surfaces. 'Pseudobedding', by contrast, has been the term introduced to describe separation along rather closely spaced (a few to a few tens of centimetres) planar partings parallel to themselves and to the outer surface of an outcrop (Waters, 1954) (Plate 2.5). The third source of confusion is the relationship between exfoliation and spheroidal, or onion-skin weathering. Some authors use these terms interchangeably, whereas others

Plate 2.5 Pseudobedding on a granite outcrop, Karkonosze, south-west Poland

argue that exfoliation is a subaerial phenomenon and it is only the exposed rock surface that is acted upon. Spheroidal weathering, by contrast, would affect the whole rock mass, both at the surface and in the subsurface, or even entirely in the subsurface, such as around unweathered compartments in deep weathering profiles (Chapman and Greenfield, 1949; Ollier, 1967). Onion-skin weathering denotes the presence of concentric spalls around a core. A very large-scale exfoliation, involving separation of rock layers in excess of a few metres thick and tens of metres long, tends to be called 'sheeting' and may occur along pre-existing fractures (Dale, 1923; Jahns, 1943; Twidale *et al.*, 1996; see also Chapter 2). A final problem with exfoliation is semantic. It is difficult to name the products of exfoliation in the way that we can say that flakes are produced by flaking and sheeting leads to detachment of individual sheets.

Schattner (1961), working on granite weathering in the Sinai Peninsula, was clearly aware of this multiple meaning of 'exfoliation' and the inherent potential for confusion. Therefore he distinguished its three forms:

(1) large-scale exfoliation, involving spalls several square metres in surface area and a few tens of centimetres thick;
(2) concentric boulder exfoliation, understood as an equivalent of spheroidal weathering;

Table 2.1 Suggested terminology for surface-parallel rock detachment

Phenomenon	Products of detachment	Surface area of detached fragments	Thickness of detached fragments
Flaking	Flakes	10–500 cm^2	up to 1 cm
Spalling	Spalls	0.5–5 m^2	1–30 cm
Sheeting	Sheets	> 5 m^2	> 30 cm
Spheroidal weathering (= onion-skin weathering)	detachment occurs concentrically around a rock fragment within a rock mass or deep weathering profile		

(3) scaling off of small and thin pieces of rock surfaces. This conforms to the widely accepted meaning of flaking.

Thus, the phenomenon of surface-parallel rock detachment appears to take four distinct forms, each best described by its own name (Table 2.1).

A variety of mechanisms have been proposed to account for surface-parallel detachment. Given the range of spatial scales of the phenomenon, it is unlikely that there is one universally applicable explanation, as already hypothesized by Blackwelder (1925). Some textbooks continue to emphasize insolation and extreme diurnal temperature changes, but this mechanism was doubted by Blackwelder as early as in 1925. He did acknowledge the role of fires in inducing rock spalling, but thought it an unlikely general explanation and favoured hydration and other chemical changes occurring in the outer parts of a rock as the principal cause of exfoliation. This view has largely been reiterated by Chapman and Greenfield (1949) and Schattner (1961). Crystallization of salts in the near-surface voids and associated rock swelling is another possible mechanism. For spheroidal weathering occurring at depth, Ollier (1967) suggested hydrolysis as the cause of rock separation into layers. The causes of sheeting are likely to be different. At this scale, climate-driven weathering processes become subordinate to large-scale manifestations of stress conditions within rock masses.

Outcrops affected by surface-parallel separation of curved or planar plates can be found in virtually every granite landscape of the world, which reinforces the view that exfoliation in its widest sense 'is not a simple process, but several processes of diverse origin' (Blackwelder, 1925: 794). Planar exfoliation (pseudobedding) is observed on tor-like outcrops in such diverse areas as the Dartmoor upland in cool humid southwest England (Waters, 1954), the Bohemian Massif in Central Europe (Chábera and Huber, 1998), eastern Mongolia (Dzulynski and Kotarba, 1979), recently deglaciated Cairngorms Mountains, Scotland (Sugden, 1968; Glasser, 1997), and Greenland (Oen, 1965). Finally, sheeting appears most common in areas of high relief, where tensile stresses set up by offloading are particularly high, but it also typifies many isolated residual hills. Rock slopes within deeply incised fluvial and glaciated valleys, as well as dome-shaped outcrops (bornhardts), are particularly conducive to multiple sheeting. Some of the best examples are provided by granite rock slopes in the Yosemite and Sequoia National Park in the Sierra Nevada, California (Matthes, 1950, 1956) (Plate 2.6).

Block Splitting

One of the most characteristic components of granite landscapes are more or less rounded boulders which occur either in isolation or in clusters, and attain a wide range of sizes, from a few tens of centimetres to more than 20 m long. There is now ample evidence, which will be reviewed more extensively in the next chapter, that the origin of boulders is largely a subsurface phenomenon. The vast majority of boulders originate as corestones in deep weathering profiles and are subsequently exposed.

However, once exposed, boulders become subject to a variety of superficial weathering processes and continue to break down. They diminish in size by means of granular disintegration, flaking, and selective hollowing of their surfaces. Not uncommonly, many of these boulders are broken into two pieces along a single vertical crack. Broken or split boulders may be as much as a few metres across. Although frost weathering has been occasionally claimed as responsible, splitting is best seen as a response to tensile stress distribution in unsupported or poorly supported rock masses,

Plate 2.6 Widespread sheeting on granite slopes of the Tenaya Canyon, Yosemite National Park, Sierra Nevada, California, accounting for frequent slab failures

hence with no relation to environmental conditions. Indeed, split boulders occur in environments as different as warm deserts, cold deserts, and humid tropical savannas. Stress is released along a pre-existing or latent fracture, which opens up causing the boulder to break into two or more parts (Plate 2.7). Ollier (1978a) has shown that the likelihood of induced cracking and boulder splitting increases for large boulders resting on a very small contact area, the diameter of the boulder being perhaps as much as 15–30 times the length of the contact area. Pre-weathering and long-term accumulation of stress lower the critical dimensions for splitting to occur, but nevertheless split boulders less than 2 m across are extremely rare.

In other cases, it is the undersides of boulders where angular fragments progressively peel off leaving the main rock mass less and less supported. Flaking in the near-ground, moist environment may add to the efficacy of losing mass. Over time, the surface of contact between a boulder and the underlying rock may become very small, for which situations the notion of a 'perched boulder' is applied (Twidale, 1982).

In site-specific circumstances the cause of block splitting can be lightning. Although a direct hit on a boulder may be a very rare event, it is worth quoting that rock fragments of up to 1 m^3 can be detached in the consequence of a strike, whereas smaller pieces can be thrown into the air to land 20 m away from the outcrop (Swantesson, 1989). In certain mountain areas witnessing frequent convective storms, repeated lightning strikes may significantly contribute to the origin of summit blockfields composed of

Plate 2.7 Gigantic split granite boulder, marginal desert environment, Erongo, Namibia

Plate 2.8 Boulder cluster affected by various breakdown processes: granular disintegration, flaking, and splitting, near Mono Lake, California

angular rock fragments. Boulder splitting has also been recorded during earthquakes, for example in the granite Rokko Mountains near Kobe, during the powerful tremor on 17 January 1995 (Ikeda, 1998). Other instances of probable earthquake-induced block splitting have been shown by Twidale *et al.* (1991).

All the above-mentioned means of disintegration are not mutually exclusive and there are field examples of boulders with pitted outer surfaces, subject to grain-by-grain breakdown, exfoliation, and split along a major crack (Plate 2.8).

Chemical Weathering and Denudation

Chemical weathering is a very complex process, the rate of which is controlled by a multitude of factors such as the mineralogy of the rock itself, water supply, residence time of water in the weathering environment, initial pH, the presence of organic acids, and the temperature of soil solutions (Kump *et al.*, 2000). The relative influence of each factor is difficult to isolate and quantify, although some information is available. Detailed geochemical studies at the catchment scale, within carefully selected reference areas, have helped to assess the role of key climatic parameters like temperature and rainfall in the chemical weathering of granite.

Granite chemical weathering is actually a group of processes, among which hydrolysis appears the most important and most widespread, affecting all silicate minerals constituting granite. It involves the reaction of a mineral with water which results in breaking down of the compound, replacement of metal cations (e.g. potassium, sodium, or calcium) by hydrogen ions, and recombination of released cations with hydroxyl ions. For example, hydrolysis of potassium feldspar may be illustrated by the following reaction:

$$4KAlSi_3O_8 + 22H_2O \rightarrow Al_4Si_4O_{10}(OH)_8 + 8H_4SiO_4 + 4K^+ + 4OH^-$$

orthoclase kaolinite

whereas hydrolysis of sodium plagioclase (albite) in a less humid environment can be written down as follows:

$$8NaAlSi_3O_8 + H_2O \rightarrow 3Na_{0.66}Al_{2.66}Si_{3.33}O_{10}(OH)_2 + 14H_4SiO_4 + 6Na^+$$

albite smectite

The resultant solid state compounds are clay minerals, which are typical constituents of weathering mantles developed on granitoid rocks. However, various clay minerals form through hydrolysis, depending on both the composition of the primary mineral and environmental conditions, mainly the intensity of leaching (Table 2.2).

One of the implications of Table 2.2 is that the influence of the mineralogy of the parent material diminishes with increasing intensity of leaching. It should also be noted

Table 2.2 Relationships between primary rock-forming minerals, secondary clays, and leaching intensity

Primary mineral	Secondary minerals		
	Weak to moderate leaching	Intense leaching	Very intense leaching
Albite	Smectite	Kaolinite	Gibbsite
Orthoclase	Illite	Kaolinite	Gibbsite
Muscovite	Illite	Kaolinite	Gibbsite
Biotite	Vermiculite	Kaolinite, goethite	Gibbsite
Hornblende	Smectite	Kaolinite, goethite	Gibbsite

Source: Simplified from Summerfield (1991), fig. 6.14, p. 140.

that the order of products may approximate changing mineralogy with the depth of the weathering zone, and is to a certain extent time-dependent.

Another mechanism of mineral alteration is hydration, in the course of which a mineral takes in an entire molecule of water. Among minerals forming granite, biotite is susceptible to hydration and alters to hydrobiotite through partial loss of potassium and its replacement by water molecules (Isherwood and Street, 1976). Progressive change of this sort, accompanied by oxidation of Fe^{+2} to Fe^{+3}, will lead to the formation of vermiculite at the expense of biotite.

Chemical composition and structure of minerals govern their resistance to chemical weathering. The silicates forming granitoid rocks are chemically varied and present a range of structures, hence their contrasting susceptibilities to hydrolysis. In most environments quartz is virtually immune to alteration, whereas some plagioclases and biotite change rather easily and become replaced by secondary clays such as vermiculite, smectite, and kaolinite. Potassium feldspar occupies an intermediate position in respect of the ease of its weathering. The idea of unequal resistance of minerals was formalized by Goldich (1938) in the weathering stability series, established after a study of granite gneiss in Minnesota and two other lithologies (Fig. 2.1). Referring specifically to granite, calcium plagioclase appears as the least resistant and is followed by other plagioclases. It is in turn followed by biotite and potassium feldspar. Muscovite and quartz are the most stable. Less common mafic constituents of granitoid rocks such as hornblende show rather limited resistance to chemical weathering.

The scheme is built on the Bowen's reaction series (Fig. 1.2), which ranks minerals in the order of their crystallization at consecutively lower temperatures, being essentially its reverse. The explanation of the order of stability lies in the internal arrangement of silicon tetrahedra (SiO_4) in various minerals and the strength of bonds between metal

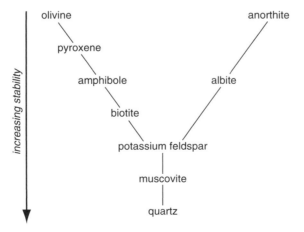

Fig. 2.1 Mineral-stability series, according to Goldich (1938)

cations and the silica structural units. Muscovite and biotite belong to the group of phyllosilicates, in which three oxygen atoms out of four are shared with adjacent tetrahedra, whereas feldspars and quartz, collectively called tectosilicates, have all their four oxygens shared. By contrast, hornblende, which is present in some granites, is an inosilicate with two/three oxygens shared and therefore breaks down easily, disrupting the whole structure.

Goldich's scheme agrees well with many real-world situations, in which alteration of plagioclase is reported as having occurred first, followed by biotite. However, if processes other than hydrolysis are involved, biotite may become the forerunner of weathering change and it is its expansion upon hydration that leads to the disruption of the entire rock fabric (Isherwood and Street, 1976). The actual role of biotite in the initiation of granite weathering is the subject of discussion. Some workers confirm its role as the weakest link in the chain (Wahrhaftig, 1965; Eggler et al., 1969; Bustin and Mathews, 1979), whereas others found little evidence that its role is significant (Williams et al., 1986). It is possible that environmental factors play a decisive part. In humid areas with efficient hydrolysis, plagioclase yields first, but in drier and/or cooler regions hydrolysis is suppressed and the effects of biotite expansion are more pronounced.

The composition of granite is thus expected to have a profound influence on the effectiveness of chemical weathering. Given other controlling factors being equal, a higher proportion of weatherable minerals such as Na- and Ca-feldspars and biotite would result in higher export fluxes of dissolved components, faster formation of weathering products, a higher proportion of secondary to primary minerals in the weathering mantle, and a faster downward advance of the weathering front. However, an equally important variable is the presence of access routes for water, which is both a reactant to and the carrying medium of decomposition products away from the minerals undergoing chemical change. Although the presence of dense jointing is emphasized most often in this context, it is in fact the occurrence of microcracks that enhances weathering at the initial stages (Pye, 1986).

Chemical weathering and denudation at the catchment scale are usually investigated via the construction of geochemical budgets. Geochemical budget data, in turn, may be used to calculate the rate of rock-to-weathering mantle-to-soil conversion, the amount of total mass and volume loss within the catchment (Pavich, 1986), as well as mass loss for particular minerals (Williams et al., 1986). In the geochemical approach elemental export fluxes are calculated from analyses of stream water chemistry and then compared with relevant precipitation fluxes to receive a net export. Fluxes are expressed in matter quantity units per unit area per time, such as $kg \cdot ha^{-1} \cdot yr^{-1}$ (Williams et al., 1986), $mol \cdot m^{-2} \cdot s^{-1}$ (White and Blum, 1995; White, 2002) or $mol \cdot ha^{-1} \cdot yr^{-1}$ (Oliva et al., 2003). Systematic geochemical studies of granite drainage basins were pioneered in the 1970s and 1980s (e.g. Cleaves et al., 1970; Ternan and Williams, 1979; Velbel,

1985; Pavich, 1986; Williams *et al.*, 1986). Most recent databases include geochemical data for well over 100 catchments of various sizes (Oliva *et al.*, 2003), although the geographical coverage is highly biased towards the USA and western Europe. Both silica and total cation fluxes differ hugely between the analysed catchments, by one to two orders of magnitude (Oliva *et al.*, 2003; Table 2.3).

In the light of the above data set climate does appear as a significant factor controlling chemical weathering and denudation rates, but correlation is not as good as one would expect. In particular, and somehow contrary to the previous albeit less exhaustive data set used by White and Blum (1995) and White *et al.* (1999), the role of temperature is not evident. An increase in precipitation, by contrast, generally correlates with increasing rates of chemical weathering and export fluxes. The non-climatic

Table 2.3 Granite chemical weathering and denudation rates in selected catchments located in different climatic zones

Catchment location	Annual precipitation (mm)	Mean annual temperature (°C)	Annual runoff (mm)	Total silica flux (mol · ha^{-1} · yr^{-1})	Total cation flux (mol · ha^{-1} · yr^{-1})	Total dissolved solid flux (kg · ha^{-1} · yr^{-1})
Experimental Lake, Canada	583	2.4	252	243	34	7
Sogndal, Sweden	984	n/a	875	141	145	8
Nsimi, Cameroon	1,751	24.0	380	479	268	20
Panola, USA	1,149	15.3	338	676	425	30
Juiadale, Zimbabwe	1,220	13.8	385	942	565	41
White Laggan, Great Britain	2,822	6.0	2,185	1,004	1,512	73
Jamieson Creek, Canada	4,541	3.4	3,668	1,534	1,714	98
Tsukuba, Japan	1,587	13.1	721	2,580	2,110	128
Anamalai Hills, India	2,300	27.0	822	1,384	3,681	142
Ilambalari, India	3,800	27.0	2,019	3,875	6,021	281
Rio Icacos, Puerto Rico	4,300	22.0	3,680	8,066	7,552	443

Source: Oliva et al., 2003.

reasons for data dispersion include minor differences in granite geochemistry, the variable thickness of the weathering mantle between catchments which affects the hydrology of the weathered rock/solid rock interface, drainage basin topography, and related denudation rate. Highest fluxes have been recorded in the subtropical mountainous basin of Rio Icacos, Puerto Rico, typified by steep topography, humid tropical climate, and high physical erosion rates, and underlain by easily weathered quartz diorite, rich in calcic plagioclases and depleted in potassium feldspar (White et al., 1998). By contrast, shield areas appear to weather at very low rates, in subpolar and equatorial settings alike.

Geomorphological interpretation of chemical denudation data collected at the catchment scale is not simple, although they do give a reasonable estimate of long-term chemical denudation and allow for comparative analysis. One obvious limitation is that export flux data are averaged over an entire drainage basin and as such, do not allow for identification of zones of localized weathering, whereas geomorphic and pedologic evidence shows that weathering is hardly ever uniformly distributed. Second, the weathering record inferred from averaged cation flux data is difficult to relate to weathering profile behaviour, whether it signifies progressive alteration of rock-forming minerals within the weathering mantle of a constant thickness, or a downward advance of the weathered rock/solid rock interface. However, White et al. (1998) do distinguish these different aspects of weathering and conclude that the rate of weathering front descent in the humid tropical mountains of Puerto Rico is about 58 m $\cdot$ Ma^{-1}. They also indicate that plagioclases and hornblende are the first minerals to alter and this occurs at the weathering front, whereas breakdown of biotite and quartz typifies the upper saprolite. Third, chemical weathering is not necessarily associated with surface lowering of a granite surface, because granite dissolves incongruently (Thomas, 1994b). Thus, chemically weathered granite is composed of both residual and neoformed minerals, which have a similar volume but decreased density, and has lower density than the original rock. In this case a significant export of solutes via throughflow and streams is not matched by any appreciable change at the topographic surface.

Weathering Indices

Another approach to assessing the extent of rock chemical weathering has been through various quantitative and semi-quantitative indices. They are applicable to *in situ* weathering, where disintegrated and decomposed rock remains in place. These indices may be based on the chemistry of weathered rock, its physical properties, or proportions between constituent minerals (Irfan and Dearman, 1978; Nesbitt and Young, 1989; Irfan, 1996). Recent applications of weathering indices to characterizing weathered granite include those by Irfan (1996), Guan et al. (2001), Gupta and Rao (2001), and Kim and

Park (2003) for different Asian granites. Some of the most commonly employed indexes are presented in Table 2.4.

The advantage of using weathering indices is that they emphasize a continuum of change, from fresh to highly altered rock, and give numerical expression to the chemical or mineralogical changes taking place during weathering, allowing for intra-profile and site-to-site comparisons. For example, Migoń and Thomas (2002) have showed how *in situ* grus weathering mantles may be characterized by indices such as the Chemical Weathering Index (CWI) (Sueoka, 1988) and the Chemical Index of Alteration (CIA) (Nesbitt and Young, 1989). Some weathering indices show reasonable correlation with weathering grades commonly used by engineering geologists. The CWI for fresh rock is 13–15 per cent which increases to 15–20 per cent for moderately weathered and to 20–40 per cent for highly weathered rock. Residual soil has a CWI in excess of 40 per cent (Sueoka, 1988). Guan *et al.* (2001) employed mobility indexes for major oxides normalized to titanium and alumina and a volume index for the entire rock, and found a systematic change of the values obtained with increasing decomposition grade. In

Table 2.4 Selected weathering indices used to characterize the degree of weathering of granitoid rocks

Index name	Formula	Source
Modified weathering potential index (MWPI)	$\dfrac{(Na_2O + K_2O + CaO + MgO) \times 100}{Na_2O + K_2O + CaO + MgO + SiO2 + Al_2O_3 + Fe_2O_3}$	Reiche, 1943
Decomposition index (X_d)	$\dfrac{N_q - N_{qo}}{1 - N_{qo}}$ $\quad$ N = quartz/quartz + feldspar $_{(q)}$ — in soil; $_{(qo)}$ — in solid rock	Lumb, 1962
Micropetrographic index (I_p)	$\dfrac{\%(\text{quartz} + \text{unaltered feldspar} + \text{unaltered biotite})}{\%(\text{altered minerals} + \text{voids} + \text{microcracks})}$	Irfan and Dearman, 1978
Chemical weathering index (CWI)	$\dfrac{Al_2O_3 + Fe_2O_3 + TiO2 + H2O(\pm)}{\text{all chemical components}} \text{mol} \times 100$	Sueoka, 1988
Chemical alteration index (CIA)	$\dfrac{Al_2O_3 \times 100}{Al_2O_3 + CaO + Na_2O + K_2O}$	Nesbitt and Young, 1989
Mobility index (MI)	$\dfrac{R^i_w/R_w}{R^i_p/R_p}$	Guan *et al.*, 2001

Notes:
R^i_w—percentage by weight of a non-stable *i* constituent in weathering mantle
R_w—percentage by weight of a stable constituent in weathering mantle
R_p—percentage by weight of a stable constituent in parent rock
R^i_p—percentage by weight of a non-stable *i* constituent in parent rock

another recent paper, Gupta and Rao (2001) recognize the good correlation between the micropetrographic index and engineering properties.

Lithological and Structural Controls on Granite Breakdown

Granitoid rocks vary in terms of their mineral and chemical composition, texture and fabric, degree of microcracking and fracturing, and all these characteristics significantly influence the course and rates of rock weathering. However, there are not too many studies that address this issue directly, for example through systematic measurements of solute exports from catchments underlain by different granites. Much more often, lithological and structural control is inferred from the analysis of landforms and landscapes, characteristics of weathering mantles, or even from very general considerations of variables controlling weathering processes.

The factor of lithology exerts evident control over the patterns of physical breakdown of granite. Granular disintegration typifies areas built of coarse granites and proceeds both at the surface, through continuous grain release from rock outcrops, and in the subsurface, where laterally continuous mantles of disintegrated granite up to a few metres thick may occur. By contrast, fine-grained granite tends to disintegrate mainly through separation and detachment of small angular pieces, with little granular material being produced. A good example might be the Serra da Estrela in Portugal, where thin aprons of granular material are very common in the parts of the massif built of coarse Seia and Covilha granite, but are virtually absent on the high plateau built of fine-grained Pedrice granite where small angular blocks dominate the surface (Vieira, 1998, 1999). Similar observations have been forwarded from the Karkonosze in Poland (Dumanowski, 1961). At the scale of individual clasts, Bustin and Mathews (1979) noted that coarse granite fragments in tills in British Columbia disintegrate more readily than those made of fine-grained granite.

Field observations are supported by laboratory experiments, although one needs to realize that debris production from granites subject to freeze-thaw or salt immersion cycles is invariably very low, irrespective of the detailed petrological characteristics of the rock. Martini (1967) tested both coarse porphyritic and fine equigranular variants and was able to demonstrate that significantly more debris was produced from coarse granites, especially if the samples being compared were not pre-weathered. Likewise, Swantesson (1989) found variants with large (up to 5 cm long) megacrysts more susceptible than more equigranular and fine-grained variants.

That coarse granites yield to granular disintegration more easily is explained through recourse to their fabric. Widely different mineral sizes, up to a few centimetres for potassium feldspar, related higher porosity and wider dimensions of grain-boundary cracks, and internal cleavage of larger crystals due to rock de-stressing upon cooling and long-term denudation, provide a good work environment for various weathering agents acting near the rock surface: salts, freezing and thawing water, insolation, and living organisms. The much tighter fabric of fine granites makes these lithologies more resistant to mechanical disintegration into individual grains or their small aggregates. Preliminary results of Schmidt hammer tests carried out on granite outcrops in the Serra da Estrela have shown that the average mechanical strength of medium and fine granite is higher than that of coarse granite and more consistent by 5–10 reading points (Migoń and Vieira, in preparation). Similar conclusions have been reached by Tuğrul and Zarif (1999), who found statistically significant inverse relationships between the mean grain size quartz, K-feldspar, and plagioclase, and uniaxial compressive strength in their sample of Turkish granites (Fig. 2.2).

The differences in fabric and strength between coarse and fine granites account for the preferential production of large angular fragments from the latter. Tight fabric and low porosity increase intergranular strength, which can be overcome along more widely

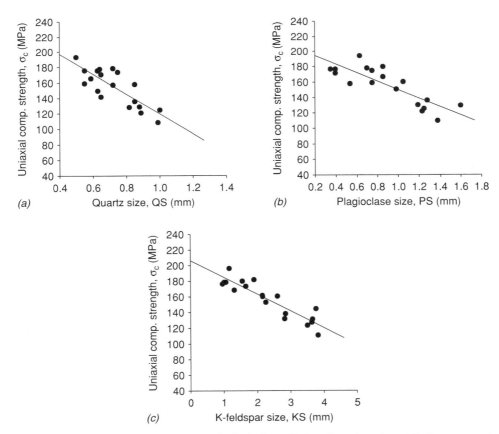

Fig. 2.2 Relationships between the size of principal granite-forming minerals and their compressive strength: (a) quartz, (b) plagioclase, (c) K-feldspar (after Tuğrul and Zarif, 1999, fig. 7)

spaced transgranular lines of weakness, set up during rock mass cooling and subsequent removal of overlying country rock. Hence, at the surface stress is released along these longer discontinuities, causing the separation of outcrops into joint-bounded blocks. Likewise, weathering agents have difficulty penetrating grain boundary cracks and voids and focus on transgranular cracks instead.

Weathering Susceptibility of Granite

Granite weathering proceeds via interactions between the rock and external environmental agents such as freezing water, salt, temperature change, including fire, living organisms, and products of their metabolism. They act upon rock surfaces simultaneously, not uncommonly reinforcing each other, and therefore it is difficult to isolate the effects of each of these interactions. In the field, relationships between observed weathering products and weathering mechanisms are particularly difficult to establish. To overcome this problem at least in part, experimental studies are attempted to quantify weathering effects under the influence of a particular weathering agent.

Frost (Freeze-Thaw) Weathering

Frost weathering, or frost shattering, denotes rock breakdown in sub-, or near-zero temperatures and occurs through two principal mechanisms (Hall *et al.*, 2002). One is the freezing of water in cracks and pores which causes volume expansion of about 9 per cent. However, critical pressures able to disrupt the rock develop only if the system is closed and the moisture content in the rock is very high, which is not a very common situation in natural environments. Therefore, a segregation ice model (Hallet *et al.*, 1991), according to which stress develops as water migrates to growing ice lenses, seems more realistic and does not require a very high level of rock saturation or a closed system. The efficacy of frost weathering is controlled by a range of climatic factors, including the number of freeze-thaw cycles, the rate of freezing, and moisture availability.

The conventional view is that frost weathering operates very efficiently in cold environments and invariably produces a mantle of angular debris, not uncommonly to form extensive block fields (*Felsenmeer*). Indeed, granite uplands in Scotland, Arctic Canada, the Rocky Mountains, Khentei in Mongolia, and in many other mountain ranges are littered with blocks and debris which are angular rather than rounded (Plate 2.3). However, that these debris mantles are really products of frost weathering is far from clear and the whole idea of the dominance of freeze-thaw weathering in cold climates has recently been seriously challenged (Hall, 1995; Hall and André, 2001; Hall *et al.*, 2002). It now seems clear that alternative mechanisms of physical breakdown such as wetting and drying and thermal shock fatigue have not been given due consideration, nor are the relationships between process and form as straightforward as once assumed. Here it is sufficient to say that block fields in high latitudes have more than one origin, but this issue will be explored further in Chapter 8.

Further light on the contribution of frost weathering to rock disintegration has been shed by experimental studies. Although porous soft rocks are more favoured in the experiments because more debris liberation is expected, various granites have also been tested (Martini, 1967; Swantesson, 1989). Overall, the efficacy of frost shattering of granite appears low, at least in comparison with other common rock types. For example, in the experiments carried out by Swantesson (1989) the largest weight loss during 864 cycles was 0.02 per cent, but most tested granite samples yielded less than 0.005 per cent. By comparison, debris liberation from sandstone and limestone was higher by one order of magnitude. Similar percentages of liberated debris, between 0.02 and 0.06 per cent, have been obtained from the rapakivi granites from Finland (Lautridou and Seppälä, 1986).

However, increased resistance of granite against freeze-thaw weathering must not be automatically assumed as the outcome of these simulations. Experimental weathering studies have numerous limitations, including a finite number of cycles applied to tested samples, but one deemed particularly important is in the context of granite weathering. Frost shattering investigated in environmental cabinets is microgelivation, which is characteristic of porous rocks such as sandstone or chalk. However, the very low porosity of granite precludes the mechanism of ice growth in pores as an efficient means of breakdown. Instead, macrogelivation accomplished by wedging in cracks is probably far more significant, but this process is very difficult to replicate in experiments (Matsuoka, 2001*a*). In addition, several studies have indicated that even slight pre-weathering of granites increases their susceptibility to physical breakdown significantly (Martini, 1967) and it needs to be recalled that periods of Pleistocene cold climate on European and North American uplands were preceded by long time spans of warm conditions, conducive to a variety of chemical weathering processes.

Salt Weathering

Salt weathering is a collective term to describe many different ways in which salt causes rock to break down, but three assume major importance (Evans, 1970; Goudie and Viles, 1997). These are salt crystal growth from solutions in pores and cracks, salt expansion due to hydration, and salt expansion due to temperature changes.

The efficacy of salt weathering of sandstone and limestone is impressive (Goudie and Viles, 1997), but the results of experimental work on weathering of granite are inconclusive. One simulation using sodium sulphate, which is often most effective in rock breakdown, showed virtually no weight loss after 30 cycles (Goudie, 1974). Another test combined freeze-thaw and salt immersion cycles, and again very little breakdown has been recorded after 35 cycles (Wessman, 1996). But field evidence of

46 Granite Weathering

Plate 2.9 Granite sculptures in Mahabalipuram, southern India, severely affected by salt weathering (Photo courtesy of Yanni Gunnell)

salt attack on granite outcrops in the form of efflorescences is available (Bradley *et al.*, 1978; Goudie, 1984; Lageat, 1994), and salt weathering of granite building stones is recognized (Matias and Alves, 2002; Schiavon, 2002). The most likely explanation for this discrepancy is that the duration of the experiments was of insufficient length to produce a more significant breakdown. Natural outcrops and architectural monuments which remain under the influence of salt attack for much longer do show an appreciable amount of mass loss. A detailed study of granite historical buildings in Braga, northern Portugal, has revealed that granite breaks down in many ways in the presence of sulphates, through granular disintegration, flaking, and scaling (Matias and Alves, 2002). Coastal environments are particularly affected by salt weathering, as will be discussed in Chapter 6.

Thermal Weathering and Fire Effects

Thermal weathering, otherwise known as insolation weathering or thermoclasty, is the breakdown of rock due to temperature changes within the rock induced by repeating heating and cooling. Temperature changes are usually considered by referring to daily cycles. Diurnal rock temperature changes reported from warm deserts are indeed high and may well exceed 40° (Cooke *et al.*, 1993), and those from polar deserts and high-mountain environments are of a similar order (Hall, 1999; Waragai, 1999), but how this affects weathering is not entirely clear. Once favoured as a key weathering mechanism in dry environments, thermoclasty has subsequently been dismissed because laboratory experiments such as those carried out by Blackwelder (1933) failed to demonstrate its effectiveness. However, it has been acknowledged more recently that short-term thermal stress events caused by the passage of cloud or the onset of rainfall may have a more significant effect than long-term (daily) temperature changes (e.g. Warke and Smith, 1994; Hall, 1999).

Recent studies carried out on a granodiorite outcrop in Antarctica (Hall and André, 2001) have demonstrated that heating and cooling may occur at rates of $2°C \cdot min^{-1}$, which is sufficient to cause thermal shock within the rock. Striking similarities between superficial crack patterns produced experimentally and those occurring in natural environments have also been noted. Granite in fact appears particularly susceptible to thermal effects because its mineral constituents have different coefficients of thermal expansion and the dark ones should absorb solar radiation rather easily, but conclusive field evidence is not yet available. Nevertheless, the occurrence of cracked boulders in desert and semi-desert areas, difficult to explain otherwise, is consistent with the hypothesis of thermal weathering (Ollier, 1963).

Much more effective, though a rather localized means of rock disintegration, is fire. Many granite terrains in areas characterized by seasonal drought are frequently swept by fire. In semi-arid grasslands the return period of fires may be less than five years, and in savanna and Mediterranean (chaparral, maquis) environments it is in the range of 5–15 years. In addition, these areas may be affected by fires induced by humans for various purposes. Despite some early observations (Blackwelder, 1926), fire used to be a rather underestimated geomorphic agent, but this view has changed in the last decade or so (Dorn, 2004). Nowadays fire is considered an important trigger of rock disintegration, which breaks down because of being subjected to an extreme range of temperatures, approaching 1,000°C in less than 10 minutes in certain instances. Fire effects may be particularly pronounced in granite and granodiorite because of the high expansion of quartz if heated and its poor thermal conductivity (Ollier and Ash, 1983). Namely, quartz expands by almost 4 per cent if heated from 15°C to 570°C, that is, four times more than feldspar, which is the other principal constituent of granite (Winkler, 1973). Experimental studies (Goudie *et al.*, 1992; Allison and Goudie, 1994) have confirmed that granite is particularly susceptible to fire-induced disintegration, as demonstrated by its 80–95 per cent reduction in the modulus of elasticity value and noted growth of intragranular cracks.

Field observations in granite areas subject to frequent fires indeed indicate the occurrence of a range of surface weathering features which are best attributed to fire (Plate 2.10) (Ollier and Ash, 1983). Detachment of thin outer layers of rock from exposed surfaces is most common in medium-grained granite. Flakes fall and may additionally break down upon impact, to form miniature grus aprons. These in turn are susceptible to wash and further re-distribution of fine debris across the slope. Splitting

Plate 2.10 The upland of the Serra da Estrela, Portugal, is frequently swept by intentionally set fires, which contribute to the widespread flaking of granite boulders

of boulders has been reported by Ollier and Ash (1983), who have also suggested that spalling may be characteristic of the heating phase, whereas splitting would typify the cooling phase. In fine-grained granites boulders mainly disintegrate into smaller pieces along straight and curved fractures, whereas further spalling leads to rounding of boulder edges (Dragovich, 1993).

Dragovich (1993), using an example from the Pilbara region in Western Australia, highlighted the consequences of repeated fires for the appearance of arid to semi-arid granite landscapes. The junction of rock outcrops and lower vegetation-covered slopes is often an effective fire boundary, the former being much less affected. In consequence, an abrupt transition from a superficial mantle of small loose stones (fire spalls) to rock outcrops or boulder slopes develops. Furthermore, a tight cover of angular fire spalls on lower slopes offers protection against erosion.

Biological and Biochemical Weathering

Biological weathering of granite is a poorly documented phenomenon and there are very few studies addressing this issue in detail. As with other weathering process studies, the common perception of granite as a very durable rock appears to discourage researchers who choose sedimentary rocks instead. That granites suffer from the action of living organisms growing on and inside the rock is evident in the light of field observations. In humid environments such as western Europe, or foggy places such as the coastal strip of the Namib Desert, granite outcrops are extensively overgrown by lichens, to the extent that almost no bare rock surface is visible (Plate. 2.11). Algae and mosses may also thrive on granite surfaces. Both natural outcrops and historical monuments are affected.

The destructive role of organisms includes their mechanical and biochemical action, perhaps mutually reinforced. The expansion of epilithic lichens on the surfaces of Neolithic megaliths in Alentejo, Portugal, has brought the following consequences to the rock (Romão and Rattazzi, 1996). Hyphae bands, as thick as 10 μm, penetrate very deeply, up to 10–20 mm, into the granite substratum, crossing and contouring constituent minerals. In effect, individual quartz crystals may become detached, while those of feldspar and mica are fragmented into smaller pieces. A closer inspection of lichen mats reveals the presence of individual crystals of up to 2–3 mm long, apparently detached from the underlying rock surface (Plate 2.11). Biogeochemical activity of sulphur and nitrogen cycle bacteria may lead to a variety of surface disintegration features, such as grain detachment and flaking (Leite Magalhães and Sequeira Braga, 2000). At the microscopic scale, Schiavon (2002) has found evidence of mineral etching and dissolution due to biochemical action on all granite-forming minerals and observed that the biological patina is largely made up of fragments of primary minerals. His study suggests that lichen mats do not protect the underlying surface but are

Plate 2.11 Lichen mats on granite boulders, Pierre de Jumaitres, Massif Central, France

active contributors to mass loss. In another study, carried out on granite nunataks in the Juneau Icefield in Alaska, Hall and Otte (1990) documented how green algae grow in microcracks parallel to the outer surface and expand upon absorption of water, contributing to the detachment of 2–4 mm thick flakes. However, there remains a problem with relating these observations, mainly from building stones and polluted environments, to the long-term development of granite rock surfaces, as it combines several of the major scale issues in weathering studies (Viles, 2001). For example, given the significance of lichens in detaching individual grains it is probable that biological agents are important contributors to surface pitting, normally explained without invoking the role of organisms.

Granite Weathering in Comparison to other Common Rock Types

Although this book is about the geomorphology of granite, it seems useful to compare the means and efficacy of granite disintegration and decomposition with those affecting other common rock types. Such a comparison may hold some clues to the widely observed distinctiveness of granite landscapes. No weathering process or mechanism has ever been claimed as unique to granite, but lithology and the structure of granite dictate a very specific course and effects of breakdown in many instances.

First, the disintegration of solid rock into a few millimetre-sized crystals and aggregates is very distinctive to granite. Granular disintegration is particularly efficient in coarse granites and/or those affected by significant microcracking; it is less common in fine-grained granites which tend to release larger fragments, several tens of centimetres long, and silt. Many individual mechanisms conspire to effect disintegration, including salt action, freeze-thaw, thermal shock, biotite hydration, and organism activity. Because of the ubiquity of granular disintegration, superficial grus is a very characteristic product of granite weathering. Few other rocks, and perhaps none of the most common rock types on Earth except some gneisses, are able to yield such large quantities of grus due to their weathering. Products of granular disintegration may occur *in situ* or become redistributed across the topographic surface.

The relative ease with which the granular disintegration of granite proceeds is attributed to both mineralogy and the texture of the rock. The presence of different minerals, each having different mechanical, chemical, and thermal properties, favours a number of weathering processes such as volume changes due to rock temperature changes, hydration, and hydrolysis. The presence of transgranular and boundary cracks, in turn, allows salt and freeze-thaw weathering to act efficiently. On the other hand, quartz and, in many environments, potassium feldspar too, are relatively inert constituents and keep the whole structure coherent, preventing its immediate collapse despite some loss of material.

Second, a multitude of breakdown products are known to derive from granite outcrops. These can be as small as silt-sized particles, or as big as massive blocks formed due to sheet structure disintegration. The latter are not uncommonly a few tens of metres long and up to 10 m thick. All intermediate sizes are possible, although it is probably true that the 2–50 cm fraction is underrepresented. The immediate consequence of the wide dimensional range of disintegration products is an equally wide range of superficial deposits in and around granite terrains, from loess through grus, block and boulder fields, to massive taluses surrounding some of the inselbergs and lining valley sides in mountainous settings. Again, such a diversity of sizes of weathering-derived particles appears specific to granite. Sandstone breakdown, for instance, often proceeds straight from joint-bound block stage to individual sand grains. Other common rock types, limestone and basalt, easily disintegrate into medium-sized blocks (10–50 cm), but sand and gravel-sized particles are not very abundant.

Third, in many areas granite weathering releases huge blocks, more than 2 m long and not uncommonly longer than 10 m, which seems to be a unique granite phenomenon. This is related to wide spacing of fractures, which in turn is characteristic of late-kinematic intrusions in geotectonic settings which have been little disturbed tectonically afterwards. Few other rocks are similarly massive, therefore isolated rock compartments of comparable size are rare.

Fourth, the results of numerous experimental studies, notwithstanding their limitations, indicate that granite is more resistant than most sedimentary and low-grade metamorphic rocks against such processes as salt and freeze-thaw weathering. However, massive volcanic rocks, for example, basalt and rhyolite, and some monomineral metamorphic rocks, for example, quartzite, are even more resistant. Likewise, the mechanical strength of granite is higher than that of common sedimentary and metamorphic rocks, but lower than that of basalt, dolerite, or quartzite. Therefore, granites often form elevations in relation to sedimentary rocks and schistose complexes, but this does not necessarily occur if they are juxtaposed with gneiss or quartzite. If contact metamorphism occurred during the intrusion of a granite body, the thermally altered rocks usually occupy higher ground than granite (Fig. 2.3). Initial alteration of granite increases its susceptibility to physical breakdown considerably.

Last but not least, one of the most distinctive features of granite weathering systems is the extent to which deep weathering occurs and the protracted residence times of its products. Many granite terrains are typified by the presence of thick and spatially extensive mantles of weathered rock *in situ*, formed primarily due to chemical weathering operating within the entire rock mass. This phenomenon has huge implications for landform development and will therefore be considered separately in the following section.

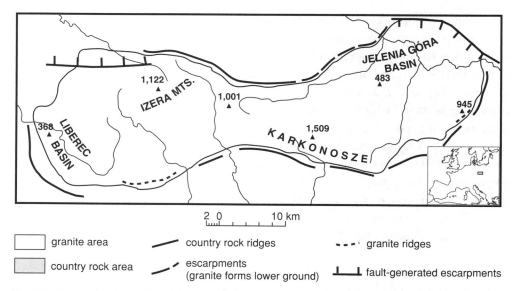

Fig. 2.3 Geomorphology of contact zone between the Karkonosze granite and the country rock indicates decreased resistance of granite against long-term weathering. Along most of the perimeter of the granite massif higher ground is associated with the thermally altered country rock (hornfels)

Deep Weathering

The phenomenon of deep weathering is ubiquitous on granites (Plate 2.12; Plate II), to the extent that it is probably the absence of a thick near-surface mantle of weathered rock that requires special explanation, rather than the presence of such materials. The majority of early descriptions of thick mantles of decomposed rock, including those derived from granite, came from humid low latitudes (Branner, 1896; Falconer, 1911), creating an impression that deep weathering is a peculiar characteristic of the humid tropics. However, in the last few decades evidence has steadily accumulated to show that deeply weathered granites are present all around the globe, including high latitudes of northern Europe and North America, much of the temperate zone of the northern hemisphere, and semi-arid areas, and such a view is no longer tenable. Some of these high latitude weathered granites may be relicts from past tropical climates, but in other cases deep weathering is evidently late Cainozoic, thus not tropical at all.

In a similar manner, many early studies focused on thick weathering mantles in topographical terrains of rather low relief, but Thomas (1994b: 19) remarked that 'any suggestion that they [weathering mantles] underlie only the ancient planation surfaces of continental interiors is misleading'. Indeed, excellent exposures of thick weathering mantles have been documented in highly dissected and mountainous terrains such as Hong Kong, Serra do Mar in south-east Brazil, or Sierra Nevada in western USA. It has

Plate 2.12 Deeply weathered granite, disintegrated into a sandy residuum, southern Sierra Nevada, California

also been revealed that products and patterns of deep weathering on granite are highly variable, controlled by many structural and environmental factors, and escape a simple explanation.

Deep Weathering Profiles on Granite

In general, near-surface mantles of weathered rock are transitions from fresh, solid rock at depth to *in situ* surface soil, subdivided in the vertical column into a number of zones showing different mineralogical, physical, textural, and engineering properties. Their presence reflects gradual change from the surface downward, resulting from a decreasing influence of weathering agents with depth and from the different time spans of weathering involved. Because of this zoning, a term 'weathering profile' is often used in a way analogous to the soil profile. The arrangement and properties of constituting zones vary widely from one rock type to another, and from one area to another; therefore no 'standard' weathering profile exists which would be applicable to all rock types and settings. However, it has long been observed that zoning of weathering mantles developed from granitoid rocks is particularly well-ordered, at least in comparison to other rocks, and that hence an opportunity for classification arises.

One of the first classifications was proposed by Ruxton and Berry (1957), working on the geomorphology of weathered granite terrain in Hong Kong. Their fourfold (in fact, fivefold) scheme included the subdivisions shown in (Fig. 2.4 see also Table 2.5).

54 Granite Weathering

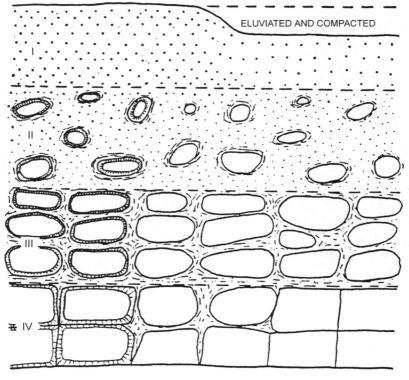

Fig. 2.4 Partition of a typical weathering profile on granite (after Ruxton and Berry, 1957, Fig. 2). For explanation of I–IV codes see Table 2.5.

Table 2.5 A weathering profile on granite according to Ruxton and Berry (1957)

Weathering zone code	Name	Description
I	Residual debris	Structureless sandy clay or clayey sand, up to 30% of clay, dominance of quartz and kaolin
IIa		Few rounded corestones, less than 10% of the section area, less than 5% of solid rock
	Residual debris with corestones	
IIb		Much of the original rock structure preserved, corestones occupy 10–50% of the section area
III	Corestones with residual soil	Dominance of rectangular corestones (50–90%), surrounded by residual matrix
IV	Partially weathered rock	More than 90% is solid rock, minor amounts of residual debris along fractures, iron staining may be present
	Bedrock	No visible signs of rock weathering

This classification has since formed the basis for various more elaborate classifications, designed chiefly for engineering purposes (see reviews in Gerrard, 1988*b*, and Irfan, 1996). Dearman *et al*. (1978), following some previous work, employed simple descriptive terms for each zone and, in contrast to Ruxton and Berry, numbered the zones from the bottom up (Table 2.6). An important novelty has been an attempt to provide numerical values of selected rock mass properties for the zones and an engineering assessment for each zone. More recently, Irfan (1996) presented a comprehensive data set relating weathering grades by Dearman *et al*. (1978) to fabric properties and the mineralogical composition of the weathered rock.

The concept of weathering grading, although relatively simple and easy to use, is seldom used outside engineering geology. In geomorphology, where the studies undertaken do not usually have a direct engineering application, the emphasis is on describing the complexity of weathering mantles, viewed three-dimensionally rather than in one-dimensional sections. A general term is 'saprolite', derived from the Greek *sapros* meaning 'rotten' and introduced by Becker in 1895 (Taylor and Eggleton, 2001), but even this is variously defined by different authorities. *The Encyclopedic Dictionary of Physical Geography* (Goudie, 1994: 439) states simply that saprolite is 'weathered or partially weathered bedrock which is *in situ*', and this broad definition has been adopted in many detailed studies and reviews (e.g. Wang and Ross, 1989; Lidmar-Bergström *et al*., 1997; Migoń and Lidmar-Bergström, 2001). Saprolite defined in this way constitutes a part of 'regolith', which is the term encompassing all fragmental and unconsolidated material between fresh rock and the surface. Hence, regolith cover usually consists of two distinct units: weathered rock *in situ* below and an overlying layer of transported material of whatever origin (Taylor and Eggleton, 2001).

However, another approach is to restrict the usage of the term saprolite to situations in which significant chemical alteration of parent rock is recorded but the weathered rock still retains the original fabric (Taylor and Eggleton, 2001). Consequently, a saprolite would be a middle part of a weathering mantle (Fig. 2.5). For the material below it, which is no longer solid but not yet friable either, and in which some minerals show initial alteration, a term 'saprock' is proposed. In the context of granite though, such a definition poses problems. One of the most characteristic products of deep granite weathering is grus, described more extensively in the next section, which is typically little altered chemically but thoroughly disintegrated, and individual mineral aggregates can be picked up by hand. Hence, it satisfies one requirement for saprock (little mineralogical change) but also one for saprolite (much weakened fabric).

On the other hand, if the requirement of retaining the original fabric is accepted, the uppermost part of a weathering profile does not conform to the definition of saprolite either. Taylor and Eggleton (2001: 160–3) discussed the terminology for this zone and eventually favoured the term 'pedolith'. Indurated veneers, or duricrusts, may occur with the pedolith part of a weathering profile.

Table 2.6 The scale of weathering grades of rock mass by Dearman et al. (1978)

Term	Grade	Description	Dry density	Percentage of altered minerals	Percentage of microcracks and voids	Micropetrographic index
Residual soil	VI	all rock material converted to soil; original structure and fabric destroyed; large change in volume	1.25–1.60	50.1–59.6	13.2–24.8	0.33–0.41
Completely weathered	V	all rock decomposed and disintegrated but original structure and fabric still largely intact	1.20–1.80	29.5–48.8	15.6–38.0	0.39–1.21
Highly weathered	IV	some rock decomposed and disintegrated; fresh or weakened rock locally present	1.70–2.40	38.0	15.6	0.87
Moderately weathered	III	rock discoloured and partly weakened; solid rock locally present, including corestones	2.30–2.58	7.0–12.6	2.4–3.0	5.41–9.18
Slightly weathered	II	discolouration as the main sign of change, mainly along discontinuities; no clear weakening of fabric	2.55–2.60	5.9–9.7	0.5–3.1	6.81–14.59
Fresh	I	no visible signs of weathering; slight discolouration along major discontinuities	2.58–2.63	2.3–3.9	0.4–0.6	24.0–33.48

Source: Quantitative data (cols. 4–7) from Irfan (1996).

The two distinctive features of weathering mantles developed on granitoid rocks are a weathering front and corestones. They are not endemic to granite but are either rare or less distinct, or even absent, in other lithologies.

The Weathering Front

A weathering front is the boundary separating solid, unweathered rock and rock that has already been weathered but remains still *in situ* (Mabbutt, 1961b). In reality, the transition between weathered and fresh rock compartments is rarely sharp and assumes the form of a zone within which the change from unweathered mass to disintegrated or decomposed rock occurs. In some lithologies, especially in foliated metamorphic rocks and clastic sedimentary rocks, the transition is highly gradational and a well-defined

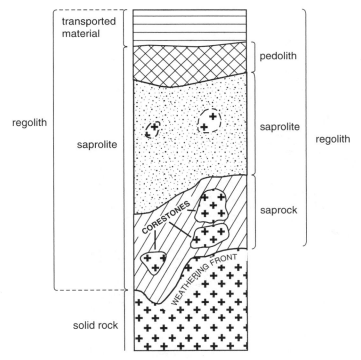

Fig. 2.5 Diagram to explain the terminology of weathering mantles and profiles

weathering front may not exist, but in granites this zone is not uncommonly only a few centimetres thick (Plate 2.13). Schnütgen (1991) suggested that sealing effects produced by the precipitation of iron oxides and silica gel might be responsible for the very sharp interface, but the evidence presented is actually from a dolerite boulder.

A weathering front should not be visualized in the form of a single, continuous rock/weathering mantle interface present at some depth, as implied by the predecessors of the term, 'basal platform' (Linton, 1955) and 'basal surface of weathering' (Ruxton and Berry, 1957). A weathering front is a dynamic feature which migrates downwards and sideways over time, as more and more rock disintegrates or decomposes. Its geometry can be highly irregular, reflecting preferential weathering along pre-existing zones of weakness. Moreover, if there are unweathered compartments within an otherwise altered rock, the boundary of each of them is a weathering front.

Corestones

Corestones are unweathered, or significantly less weathered, compartments of parent rock, roughly of the size of a boulder (from a few tens of centimetres to a few metres), within an otherwise weathered mass (Plate 2.14). Many are rounded and may exhibit

Plate 2.13 Very sharp weathering front in granite, Valley of the Thousand Hills, South Africa

onion-skin weathering, but others are more cubic and their edges are yet to be smoothed. Corestones are usually cited as products of selective deep weathering, which is initially concentrated along orthogonal fractures and progresses from fracture planes inward to the host rock. In this way, the size of solid rock cores diminishes and their edges become rounded with the passage of time (Fig. 2.6). Assuming no erosion over a longer time, weathering should be able to consume a corestone, and indeed, 'ghost' corestones can be observed in sections through completely weathered rocks. Therefore, corestones are usually concentrated, and attain larger dimensions in the lower parts of deep weathering profiles, as envisaged in the scheme of Ruxton and Berry (1957). There are instances, though, in which corestones are observed in the upper parts of profiles and the corestone zone is underlain by a thick mass of thoroughly decomposed granite (Plate 2.15). This is probably indicative of a certain structural predisposition in the parent rock, such as wider joint spacing or minor mineralogical differences.

Granite Weathering 59

Plate 2.14 Rounded corestones within a deep weathering profile on coarse granite, Miłków, Karkonosze, south-west Poland

However, some corestones are difficult to explain by progressive rounding of a formerly cubic block. First, they occur at a significant depth within profiles and might be expected to be more angular, yet they show almost perfect spherical shapes. Second, they are surrounded by a series of concentric bands (layers) of rock, which may even be

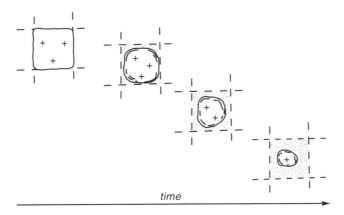

Fig. 2.6 Corestone development through progressive weathering of joint-bounded compartments within saprolite

60 Granite Weathering

Plate 2.15 Less weathered compartments concentrated in the upper part of the weathering profile, Kings Canyon National Park, Sierra Nevada, California

wrapped around a few adjacent corestones. Third, these outer layers are not necessarily very much weathered. Farmin (1937) suggested unloading at shallow depth, but this does not explain the localized nature and geometric pattern of the phenomenon. Ollier (1967) noted no unequivocal evidence for volume change in the course of spheroidal weathering and considered the development of outer bands to be the result of hydrolysis and migration of mineral elements. On the other hand, spherical structures may be related to differential crystallization of magma and are revealed at shallow crustal depth by a combination of rock dilatation and incipient weathering changes.

The scheme by Ruxton and Berry (1957), the engineering scheme by Dearman *et al.* (1978), as well as the scheme by Taylor and Eggleton (2001), all imply a rather gradual progression from highly altered granite at the top of the profile to fresh parent rock at the bottom. In reality, granite weathering profiles can be complex and weathering grades as listed in Table 2.6 do not necessarily occur in sequence. There are numerous field examples in which the occurrence of a less weathered rock mass above a more weathered one is recorded (Plate 2.15). Nor do corestone percentage and shape change regularly with depth. In the wider spatial context, Shaw (1997: 1084) concluded, after analysing 4,730 borehole logs from the Kowloon peninsula in Hong Kong, that 'the concept of a regular vertical zonation of weathering profiles is not supported by the results of this study'.

Equally important are lateral variations within granite-derived deep weathering mantles. Not only may the thickness change significantly over short distances (Thomas,

Plate 2.16 Lateral contact of solid and weathered granite, Kings Canyon National Park, Sierra Nevada, California

1966; Shaw, 1997), but there are lateral contacts between weathered masses showing contrasting degrees of fabric preservation, mineralogical composition, engineering properties, and corestone content. The latter may be observed in countless outcrops around the world, in all possible environmental and topographic settings (Plate 2.16).

Lithology and Mineralogy of Granite-Derived Weathering Mantles

As deep weathering proceeds, there is a progressive change in the mineralogy of the weathering mantle which in turn induces further change in fabric, strength, and other engineering properties. The weathering grades and their descriptions introduced earlier indicate the nature of change rather clearly, but there is nevertheless a tendency to use specific names for various types of weathering mantles. Thus, the name *grus* is used to describe products of *in situ* granular disintegration of coarse-grained rocks. Grus is characterized by its specific grain size distribution, where the sand (0.1–2.0 mm) and gravel (>2.0 mm) fraction predominates and may constitute up to 100 per cent of the total. The percentage of finer particles (silt plus clay) liberated by weathering is often negligible, and does not exceed 25 per cent. Accordingly, neoformed minerals, usually smectite and vermicullite, occur in minor quantities (Migoń, 2004b).

In turn, *kaolin* is a weathering mantle that contains predominantly kaolinite as a neoformed clay mineral. Since the mineral kaolinite comprises the more resistant elements of aluminium and silica and lacks any more soluble components such as magnesium or sodium, it indicates quite an advanced stage of weathering (Trudgill, 2004). Therefore, the abundance of kaolinite in the weathering mantle is usually associated with significant loosening of the rock fabric and an increasing percentage of the clay fraction. Saprolites, as defined by Taylor and Eggleton (2001), if developed on parent rock rich in aluminosilicates, are usually kaolins.

Despite frequent usage, this bipartite typology lacks consistency, emphasizing predominant grain size in grus and predominant neoformed minerals in kaolins. Therefore, some authors prefer to speak about clayey weathering mantles instead of kaolins, which would be a term consistent with the principle behind the term 'grus'. Differentiation between arenaceous (i.e. rich in sand-size fraction) and argillaceous (i.e. rich in clay) weathering (Power and Smith, 1994) follows exactly the same principle and may be recommended. However, in weathering mantles with a high clay content, kaolinite is indeed the dominant mineral formed through weathering. In many regional accounts a distinction is made between grus and kaolins (or clayey saprolites) as if they formed two fundamentally different products of separate weathering types. In reality, there is a continuum of weathering products from incipient disintegration to the most thoroughly decomposed material and any boundary lines between them will inevitably be arbitrary.

Clayey mantles dominated by kaolinite and residual quartz are typical products of very advanced weathering of granites. Iron-rich, indurated mantles, commonly described as *laterites*, are comparatively rare in granite areas because of the low iron content in the parent rock. In addition, there is an ongoing debate about the meaning and usage of the term, which it is impossible to review properly here. For example, Bourman and Ollier (2002) have recently recommended using a less ambiguous term *ferricrete*, which simply denotes an enrichment in iron compounds, but Widdowson (2004) urges us to distinguish between *in situ* formed laterite and ferricrete as two fundamentally different types of iron-rich duricrusts.

An alternative approach to deep weathering products is based on the geochemistry of weathered rock and has been championed mainly in France. For example, Tardy (1971) distinguishes between the following processes:

- allitization, in which all basic cations and most of the silica is removed. Hence the product is dominated by Fe and Al hydroxides. The proportion $SiO_2:Al_2O_3$ is below 2;
- monosiallitization, in which silica is partially removed and kaolinite is formed ($SiO_2:Al_2O_3 = 2$);
- bisiallitization occurs when most of the silica remains in weathering products, 2:1 clay minerals originate, and some of the cations are retained. The proportion $SiO_2:Al_2O_3$ is above 2.

Weathering mantles dominated by the effects of allitization would be equivalents of laterites, whereas kaolins are products of prevailing monosiallitization. In bisiallitization the original fabric is usually retained and the proportion of altered minerals is minor; hence a transitional nature between grus and clayey saprolite may be expected.

Grus

Grus (Fig. 2.4; Plate 2.14; Plate II) is a product of rock disintegration, which is very characteristic of granite, especially of its coarse textural variants. In fact, a number of authors tend to restrict the usage of this term to granite, for example Bates and Jackson (1987: 294), who propose that grus is 'the fragmental products of in-situ granular disintegration of granite and granitic rocks'. Other descriptive terms present in the literature such as 'gravelly weathering (mantles)' (Lidmar-Bergström *et al.*, 1997), 'sandy weathering' (Bakker, 1967) and 'arenaceous weathering' (Power and Smith, 1994) fulfil the criteria for grus. The French term *arène* is considered as an equivalent of grus (Godard *et al.*, 1993, 2001) and older English literature frequently refers to 'growan', which is a local name to describe grus-type weathering products in the granite uplands of south-west England (Brunsden, 1964). However, it needs to be noted that the term is also used by sedimentologists to describe products of accumulation of weathering-derived, poorly sorted angular quartz and feldspar grains, subjected to very limited transport and forming a sedimentary veneer around an outcrop (Hoskin and Sundeen, 1985).

Although most early reports about the occurrence of grus come from the mid-latitudes of western and central Europe and North America, grus is much more widespread (Migoń and Thomas, 2002; Fig. 2.7). Its presence in arid and semi-arid areas of middle latitudes, as well as in semi-arid to humid subtropical and tropical environments, demonstrates that it is very difficult to relate grus to any narrow range of climatic conditions, or to any particular climatic zone, as once suggested by Bakker (1967) in the context of the interpretation of palaeoweathering remnants. Sustained claims that grus is a product of deep weathering occurring in temperate climatic conditions, and hence characteristic of mid-latitudes (Sequeira-Braga *et al.*, 1990) oversimplify, and in fact contradict the reality.

It appears that the core of the problem with *in situ* grus is that it escapes simple explanation if characterized solely by grain size and textural features. To understand this phenomenon, lithological characteristics need to be considered alongside its relationships to other weathering products and local and regional topography. Evidently, grus forms in an early stage of bedrock weathering. Hence it may be expected to occur both below a topographic surface of young age, and near the weathering front. Indeed, deep weathering profiles derived from granite usually have grus of variable thickness at their base. This lower part of a profile, with limited mineralogical and geochemical change, is called 'saprock' by some authors (Robertson and Butt, 1997; Taylor and Eggleton, 2001). Likewise, a grus layer of variable thickness occurs around corestones, notwithstanding the climatic zone and general characteristics of a weathering mantle

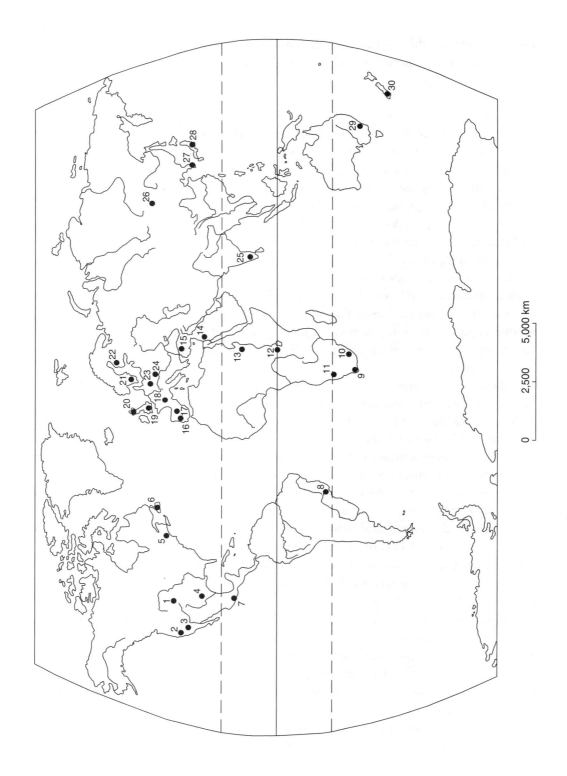

(Schnütgen, 1991). However, grus can also be envisaged as an end-product of weathering in certain environmental conditions unfavourable to advanced alteration. The latter hypothesis seems to reflect a view presented by many authors (e.g. Bakker, 1967; Eden and Green, 1971; Sequeira-Braga et al., 1990) who claim that temperate conditions inhibit more advanced geochemical changes within weathering profiles. Logically, the duration of weathering in such environments should be largely immaterial for grus.

Consequently, and theoretically, at any one site the presence of grus may be interpreted in one of three ways (Fig. 2.8). Grus could be (a) a geologically young material formed after the relevant topographic surface has developed, (b) an exposed basal part of a profile that suffered from significant erosional truncation, or (c) a remnant of palaeoweathering under temperate conditions. Therefore, its occurrence would be of geomorphic (cases (a) and (b)) rather than of climatic, or palaeoclimatic significance (case (c)). Most tropical to warm temperate occurrences of grus (Power and Smith, 1994; Taylor and Howard, 1999; Le Pera and Sorriso-Valvo, 2000) are recorded in uplifted and dissected areas with high rates of denudation, and hence they are consistent with case (a).

Weathering versus Hydrothermal Origin of Deeply Weathered Granite

It has long been recognized that alteration of granite can also be accomplished by interactions between rock and hot hydrothermal fluids, and that a hydrothermal origin of decomposed granite is an alternative to a weathering origin. Unequivocally diagnosing hydrothermal alteration is not however easy, because the effects of weathering may be overprinted on the effects of early hydrothermal alteration. It appears that isotope

Fig. 2.7 (opposite) Geographical distribution of deep grus weathering (selected examples): (1) Rocky Mountains (Eggler et al., 1969), (2) Sierra Nevada (Wahrhaftig, 1965), (3) Mojave Desert (Oberlander, 1972, 1974), (4) Texas (Folk and Patton, 1982), (5) northern Appalachians (Bouchard and Godard, 1984; Wang and Ross, 1989; Bouchard et al., 1995; O'Beirne-Ryan and Zentilli, 2003), (6) Newfoundland (LaSalle and De Kimpe, 1989), (7) Sierra Madre del Sur (author's observations), (8) Serra do Mar (Power and Smith, 1994; Thomas, 1995), (9) Cape Mountains (Thomas, pers. comm.), (10) southern Natal (author's observations), (11) central Namibia (author's observations), (12) Uganda (Taylor and Howard, 1999), (13) Sudan (Ruxton and Berry, 1961; Mensching, 1978); (14) south-west Jordan (author's observations), (15) western Turkey (Arel and Tuğrul, 2001), (16) Serra da Estrela (author's observations), (17) Serra da Guadarrama (Pedraza et al., 1989), (18) Massif Central (Flageollet, 1977; Coque-Delhuille, 1978, 1979), (19) Dartmoor (Eden and Green, 1971; Doornkamp, 1974), (20) Cairngorms (Hall, 1991, 1996), (21) southern Sweden (Lidmar-Bergström et al., 1997, 1999), (22) southern Finland (Kejonen, 1985; Lahti, 1985), (23) Harz (Hövermann, 1953; Wilhelmy, 1958), (24) Bohemian Massif (Jahn, 1962; Migoń, 1997a), (25) southern India (Brunner, 1969), (26) Mongolia (Kotarba, 1986), (27) Korean Peninsula (Oh and Kee, 2001), (28) Japan (Ikeda, 1998; Chigira, 2001), (29) Bega Basin (Dixon and Young, 1981; Ollier, 1983), (30) Southern Island (Thomas, 1974)

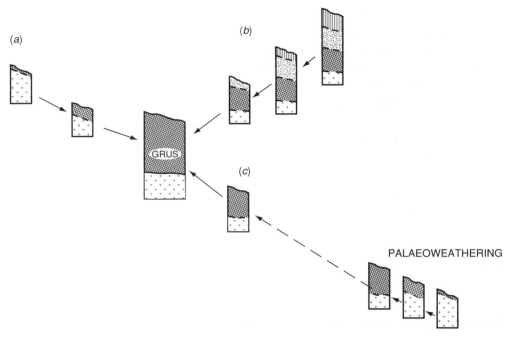

Fig. 2.8 Diversity of evolutionary pathways and interpretations of grus weathering profiles

geochemistry offers the only reliable clue to the origin of kaolinite (Gilg *et al.*, 1999; Table 2.7).

One of the much debated areas is south-west England, where extensive kaolinization of granite occurs. An exceptional thickness of kaolin deposits and their localized occurrence within the granite massifs of Dartmoor, Bodmin Moor, and St Austell has led to the widespread assumption that the kaolins are hydrothermal (Bristow, 1969). Later isotopic research demonstrated that a weathering origin of kaolinite is more likely (Sheppard, 1977), but initial hydrothermal changes (Bristow and Exley, 1994) or the contribution of deeply circulating groundwater, with temperatures around 70°C (Alderton and Rankin, 1983; Psyrillos *et al.*, 1998), are not totally excluded. Bristow (1998) has recently summarized the debate by proposing a multi-stage history of alteration processes as shown in Table 2.8. Another debate, about the origin of deeply grusified granite of the Bega Batholith in New South Wales, Australia, has not been satisfactorily solved (Dixon and Young, 1981; Ollier, 1983).

The overwhelming majority of studies on hydrothermal effects in weathered granite have little relevance to geomorphology, but the contribution of Coque-Delhuille (1988) is exceptional in this respect. She argues that the basin-and-hill topography of granite massifs in south-west England is a direct reflection of the extent of hydrothermal alteration affecting the intrusions. Topographic basins, such as the lower ground

Table 2.7 Criteria to distinguish between weathering and hydrothermal kaolinitic mantles

Diagnostic feature	Weathering origin	Hydrothermal origin
Isotope geochemistry	use of stable oxygen and hydrogen isotopes to construct $\delta D - \delta^{18}O$ diagrams	
	values fall near kaolinite line, indicating equilibrium with meteoric waters at 20°C	values fall significantly left from the supergene/hypogene (S/H) line
Mineralogical composition and zoning	occurrence of gibbsite	absence of gibbsite
	absence of high-temperature minerals such as pyrophyllite, diaspor and dickite	presence of high-temperature minerals such as pyrophyllite, diaspor, and dickite
		characteristic mineralogical zonation: inner kaolinite—alunite—pyrophyllite zone; outer illite—smectite zone
Clay textures	ambiguous: very fine-textured clays may form from fine-grained parent rock	
Fluid inclusions	scarce	abundant
	ambiguous: fluid inclusions occasionally exist in solid rock	
Chemical composition	ambiguous: trace elements do reflect temperatures of their formation, but also depend on the composition of parent rock	

Source: Modified after Gilg *et al.*, (1999).

south-east of Postbridge in central Dartmoor, are believed to have been excavated in areas where granite had been intensely fractured and weakened under the influence of hot fluids. But the situation is not entirely clear. Some basins have solid rock exposed within them, whereas some china clay deposits are located in elevated rather than low-lying areas, for instance in the western part of the St Austell upland. The advocates of a hydrothermal contribution to the origin of grus (Eggler *et al.*, 1969; Dixon and Young, 1981; Evans and Bothner, 1993) also point out that low-lying ground and topographic basins preferentially form if the rock is affected. Locally, however, interaction with hot fluids may lead to the development of quartz veins and replacement of some feldspar and mica by tourmaline, which strengthen rather than weaken the rock. Linear rock walls and angular rock spurs, such as the Roche Rock in St Austell granite, may form in such geological circumstances.

Thickness and Rate of Deepening of Weathering Profiles on Granite

It is very difficult to generalize about the thickness of weathering mantles derived from granites as the range is extremely wide. Selected data are presented in Table 2.9.

A few important observations emerge. Most significantly perhaps, it is clear that deep weathering of granite is not a phenomenon restricted to any particular climatic

Table 2.8 Alteration processes in the St Austell granite massif, south-west England

Phase	Process	Age (Ma)	Depth (km)	Temperature (°)	Heat source	Feldspar alteration
I	emplacement of biotite granite	290–285	3 (?)	500–600	magmatic	
II	1st phase of post-magmatic alteration and mineralization	285–275	2–3	200–500	magmatic	limited greisenization
IIIA	emplacement of lithium-rich and biotite granite in the W part of the massif	275–270	2–3	500–600	magmatic	
IIIB	2nd phase of post-magmatic alteration and mineralization	275–270	2 (?)	380–450	magmatic, some radiogenic	greisenization and tourmalinization
IIIC	emplacement of felsitic dykes	275–270	2 (?)	500–600	magmatic	
IV	1st phase of argillic alteration, origin of quartz veins	c. 240	1–2 (?)	250–350	radiogenic	alteration to smectite (Na) and illite (K)
V	2nd phase of argillic alteration, deep supergene alteration	180 to present	0.2–1.5	30–200	radiogenic	alteration to kaolinite, more efficient for Na-feldspar, and smectite to kaolinite
VI	chemical weathering	180 to present	0–0.5	20–50	high surface temperature	alteration to kaolinite

Source: After Bristow, (1998).

environment. Although humid low-latitude settings do host abnormally thick mantles of alteration (Thomas, 1994*a*), profiles in excess of 30–50 m are not totally absent from semi-arid areas, such as the south-west USA, or cool temperate areas of Scotland. Maximum depths to solid rock appear to typify the stable continental interiors of Australia and Africa, but highly dissected terrains may also exhibit very deep weathering, up to 80 m (Plate 2.17).

The thicknesses of weathering mantles formed at the expense of solid granite vary within a very broad range, from a mere few metres to more than 100 m. These differences reflect the influence of four key factors controlling the progress of weathering and the development of a saprolite: parent rock lithology, climatic conditions, particularly water availability, local relief, and associated relations between the rate of

Granite Weathering

Table 2.9 Thickness of weathering mantles recorded on granitoid rocks (selected examples)

Area	Type of weathering mantle	Maximum thickness (m)	Source
Europe			
Mourne Mountains (N Ireland)	arenaceous	10 m	Smith and McAlister, 1987
Scottish Highlands	arenaceous and clayey grus	10–20 m (locally > 50 m)	Hall, 1986
Dartmoor (SW England)	argillaceous	200 m[1]	Bristow, 1969; Smith and McAlister, 1987
	arenaceous	8–10 m	Eden and Green, 1971
Brittany (France)	argillaceous	60 m	Esteoule-Choux, 1983
Massif Central (France)	arenaceous and clayey grus	20–30 m	Flageollet, 1977; Coque-Delhuille, 1979; Pierre, 1990
Bohemian Massif (Czech Republic, Germany, Austria, Poland)	argillaceous	60 m (up to 100 m in fault zones)	Kužvart, 1969; Lippert et al., 1969; Kural, 1979; Störr, 1983
	arenaceous	15–20 m (locally up to 40 m)	Körber and Zech, 1984; Kubiniok, 1988; Migoń, 1997a
Southern Sweden	argillaceous	30–50 m	Lidmar-Bergström, 1989; Lidmar-Bergström et al., 1997, 1999
	arenaceous	10–15 m	Elvhage and Lidmar-Bergström, 1987; Lidmar-Bergström et al., 1997
Southern Finland	argillaceous	100 m	Sarapää, 1996
Iberian Range (Spain)	arenaceous and clayey grus	25 m	Martínez Lope et al., 1995
Serra da Estrela (Portugal)	arenaceous and clayey grus	10 m	Migoń and Vieira (in preparation)
Sila Massif, southern Italy	arenaceous and clayey grus	50–60 m	Le Pera and Sorriso-Valvo, 2000
Asia			
North-western Turkey	arenaceous and clayey grus	30 m	Arel and Tuğrul, 2001
South-east China	argillaceous	60 m	Lan et al., 2003
Hong Kong	argillaceous	40–50 m (exceptionally 90 m)	Ruxton and Berry, 1957; Shaw, 1997
Malay Peninsula	argillaceous	30 m	Hamdan and Burnham, 1996
Singapore Island	argillaceous	30 m	Pitts, 1984
Japanese Alps	arenaceous	30 m	Ikeda, 1998

(Continued)

Table 2.9 (Cont'd)

Area	Type of weathering mantle	Maximum thickness (m)	Source
Africa			
Sierra Leone	argillaceous	15 m	Thomas, 1980
Nigeria	argillaceous	50 m	Thomas, 1966
Uganda	argillaceous and arenaceous	60 m	Taylor and Howard, 1999
Eastern Sudan	arenaceous and clayey grus	10 m	Ruxton and Berry, 1961
South Africa	n/a	60 m	Brook, 1978
North America			
Laramie Range, Wyoming	arenaceous	60 m	Eggler et al., 1969
Sierra Nevada, California	arenaceous	20 m	Wahrhaftig, 1965; author's observations
Mojave Desert	arenaceous	30 m	Oberlander, 1972
Appalachian Piedmont	argillaceous and arenaceous	30 m	Pavich, 1985; Pavich and Obermeier, 1985
Northern Appalachians	argillaceous	30 m	O'Beirne-Ryan and Zentilli, 2003
	arenaceous	10 m	LaSalle and De Kimpe, 1989; Bouchard et al., 1995; O'Beirne-Ryan and Zentilli, 2003
Acapulco region, Mexico	arenaceous	10 m	author's observations
Puerto Rico	argillaceous	10–15 m	White et al., 1998
South America			
Guyana Shield	argillaceous	30–45 m	Eden, 1971; Kroonenberg and Melitz, 1983
South-east Brazil	argillaceous and arenaceous	120 m	Branner, 1896; Power and Smith, 1994; Thomas, 1994b, 1995
Australia			
Bega Basin	arenaceous	15 m	Dixon and Young, 1981
Yilgarn, western Australia	argillaceous	40 m	Mabbutt, 1961a

Note: [1] Initial rock weakening by hydrothermal alteration is envisaged.

profile deepening and the rate of surface denudation, and duration of weathering. If all these factors favour the development of a weathering mantle, then the resultant saprolites may approach 100 m thick and show thorough decomposition.

Weathering mantles which have a clay-rich horizon in their upper parts are generally thicker than grus mantles, but the latter may approach 50 m thick in specific circumstances. Such a figure is not uncommon in the Serra do Mar range in south-east Brazil

Granite Weathering 71

Plate 2.17 Deeply weathered granite in a mountainous subtropical environment, south-east Brazil (Photo courtesy of Michael F. Thomas)

(Thomas, 1994b, 1995). The wide presence of thick grus in this very humid (annual rainfall > 1,800 mm, up to 3,500 mm in places) environment is explained by the combination of very fast advance of weathering front favoured by structural circumstances on the one hand, but very efficient surface lowering by landsliding, gullying, and sheet erosion on the other, which causes geomorphic instability and inhibits more thorough geochemical and mineralogical changes. It is likely that thick grus present on the western slope of the Sierra Nevada, California, in the New Zealand Alps (Thomas, 1974), and in the Japanese Alps is similarly a specific response of the weathering system to the combination of high relief, sufficient water availability, appropriate lithology, and efficacy of sediment transfer across the topographic surface.

Table 2.9 shows also that weathering mantles occur in formerly glaciated areas of North America and northern Europe, contradicting a view that high latitudes equate with extensive tracts of bare rock, stripped of any pre-existing sediments or saprolites by ice sheets acting as giant bulldozers over hundreds of kilometres. In fact, weathering mantles can be very widespread there (Bouchard and Godard, 1984; Lahti, 1985; Lundqvist, 1985; Hall, 1986; Le Coeur, 1989; Wang and Ross, 1989; Lidmar-Bergström *et al.*, 1997, 1999; Bouchard and Jolicoeur, 2000; Fig. 2.9). The presence of a loose superficial cover of disintegrated rock was difficult to reconcile with glacial history and purported deep erosion brought about by ice sheets. Consequently, as reported by Kejonen (1985) for example, patches of grus used to be considered as products of postglacial Holocene

72 Granite Weathering

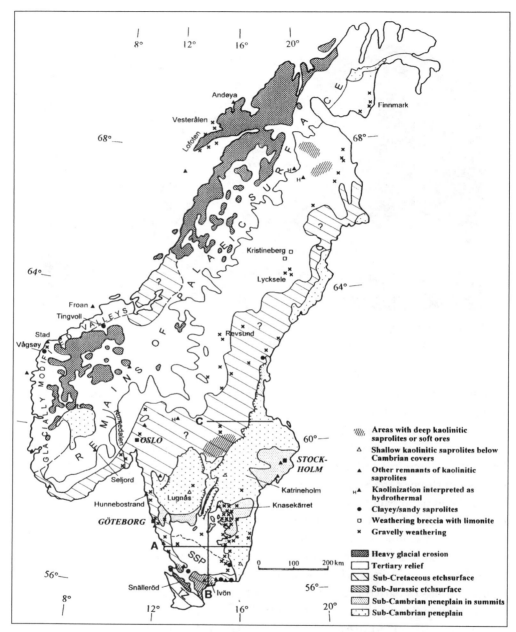

Fig. 2.9 Distribution of deep weathering sites in Scandinavia (after Lidmar-Bergström et al., 1999 Fig. 2)

weathering. Conceptual developments in glaciology and glacial geomorphology since the 1960s, and the introduction of the selective glacial erosion model in particular (Sugden, 1968, 1978), have led to a reappraisal of this view. Nowadays it is widely accepted that prior to the Quaternary high latitude shields were blanketed by a weathering mantle of variable thickness and the present-day saprolites are all but its remnants, truncated to various extents. Saprolites exceeding 15–20 m thick testify to the depth to solid rock that may have typified most of the shield terrains before ice sheets developed.

Unfortunately, not very much is known about the rates of deep weathering profile deepening on granite bedrock. The age of parent rock, although not uncommonly established with sufficient confidence, has little relevance to the onset of deep weathering, whereas the more meaningful age of batholith unroofing is usually either poorly specified or too distant to provide any reasonable time constraint for weathering studies. Mesozoic and Cainozoic marine transgressions may have trimmed weathering profiles but they failed to expose solid rock, hence these events do not zero the weathering system either. The survival of ancient saprolitic cover beneath marine sandstones and limestones is a well-documented phenomenon (e.g. Thomas, 1978; Lidmar-Bergström, 1989).

Mass balance calculations offer some insights, although the geographical coverage of such studies is highly biased towards the humid temperate eastern part of the USA (Cleaves et al., 1970; Velbel, 1985; Pavich, 1986, 1989). The results vary by one order of magnitude, from 4 m · Ma^{-1} to almost 40 m · Ma^{-1}. The highest rate of weathering front descent obtained so far has been found in the warm and humid subtropical environment of Puerto Rico and is 58 m · Ma^{-1} (White et al., 1998). Using these figures as guidelines, Thomas (1994a) suggests that the rates in the humid tropics may be even higher, and if this is correct, then many thick weathering mantles may have formed entirely during the Quaternary. However, the advance of a weathering front is likely to slow down with time, alongside increasing rock alteration, relief decline, and saprolite thickening.

Spatial Patterns of Deep Weathering

Regional Scale

The selectivity of granite weathering has already been emphasized in the context of surface weathering. This phenomenon also has important implications for spatial patterns of deep weathering and landform evolution in deeply weathered areas.

The downward advance of a weathering front is controlled by a number of factors, among which the mineralogy of the parent rock, the presence of structural discontinuities, and water availability are most important. Since the spatial distribution of each of these characteristics within any given landscape is far from uniform, the rates of weathering front advance may logically be expected not to be uniform either. Consequently, the thickness of a weathering mantle will be uneven and the depth to solid rock different. This situation is typical for many outcrops of weathered granite (Plate 2.18), but geological prospecting has revealed that thickness variability is a regional feature as well.

Plate 2.18 Uneven weathering front exposed in a former kaolin mine, Ivön, southern Sweden. The massive compartment on the right retarded the progress of weathering, which proceeded with ease in the adjacent rock, causing a rapid increase in thickness of the weathering mantle

A study near Jos in the Jos Plateau (Nigeria), underlain by fine-grained biotite granite of Jurassic age, has showed that the thickness of the weathering mantle varies from a few to more than 50 m over distances of a few hundred metres (Thomas, 1966; Fig. 2.10). The topography of the weathering front is thus very varied and includes closed and semi-closed basin-like features as well as localized highs where solid rock occurs at very shallow depth. More significantly, the relief of the weathering front does not match the contemporary surface relief developed across the weathering mantle, alluvial cover, and isolated outcrops of solid rock. In particular, there is no consistent relationship between the spatial patterns of weathering and drainage lines.

Hilly terrains appear to show even more pronounced variability in the thickness of weathering mantles. This has been inferred from observations in road cuts and at construction sites in areas such as the Serra do Mar range and the coastal strip in south-east Brazil (Thomas, 1995), but documented beyond any doubt in Hong Kong by Shaw (1997). His study area encompasses the Kowloon Peninsula and adjacent hills, 22 km² in total, and is based on more than 4,700 borehole records. Within this relatively limited area the thickness of weathering profiles varies from 1 m to more than 90 m, although thicknesses within the range 20–40 m are typical (Fig. 2.11). Again, dome-like rises and enclosed depressions typify the topography of the weath-

Granite Weathering 75

ering front. Furthermore, increasing thickness of the saprolite is associated with hill tops and low ground between the topographic highs, whereas the depth to solid rock below side slopes is smaller.

Grus mantles in higher latitudes may have variable thickness as well. In north-east Scotland 60 m thick saprolites co-exist with surface outcrops of solid granite (Hall, 1986). In the Fichtelgebirge, south Germany, grus is typically less than 10 m thick, but locally a weathering front has been encountered as deep as 40 m beneath the surface

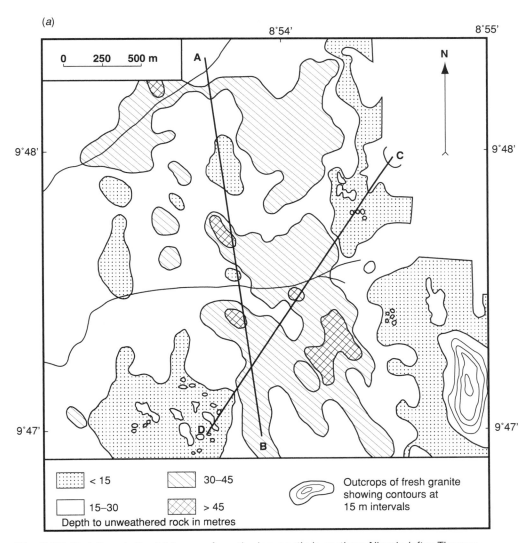

Fig. 2.10 Variations in the thickness of weathering mantle in northern Nigeria (after Thomas, 1966, figs. 2 and 4): (a) spatial pattern; (b) cross-sections (A–B, C–D). Saprolite is marked by dots, alluvial–colluvial fill is marked by lines. Solid bedrock beneath is left blank

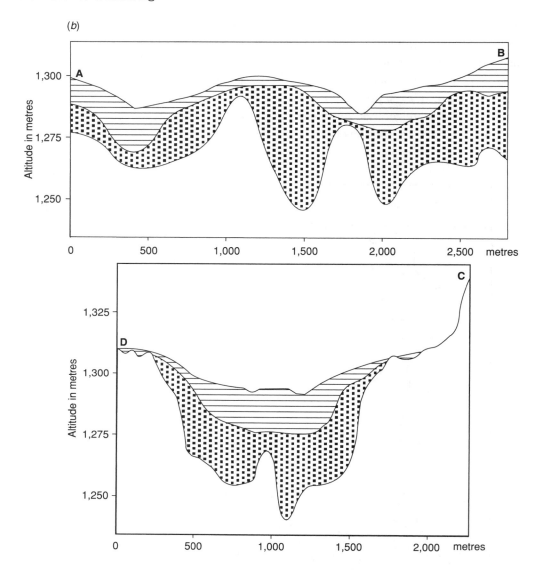

Fig. 2.10 (cont'd)

(Stettner, 1958). Wahrhaftig (1965) noted that in the Sierra Nevada, California, thick grus up to 20 m occurs within level surfaces separated by topographic steps cut in solid granite (Plate. 2.12).

The highest gradients in the thickness of weathering mantle are associated with fault lines, because rock is usually more fractured within fault zones and fluid circulation promoting alteration is therefore enhanced. For example, in the south-east part of the Bohemian Massif, the Czech Republic, the typical thickness of saprolite is 30–40 m,

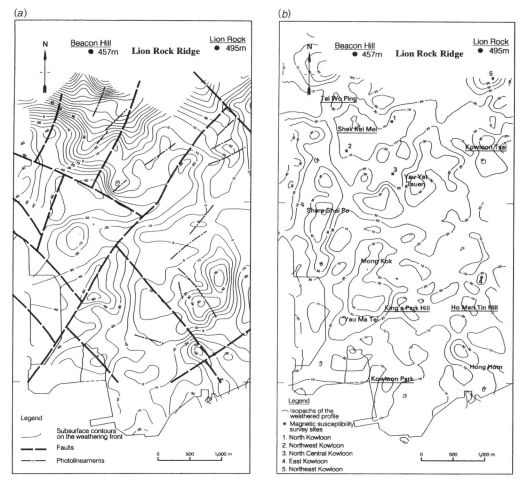

Fig. 2.11 Deep weathering patterns in Hong Kong (after Shaw, 1997, figs. 5 and 7): (a) contours of the weathering front; (b) thickness of the weathering mantle in metres

but values exceeding 100 m have been recorded along regionally important faults (Kužvart, 1969; Krýstková, 1971). In Dartmoor, there occur funnel-like depressions within the weathering front, in which the total thickness of china clay approaches 200 m (Bristow and Exley, 1994). Further examples are quoted by Ollier (1965) from south-east Australia.

Two important conclusions arise from these findings:

(1) The thickness of a weathering mantle may be highly variable and may change by tens of metres over distances of a few hundreds, or even a few tens, of metres.
(2) Current topography is not a reliable key to the topography of the weathering front. A gently rolling plain may conceal a very uneven weathering front.

Consequently, the removal of the saprolite would expose a solid rock relief which would be considerably different from the relief seen today.

Slope Scale

At the more localized slope scale, the thickness of a weathering profile may vary as well. One of the characteristic patterns involves the increase in depth to solid rock beneath the lower slope or in the footslope. This feature is explained by greater availability of moisture, as Wahrhaftig put it (1965: 1166): 'Where buried by overburden or gruss, the solid granitic rocks are moist for most of the year, and disintegrate comparatively rapidly to gruss; where exposed, the solid granitic rocks dry after each rain and therefore weather slowly'. The phenomenon of enhanced scarp-foot weathering has been recognized in different geographical settings, from the cool temperate zones of northern and central Europe to the semi-arid interiors of Africa and Australia. It has been formalized by Bremer (1971) under the name of 'divergent weathering'.

Twidale (1962, 1982) documented a number of examples from south Australia, in which excavations for reservoirs revealed a few metres thick zone of grus juxtaposed with plunging rock slopes of isolated granite hills. He also observed that the thickness of the weathering mantle on pediments does not increase steadily with increasing distance from the rock slope, but there is a belt of thicker saprolite immediately at the rock slope/pediment junction, followed by very shallow regolith interspersed with bedrock outcrops (Twidale, 1978*b*). Likewise, Mabbutt (1966) showed that the solid rock/saprolite interface at the foot of residual hills can be vertical rather than sloping. In the Karkonosze, south-west Poland, the thickness of grus varies from 2–3 m to 15–20 m and the highest figures are noted along the footslope of a distinct topographic escarpment bordering the entire massif (Migoń, 1997*a*; Fig. 2.12).

However, middle and upper slopes should not be uncritically thought of as slope sections where solid rock occurs in a very shallow subsurface. In many hilly terrains of humid and seasonally humid tropics the actual situation is reverse. Outcrops of solid or slightly disaggregated rock are best found along valley floors and watercourses, whereas higher slopes are underlain by a thick mantle of weathered rock.

Such relationships have been recognized in south-east Brazil (Power and Smith, 1994; Thomas, 1994*b*), Hong Kong (Ruxton and Berry, 1957; Shaw, 1997), New Zealand (Thomas, 1974), and Papua New Guinea (Pain and Ollier, 1981). In the southern Sierra Nevada, California, road cuts across steep slopes (20–30°) show deeply weathered granite, at least 15 m thick, with occasional corestones (Plate 2.15), whereas streams in V-shaped valleys not uncommonly flow around big boulders indicating the proximity to the weathering front. In all these examples, geomorphic considerations suggest that deep weathering of interfluves is not antecedent to linear erosion but proceeds concurrently with stream incision.

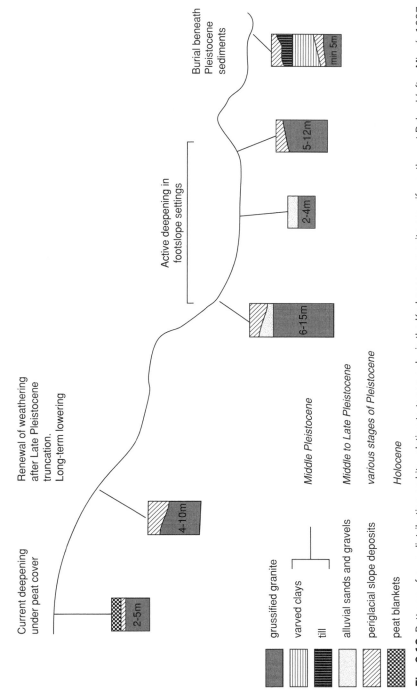

Fig. 2.12 Patterns of grus distribution and its relation to topography in the Karkonosze granite massif, south-west Poland (after Migoń, 1997a, fig. 3)

It is possible that these contrasting behaviours of slope weathering systems reflect environmental conditions crucial for the progress of deep weathering. In humid and warm areas, where abundant rainfall and high temperatures favour rock alteration but dense vegetation cover acts towards the reduction of surface denudation, rock breakdown may proceed rapidly enough to outpace erosion, even beneath moderately steep slopes. However, if water supply decreases and/or temperature falls, upper and middle slopes will no longer be suitable places for weathering profile deepening. Seasonal shortage of water also means more open vegetation, which offers less protection against erosion. Therefore, it is only less inclined footslopes receiving additional water through runoff from upper slopes which allow the weathering profile to deepen. Structural factors and human interventions (e.g. vegetation clearance) may, however, significantly alter the balance between weathering and slope denudation and change the proposed relationships.

Evolutionary Pathways of Weathering Mantle Development

Weathering mantles are not static but evolve and hence change, over time. Their evolution may be analysed from different, not mutually exclusive, points of view. Geochemical and mineralogical evolution of weathered rock is the subject of a voluminous literature, in which distinctive geochemical trends as well as transitional and end-products of weathering have been identified, depending on parent lithology, site drainage conditions, temperature, and rainfall (see Tardy, 1971; Tardy *et al.*, 1973; Gerrard, 1994*b* and Taylor and Eggleton, 2001, for more detail).

From a geomorphological point of view, the changes in thickness and position of weathering mantles over time are most important, yet these have long been given insufficient attention. This is surprising, given the long-standing interest of soil scientists in catenary relationships among soils, but is partly explained by the necessity to have recourse to much longer time scales than those involved in soil production on the one hand, and to view weathering mantle three-dimensionally on the other. Ollier (1976) suggested some relationships between relief, saprolites, and ferricretes, but long-term evolutionary pathways in the development of deep weathering mantles have been most comprehensively outlined by Thomas (1989*a*, 1989*b*, 1994*b*). The suggested categories include the following:

(a) *profile lowering* occurs if the weathering front advances downward at a rate similar to the rate of surface lowering by denudation. Over time, the weathering column descends as the topographic surface is lowered, but its thickness remains unchanged and no widespread exposure of solid rock takes place;
(b) *profile deepening* signifies the environmental conditions under which weathering penetrates into the rock at a higher rate than that of surface denudation. If this is

the case, the thickness of a weathering mantle will increase with time. This profile behaviour does not imply a static land surface, nor any particular rates for two concurrent processes. It may occur under planate tropical shields, where minimal relief prevents denudation, but also in dissected areas with a widespread duricrust cap;

(c) *profile collapse* would be characteristic of very advanced weathering, so that even relatively immobile iron, aluminium, and silica are leached and lost in solution. The thickness of the weathering mantle would decrease, but not because of efficient mechanical surface denudation;

(d) *profile thinning* occurs if the descending weathering column encounters a highly resistant rock compartment at depth, for example an extremely massive core of granite. Then, despite little erosion, the profile becomes thinner because further downward advance of the weathering front is considerably slowed down;

(e) *profile truncation* takes place if surface erosion accelerates, usually because of an environmental crisis, natural or human-induced, and upper portions of pre-existing saprolites are removed. Agents of erosion are different and include slope wash and gullying, landsliding, glacial stripping, solifluction, and waves. The presence of truncated profiles is indicative of a relatively recent change in balance between weathering and erosion, but to determine if a particular profile has been truncated is not always straightforward;

(f) *profile burial* will happen if material becomes deposited on top of the weathering mantle and where no, or little, erosion has occurred. If the overburden is sufficiently thick or impermeable, the pre-existing profile may be effectively sealed, so that no further geochemical changes take place and the mantle is entirely relict. However, sealing must not be automatically assumed (Pavich and Obermeier, 1985);

(g) *profile renewal* denotes a situation in which, after a period of truncation, weathering becomes more efficient again and the weathering column begins to thicken.

The geomorphological relevance of the above classification becomes clear if one realizes that profile behaviour is critically dependent on the balance between the descent of the weathering front, itself very much controlled by the local topographic and hydrogeological factors, and surface lowering by denudation. Linking profile behaviour and landscape evolution is one of the potentially rewarding avenues for further research.

Economic Significance of Deeply Weathered Granite

The phenomenon of deep weathering is also significant from an economic point of view. On the one hand, the occurrence of weathered granite is a challenge to engineers and planners who often have to deal with rock which is not solid but significantly weakened and therefore very unstable. Ollier (1977) has reviewed selected case studies and their implications in the context of dam and tunnel construction, road building, and foundation engineering. How pre-weathering of rock controls the patterns of mass movement in granite terrains will be shown in Chapter 5.

On the other hand, however, the phenomenon of weathering can be beneficial. Weathered granite is used as a raw material in the ceramic industry, for primitive house building if ferruginized, or as an aggregate for road building in many countries around the world. The volume published on the occasion of the Twenty-Third International Geological Congress in Prague (*Kaolin Deposits of the world*, 1969) and the later IGCP 23 'Genesis of Kaolins' programme reports (e.g. Störr *et al.*, 1977), although now rather dated and restricted to operating mines, provide a good source of reference about the extent of kaolin quarrying in the world and Europe, most of it derived from alteration of granite. The use of partially weathered granite as a building stone may also be mentioned in this context. The reddish colour of ferruginized granite blocks gives a distinct flavour to monumental architecture, for example in old medieval Spanish towns such as Avila or Toledo.

3

Boulders, Tors, and Inselbergs

Boulders, tors and inselbergs (Plates III, IV, V) are regarded as the most characteristic individual geomorphological features of granite landscapes and it is their assemblages extending over large areas that give granite terrains their unmistakable appearance. Although none of these landforms is unique to granite, nor even specific to basement rocks, it is perhaps true that the most astounding ones occur within granite areas.

Twidale (1982) in his *Granite Landforms* considered boulders and inselbergs as two key individual components of granite landscapes and devoted to them almost 100 pages, whereas the other major landforms received only 35. Likewise, rock-built residual hills figure prominently in *Klimamorphologie des Massengesteine* by Wilhelmy (1958). In the voluminous literature about inselbergs, papers focused on those developed in granite evidently prevail (see the reviews by Kesel, 1973 and Thomas, 1978). Likewise, granite tors, especially in classic areas such as Dartmoor (Gerrard, 1994a) do not cease to attract the attention of geomorphologists.

The unifying characteristic of all three landforms considered in this section is that they are essentially outcrops of solid rock rising above a surface cut across a weathering mantle, even if the thickness and lithology of the weathering mantle may be very variable. Outside arid areas there are very few examples of tors and inselbergs, surrounded by a rock-cut platform. Therefore, the discussion about their origin and significance has inevitably been tied to the increasing recognition of the significance of deep weathering. Twidale (1981a, 1982, 2002) reviewed many early accounts and concluded that selective subsurface weathering and subsequent exposure of unweathered cores to form boulders and inselbergs had been appreciated as early as the end of the eighteenth century. Nowadays, there is little doubt that the majority of individual medium-scale granite landforms are due to selective subsurface weathering.

Problems of Terminology

Before the presentation of residual granite landforms commences, a few terminological issues need to be raised. Although it may appear that the distinction between boulders, tors, and inselbergs is a simple task, it is in fact not at all straightforward. Boulders are rather easy to define, but there is a great deal of confusion in the literature concerning the definition of a tor, the boundary between a tor and an inselberg, and the criteria to define a residual hill as an inselberg.

Boulders are defined in sedimentology as detached rock masses with a diameter greater than 256 mm and somewhat rounded in the course of transport, whereas blocks are rock fragments of the same minimum size, but angular because of no or limited transport (Bates and Jackson, 1987) (Plate III). Twidale (1982: 6) offers a less strict definition which does not imply movement, stating that a boulder is 'a more-or-less rounded mass standing either in isolation or in groups or clusters on plains, in valley floors, on hillslopes or on crests of hills. Granite boulders ... vary in degree of roundness, and in diameter from 25 cm to some 33 m, though the mode is of the order of 1–2 m'. The essence of this statement appears to be the minimum size and the absence of primary joints within the rock mass.

Inselberg is a German term, which literally means an 'island hill', but has found its way into English terminology in unchanged form. It was coined by a German naturalist and traveller from the turn of the nineteenth century, Walter Bornhardt, who was impressed by isolated rock-built hills rising steeply, as if islands from the sea, from the savanna plains of East Africa. There have been many attempts to define an inselberg. According to Twidale (1968: 556), inselbergs are 'residual uplands which stand in isolation above the general level of the surrounding plains in tropical regions', whereas Young (1972: 205) states that they are 'steep-sided isolated hills rising relatively abruptly above gently sloping ground'. The geographical restriction of inselbergs to tropical environments appears unnecessary, as apparently admitted by Twidale (1982) himself some years later. Indeed, isolated hills rising steeply from a surface of low relief are as common in the Tropics, as they are in arid lands, some temperate areas, and even in subpolar environments, for example, in northern Finland (Schrepfer, 1933; Kaitanen, 1985) (Plate V).

But it remains a difficult task to define the isolation and abruptness of the slope/plain junction. A few authors have attempted to offer quantitative indices. Thus, Faniran (1974) suggested a minimal distance of 0.8 km to the nearest neighbour, a minimum height of 15 m, and 25° as the minimum angle of the steepest slope. Young (1972) is more liberal, accepting 20° as the minimum value, whereas Kaitanen (1985) distinguishes a 'pure' inselberg, which needs to have at least 95 per cent of slopes inclined 18° or more and at least 50 per cent of slopes to be inclined in excess of 27°. The piedmont angle ought to be at least 5°, and that of the upper piedmont should not exceed 12° (Kesel, 1973). The measures are all purely arbitrary and may be difficult to apply in the real world, where a distinct piedmont angle may not occur. Thomas (1967) noted great variability of form around a single, rather low granite inselberg near Oyo in Nigeria and concluded that advancing more restrictive criteria for inselbergs is not realistic. However, a large degree of freedom in deciding which hill is to be called an inselberg makes comparative analyses of inselberg landscapes difficult.

Another source of persistent confusion is the relationship between the two terms: an inselberg and a bornhardt. The latter are 'dome-shaped, steep-sided hills with bare convex slopes covered with very little talus and flattened summit surface' (Migoń, 2004a: 92). The term itself was first used by Willis (1936) with reference to particularly massive, dome-shaped inselbergs, and also to honour Walter Bornhardt who had introduced the inselberg. Subsequently, the term evolved to describe monolithic domes regardless of their isolation in the landscape (Twidale and Bourne, 1978). The German term *Glockenberg* (Bartels, 1975), meaning a bell-shaped hill, is an equivalent. Thus, whereas in 'inselberg' the emphasis is on the landscape setting, it is on the hill form in 'bornhardt'. There is partial overlap between the two and some inselbergs fulfil the criteria for a bornhardt, but there exist granite dome-shaped hills, thus bornhardts, which are not inselbergs (e.g. domes in the Sierra Nevada, California; Gilbert, 1904; Huber, 1987; Ericson and Olvmo, 2004), and inselbergs which are by no means bornhardts (e.g. conical hills in central Arizona; Kesel, 1977). The other morphological forms of inselbergs include block-strewn residuals (or 'nubbins') and castellated shapes (or 'castle koppies') (Twidale, 1981b), and other types more typical of non-granitic lithologies such as conical hills (Young, 1972).

But the most ambiguous term of all three is the 'tor'. It is of local Cornish origin but derives from the Old Welsh *twr* or *twrr*, meaning a heap or pile. It was long used to describe castellated granite outcrops rising from the moorlands of Dartmoor and Bodmin Moor in south-west England (Plate IV), but formally defined by Linton (1955: 470) as 'solid rock outcrops as big as a house rising abruptly from the smooth and gentle slopes of a rounded summit or broadly convex ridge'. There is a similarity with the definition of an inselberg in the emphasis on abrupt margins, but evidently both the setting and the size of a tor are different. Tors do not need to rise from plains, and certainly they do not do so in localities such as Dartmoor. Its upland surface does locally have a very gentle slope, but certainly it is not a plain. Tors would not exceed 10–15 m, whereas inselbergs are much higher. Also, tors are expected to show compartmentalization into joint-bounded blocks and boulders, whereas inselbergs, especially dome-like ones, are more massive. With broadly this meaning the term 'tor' has been used in subsequent studies of rock residuals from all around the world, not all of them necessarily granitic (e.g. Demek, 1964; Godard, 1966; Cunningham, 1969; Derbyshire, 1972; Jahn, 1974; Gibbons; 1981; Ballantyne, 1994; Stroeven *et al.*, 2002; André, 2004).

In the context of this discussion a Spanish terminology of granite residuals is worth presenting, as it encompasses more categories than the English language appears to allow. It emphasizes size and fracture control as mutually dependent parameters (Pedraza *et al.*, 1989). Four major types are distinguished, namely *lanchas y bloques dispersos* (= tors), *pedrizas* (= ruiniform products of dome disintegration), *berrocal* (= domes with superimposed boulders) and *domos* (= domes). Going from the former to the latter, fracture density decreases and the size of the landform increases.

But not all geomorphologists seem to agree with the above descriptive meaning of a 'tor'. For example, Thomas (1965: 64) stated that tors 'can be described simply as groups of spheroidally weathered boulders, rooted in bedrock' and, more significantly, that 'a castle kopje (angular outcrop) is a distinct landform which also has a different origin'. In his later paper (1976), he retains this distinction, but also admits that tors may assume three main morphological types: tower-like, tabular, and hemispherical. What these forms have in common, and what renders them different from castle koppies, is their origin within a deep weathering profile. Thus, we have an attempt to restrict the definition of a 'tor' to rock outcrops formed in a particular way, as envisaged by Linton (1955) for the tors of Dartmoor. However, difficulties in recognizing the formative process from residual landforms on the one hand, and similarity of the end-forms despite different formative processes on the other, rendered such a restriction unnecessary and it was eventually abandoned. Today, tors are considered as good examples of equifinality.

Perhaps more interestingly from the point of terminological discussion, Twidale (1982) in his *Granite Landforms* does not consider tors as a separate group of landforms at all. He argues that 'features similar to the tors in southwestern England occur in different and various climates in other parts of the world are commonly called inselbergs and in particular castle koppies (*sic*)' (1982: 17). Further down he claims that 'if the tors of England occurred anywhere else they would be referred to as inselbergs and most of those of Dartmoor, for example, as castle koppies'. This assertion is difficult to accept, given the definition of an inselberg given by Twidale himself and the significant number of papers about tors from outside England. In addition, the photographic examples of castle koppies provided (Twidale, 1982:

170–1) are from high mountains (the Pyrenees), medium-high mountains (the Massif Central), and uplands (Dartmoor), neither of which fulfils any criteria for a plain, from which inselbergs need to rise. Consequently, castle koppies cannot be regarded solely as derivatives of inselbergs, but can evidently occur in topographic settings other than plains. Furthermore, the list of tor examples from other areas in Britain (Linton, 1955, 1964) leaves no doubt about Linton's intentions to restrict the term to residuals of roughly this size and location and nowhere does he suggest that inselbergs and tors would be synonyms.

Therefore, there are good reasons why the term 'tor' is still in use, although with a descriptive rather than genetic meaning. It covers all those residual landforms rising from a regolith-veneered surface or rock platform, whether flat or sloping, extensive or local, low-lying or elevated, which are composed of more than one individual compartment (boulder), but are too small to be distinguished as separate hills. Hence, a tor is a part of a slope, whereas an inselberg possesses slopes itself. In addition, tors and inselbergs are bedrock-rooted, whereas boulders may be completely isolated. Some proposals of terminological clarification are presented in Table 3.1.

Table 3.1 Suggested terminology of tors and inselbergs

Dominant fracture pattern	Tors	Inselbergs
curved (sheeting) fractures	hemispherical tor	dome domed inselberg
orthogonal joints	castellated tor castle koppie	castellated inselberg
flat-lying joints	lamellar tor	—
fracture pattern obscured	bouldery tor nubbin	bouldery inselberg block-strewn inselberg

Nevertheless, in specific circumstances, for example on a plain, there remains a problem of size and acceptable threshold values between various categories of residuals. One would probably agree that boulders are generally smaller than tors, and tors are smaller than inselbergs. Therefore boulders can be superimposed on tors and inselbergs, and tors can occur on the slopes of inselbergs. However, this is not a universal relationship. Monstrous boulders in excess of 10 m are known and these are higher than many tors, which could be a mere few metres high. On the other hand, very tall ($>$ 25 m) tors may rise higher than some inselbergs. However, such very tall tors seem rare and the boundary value, although arbitrary, should probably be around 25 m.

Boulders

Distribution and Structural Control

As Twidale (1982: 89) put it, granite boulders 'are certainly the most numerous and widely distributed of the positive relief forms developed on granite'. Indeed, isolated boulders or their clusters can be found in a wide variety of geographical settings, from the cool northern latitudes of Finland and Sweden, through to the arid landscapes of the Sahara and Australian deserts, and to humid tropical hilly terrains in equatorial Africa (Plate 3.1). In addition, many granite coasts are typified by boulder accumulation along the shores. It is perhaps only areas of vigorous glacial erosion and Alpine mountainous landscapes where rounded boulders are a rarity and bare rock-cut plains and hillslopes dominate.

Boulders, despite their superficial similarities, may have different histories. Some occur *in situ* and may simply be visualized as more resistant bedrock compartments (corestones) left in place after the surrounding rock mass has been eroded away. Others have been detached from an original outcrop and subjected to transport to the place of their present rest. The distance they have travelled, and the means of movement, may have varied. In certain cases, boulders occur in footslope settings and these have probably fallen and rolled down the local slope. But erratic boulders present in formerly glaciated terrains may have travelled much longer distances, up to hundreds of kilometres.

Plate 3.1 Granite terrain dominated by boulders, Aubrac, Massif Central, France

Granite boulders show a large variety of sizes and shapes, even if they occur adjacent to each other. For *in situ* boulders, it appears that no systematic recording of their diameter for statistical purposes has ever been attempted, but Twidale (1982) is probably right in saying that the majority are 1–2 m long. However, examples of much larger boulders are known. The Leviathan boulder in the Mt Buffalo area, Victoria (Australia) is perhaps the biggest boulder of the world, being 33 m long, 21 m across, and 12 m high (Twidale, 1982). In the Erongo Massif, central Namibia, at the Bull's Parties locality, there occurs a group of massive boulders, some of them split into two or more pieces, measuring well over 10 m across (Migoń and Goudie, 2003). Boulders dotting the elevated plateau of Serra da Estrela in Portugal are up to 10–15 m long (Plate 3.2). Another example that may be cited is Huelgoat in Brittany, France, where immense boulders of up to 12–15 m long occur interlocked within a valley deeply incised into the plateau, to form the ceilings of boulder caves located beneath.

Enormous sizes can be attained by granite boulders which have been moved by glaciers. Erratics in Estonia are not uncommonly more than 10 m long and the Kabelikivi boulder near Tallinn is 19 m long, 7 m tall, and has a volume of 728 m^3 (Raukas, 1992). Erratic boulders further away from the centres of Scandinavian ice mass dispersal, in northern Poland and northern Germany, are smaller, but still impressive. The largest erratic in northern Poland, Tryglaw in the village of Tychowo, is only partially exposed but even in the exposed part it is 13.7 m long and 3.8 m tall.

Plate 3.2 One of many enormous boulders on the high plateau of the Serra da Estrela, Portugal. Note a person standing next to the boulder

An indirect indicator of the large size that granite boulders may attain is the length of rock fragments used to build prehistoric monuments, for example megaliths in Brittany, France. Boulder dimensions may approach 10 m in length. The Neolithic age of these structures, and the rounded shape of the boulders themselves, eliminate the possibility that they may have been chiselled out of a quarry face. Rather, existing boulders, perhaps corestones protruding from the ground, were used to build stone rows and dolmens, and erected as standing stones.

The variety of sizes of granite boulders reflects the variability of primary fracture spacing within granite masses, as the boulders are essentially devoid of primary discontinuities. We have seen above that fracture density tends to vary with lithology, being smaller in coarse granite variants and in post-kinematic settings, and this relationship influences the sizes of the boulders. Areas underlain by coarse, potassium-rich, post-kinematic granites are typified by the occurrence of huge boulders standing in isolation or in clusters, whereas slopes and plains developed in fine-grained granite bear a mantle of distinctly smaller and more angular boulders, rarely exceeding 1 m across. The former can be exemplified by outcrops of the Jurassic Spitzkoppe Granite in central Namibia that postdates the break-up of Gondwana. There, boulder diameters in excess of 10 m are not uncommon (Plate 2.7). By contrast, boulders built of the much older Precambrian Salem granite, which suffered from many episodes of rock deformation, are rarely more than 2 m long (Migoń and Goudie, 2003).

Origin and Significance

The origin of isolated granite boulders is usually associated with selective subsurface weathering guided by fracture patterns and concentrated along these fractures. Ongoing weathering leaves unweathered cores between the fractures, set in a mass of disintegrated and decomposed rock, whereas subsequent removal of the weathered mass through erosion leads to the exposure of the corestone at the topographic surface. Numerous outcrops of deeply weathered granite provide convincing evidence that the separation of a corestone from the saprolite can be accomplished entirely in the subsurface (Plate 3.3), hence the evolutionary pathway of a boulder is easily visualized as a two-stage process. The first stage involves subsurface weathering and the development of a corestone; the second stage is the excavation of the corestone to form a boulder.

In the modern literature, the two-stage theory of boulder origin is usually linked with the name of D. L. Linton (1955), although his seminal work focused on tors rather than on boulders. However, the essence of the story had already been captured more than 200 years ago, by a French naturalist, J.-C. Hassenfratz, who observed scattered boulders in the Massif Central and inferred their excavation from a mass of otherwise disintegrated rock in his paper published in 1791 (Twidale, 1978a, 1982). Similar

Plate 3.3 Corestones of various sizes and shapes co-existing in one outcrop, Harz, Germany. The least weathered corestones cluster in the uppermost part of the profile (Photo courtesy of Karna Lidmar-Bergström)

interpretations have subsequently been put forward by Jameson (1798) from the Isle of Arran in Scotland, MacCulloch (1814) and de la Beche (1839) from Dartmoor, England, and above all, by Logan (1849, 1851) from Malaya (all citations after Twidale, 1978a, where exact quotations can be found).

However, stripping of the saprolite may remain incomplete and the excavation of a boulder can be partial. Gently rolling terrains with half-exposed corestones, forming isolated hemispherical outcrops of solid rock rising from grasslands, typify many granite uplands in Central and Western Europe (Plate 3.4). In the Bohemian Massif these boulder fields are locally called *kamenné stado*, translated as 'stone herd'. Again, this is a climate-independent phenomenon. Grassy granite slopes in the coastal part of Sierra Madre del Sur in Mexico are dotted with partially exposed rounded rock compartments. But it is difficult to state the degree of corestone separation in the

Plate 3.4 Partly exposed boulders on hillslopes, Sierra Madre del Sur, Mexico

subsurface. Boulder-like outcrops may in fact be elements of a partially stripped weathering front and depending on the balance between weathering and erosion they may or may not evolve into true boulders.

Fracture spacing determines the maximum size and overall shape of boulders, whereas their roundness and the actual size appear to reflect the duration and/or intensity of subsurface weathering. In a conceptual model of deep weathering, an unweathered but fractured rock mass is first attacked by weathering along structural discontinuities. At this stage, unweathered compartments between fractures are almost cubic and it is only along the fractures that narrow bands of disintegrated rock exist. With the progress of weathering, the zone of disintegration extends into the solid rock and the dimensions of unweathered cores diminish. Core margins and edges suffer most from weathering attack. Therefore they become more and more rounded. Not uncommonly, the rock begins to show evidence of spheroidal weathering and rounding is aided by the development and detachment of onion-skin concentric layers (Ollier, 1967). At a very advanced stage of weathering, the core may attain an almost perfectly spherical shape (Fig. 2.6). Thus, one could assume that rounded boulders have undergone a long period of subsurface weathering, while in the case of the more cuboid boulders the progress of deep weathering has been arrested and boulder excavation occurred rather early. Similarly, a smaller size of boulders might be expected if a long period of deep weathering preceded the stripping.

This model linking form and time has to be treated with caution, though, because lithology and rock structure may significantly alter the progress of weathering. It is true that at an early stage disintegration tends to proceed preferentially along discontinuities, as numerous exposures of shallow weathering in recently deglaciated terrains show. However, in the course of further weathering various pathways and effects are possible, although the reasons are not fully understood. First, there are saprolites of an immature grus type which lack corestones, indicating that a weathering attack proceeds across the weathering profile with similar intensity. In this case, stripping of the saprolite would not leave any boulders. Second, spheroidal weathering leading to spherical shapes does not appear to be associated with any particular style or stage of weathering. Spherical corestones and onion-skin shells around them have been reported from both tropical saprolites in Brazil and Venezuela (Schnütgen, 1991) and shallow grus in the Fennoscandian Shield (Lidmar-Bergström *et al.*, 1997). Third, different corestone shapes may form in the same period, depending on fracture patterns, groundwater availability, and possibly minor mineralogical differences (Plate 3.3). Therefore, inferences about landscape evolution and its duration from boulder shape and size may prove very misleading. Boulder roundness does not indicate advanced weathering and rounded boulders in high latitudes cannot be uncritically taken as a tropical inheritance.

Referring to the classic model of deep weathering profile by Ruxton and Berry (1957), one can also assume that boulder shape and size would indicate their original position within the weathering profile. Less stripping would expose the upper part of the corestone zone, hence the boulders would be rounded and of smaller size, whereas deeper erosion would result in the excavation of larger, more cuboid compartments closer to the solid rock surface. But there are too many field examples of corestones clustering in the upper horizons of weathering profiles, apparently contradicting the scheme developed by Ruxton and Berry. The occurrence of big boulders at the topographic surface does not imply that the weathering front is close to it; they may be underlain by a many metres thick zone of uniformly decomposed rock, with no corestones at all (Plate 2.15). In fact, massive boulders, once exposed, may become very durable landforms, highly resistant against surface weathering. It may be hypothesized that in the course of long-term weathering and stripping boulders will passively settle down concurrently with general landscape lowering, to form eventually boulder blankets mantling the surface.

Although the two-stage theory appears to be the most plausible explanation of boulder origin, selective subsurface weathering is probably not the necessary precursor of boulder emergence. Arid granite landscapes are not uncommonly dominated by boulders of various size, but the evidence for deep weathering is usually lacking. In certain cases, recourse to the history of environmental change in the late Cainozoic can be made to explain the presence of boulders and a period of deep weathering under more humid conditions followed by a period of stripping after a shift towards aridity can be inferred. The south-western Mojave Desert in California is a locality where such a conceptual

94 Boulders, Tors, and Inselbergs

model is supported by field evidence (Oberlander, 1972). Lava flows dated at 8–9 Ma buried thick weathering mantles with corestones, which originated before the coastal mountains rose high and the Mojave turned into a desert. In places lacking a protective blanket of lava, loose saprolite has been eroded away, leaving boulders and tors exposed.

However, this model can hardly be transferred to the Namib Desert, which has a much longer history of aridity than North American and North African deserts, spanning at least the period since the inception of the Benguela Current in the Miocene, some 10 Ma ago (Goudie and Eckardt, 1999). Ward *et al.* (1983: 182) even say that '... the Namib tract... has not experienced climates significantly more humid than semi-arid for any length of time during the last 80 million years'. Occasional claims for a relatively recent (late Cainozoic) deep weathering have been disproved by Ollier (1978*b*), and at present the near-surface zone of rock disintegration is less than 1 m thick. Nevertheless, boulders do occur, both in the hyperarid coastal belt, such as at Gobabeb, as well as in the marginal desert, in the Spitzkoppe area (Migoń and Goudie, 2003). The granite landscape around Gobabeb is dominated by spherical and half-spherical outcrops rising from a rock-cut plain. They are 1–2 m high and 1–3 m long, hence of typical boulder size. But amidst the boulders, there are commonly bedrock convexities present, less than 0.5 m high, which are probably the top parts of boulders to be exposed. Rounding appears here to proceed concurrently with the development

Plate 3.5 Granite boulder fields near Gobabeb, Central Namib Desert, develop without any significant contribution of subsurface weathering. Flaking and exfoliation work towards rounding the boulders, whereas splitting acts in the opposite direction

of shallow grus. Exposed granite surfaces appear subject to efficient superficial weathering, probably caused by rock temperature changes and moisture change related to fog advections, and accomplished mainly by flaking and exfoliation (Plate 3.5). By these mechanisms, initial convexities become enhanced and joint-bounded cuboid granite compartments assume a classic boulder shape. Their further evolution includes case hardening, tafoni development, and splitting along vertical cracks, all leading to the destruction of boulders (Migoń and Goudie, 2003).

Boulders *in situ* may occupy different settings within a landscape. Some cluster on and around local hilltops, being by virtue of their massiveness and close juxtaposition responsible for the very existence of these hills. They may have been exposed as a group of isolated boulders, or may result from the decay of a tor, as the scheme proposed by Thomas (1965) implies (Fig. 3.1). In fact, any boundary between a boulder cluster and a boulder-type tor would be artificial and each one can evolve into the other through either disintegration or further surface lowering and corestone excavation.

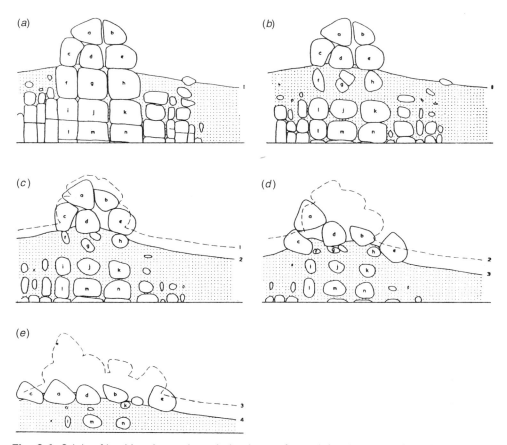

Fig. 3.1 Origin of boulder cluster through the decay of a tor (after Thomas, 1965, fig. 4)

Middle and lower slopes within a hilly landscape are another common setting for boulders. Upper slopes are often occupied by tors and domes, and the presence of boulders downslope would indicate increasing fracture density and the localized occurrence of more massive compartments. Other boulders can be found within relatively flat, or gently rolling landsurfaces, again indicating differential weathering at the small scale, but also the probable absence of very large massive compartments, which would otherwise give rise to tors and inselbergs. Boulders may also cluster in valley incisions and these should be interpreted as residuals left after the loose saprolitic cover has been washed away by running water. A very good example of an in-valley *in situ* boulder field composed of blocks as big as 10 m long is at Huelgoat in Brittany. The granite massif of Sidobre in the Montagne Noire, southern France, hosts a few block streams (*chaos de blocs* or *rivière de rochers*) within larger concave landforms which are residuals left after fine-weathered material has been eroded. Another impressive example comes from Mt Vitosha near Sofia, Bulgaria, where huge boulders assemble into a boulder stream a few hundred metres long, although some input from valley sides is likely in this case (Krygowski and Kostrzewski, 1971). Further examples are forwarded by Ikeda (1998) from Japan. Evacuation of fine saprolite from between corestones may also put them in a precarious position. Such perched blocks are presented more thoroughly while discussing the degradation of tors.

Boulder accumulations can also be found along coasts, but their genetic interpretation requires caution. They may be residuals from a corestone-rich weathering mantle which has been progressively washed away by wave action. If this is the case, one would expect marine cliffs cut in saprolite. Low latitude coasts offer relevant examples, for example in Pulau Ubin, Singapore (Swan, 1971), but boulders can also be excavated from grus mantles in higher latitudes, for example along the Pink Granite Coast in northern Brittany or along the north-western coast of Norway (Peulvast, 1989). A fossil bouldery coast of this type can be seen buried by Cretaceous deposits at Ivön, in southern Sweden (Lidmar-Bergström, 1989). An alternative explanation holds that rock fragments delivered to the shore were originally angular and derived from rock-cut cliffs and become rounded only subsequently, due to wave action. This is probably true for bouldery beaches along high-energy coasts such as those around Land's End in Cornwall, England.

Tors

Distribution and Appearance

Tors are ubiquitous landforms in granite terrains, being particularly, though not exclusively, associated with upland topography. Perhaps the most famous ones are the tors of Dartmoor in south-west England. Here, scattered across more than 500 km^2 of an upland there are more than 150 individual tors, which show a wide range of locations, shapes, and sizes (Plate IV). Likewise, the upland of Bodmin Moor further westwards, although smaller and less elevated, abounds in tors and some of them rival the classic examples from Dartmoor in the weirdness of their shapes, for example, Cheesewrings near Minions.

It is not feasible to comment about each granite area where tors occur in abundance. Selected locations and references are given in Fig. 3.2. In their light, the following generalizations about the occurrence of tors are permitted.

(1) The presence of tors appears unrelated to environmental conditions. Examples can be found in the Arctic, Antarctica, cold-climate mountain plateaux of northern Europe, as well as in Mediterranean lands, hot deserts, and tropical savanna landscapes.

(2) Tors tend to be associated with uplands and those mountain ranges which bear extensive surfaces of low relief in their most elevated parts. The latter setting is typified by the plateau of the Cairngorms in Scotland, the Massif Central in France, the Karkonosze Mountains in Poland, or the Serra da Estrela in Portugal. Tors do occur in basins, for example in the Bega Basin in New South Wales, Australia, but are less common and, on average, less impressive.

(3) Typical mountain relief does not support tor landscapes, although they may occur. Instead, juxtaposed domes and rocky crests dominate the landscape, such as is the case of the Sierra Nevada in California or the Sierra de Guadarrama in central Spain.

(4) Tors are not restricted to areas which have not been glaciated during the Pleistocene. Quite to the contrary, they do occur in those areas as well, both in North America and northern Europe. The presence of tors in the formerly glaciated areas has generated a lively discussion about their significance, as will be reviewed later.

Shapes of tors are enormously varied, but they have one common characteristic. They are controlled by fracture patterns, that is, by their orientation, density, and curvature. This close relationship between fractures and tors was noted long ago as expressed clearly by Berg: 'Everyone who knows granite landscapes is aware that tor development has been controlled, to the faintest detail, by the presence of orthogonal partings related to the fracture system that includes three joint sets perpendicular to themselves. These, as weathering begins to act, govern rock disintegration into blocks' (Berg, 1927: 5; present author's translation). Subsequent work in various environments confirmed some of these relationships quantitatively (e.g. Gerrard, 1974, 1982; Moyersons, 1977; Ehlen, 1991, 1992, 1994).

Typically, three major fracture sets are present in tors, forming together an orthogonal pattern, and this is reflected in the castellated appearance of a tor. Two are vertical or steeply dipping and join each other at a right angle. Vertical surfaces of granite tors are

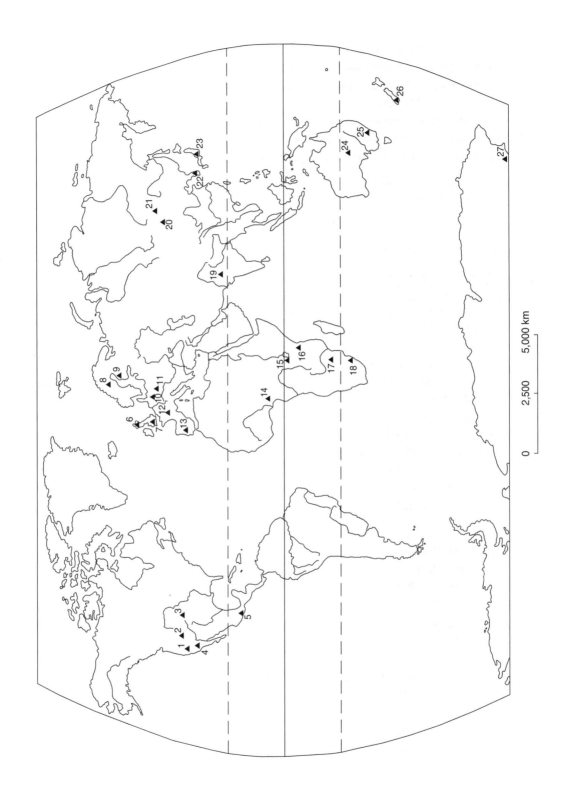

often joint faces and the sides of rock pillars and columns are defined by the two vertical sets (Plates 1.2; 3.6). The deep clefts present in many tors are either widened vertical joints or products of preferential weathering of a more jointed rock compartment. In addition, there is a third set of horizontal or gently dipping joints. This usually defines the flattened top of a tor and is responsible for the step-like appearance of their sides. Individual fractures belonging to these sets are not necessarily developed with equal frequency. If subhorizontal joints are rare, the tor will tend to rise higher and acquire a tower-like or pinnacle shape. With a very dense vertical jointing, a tor will resemble a row of tall fins, as in the Alabama Hills, California. By contrast, if the flat-lying joints prevail over the vertical ones, lamellar tor forms will develop, covering large surfaces but not being significantly tall. Many of the Dartmoor tors show this lamellar appearance due to very dense pseudobedding (Plate 3.7). The classic shape of a massive and blocky tor originates if all sets occur with similar frequency. Castellated tors show a wide range of dimensions, but the height of 25–30 m appears to be a limiting value. For example, Vixen Tor in Dartmoor is 21 m high, tors on Ben Avon plateau in the Cairngorms approach 24 m, Pielgrzymy in the Karkonosze Mountains is 25 m high, and some of the most massive tors in Serra da Estrela, Portugal, are more than 20 m high.

Other shapes may be expected if the fracture pattern is not orthogonal. Some tors resemble small cupolas and develop upon low-radius domes, deficient in vertical fractures. Various names have been proposed for these features, such as whalebacks, turtlebacks, or *ruware*. However, more pronounced dome-like tors may occur, rising to 20–30 m high (Plate 3.8). Asymmetrical ramp-like shapes form if an orthogonal pattern is replaced by moderately dipping major fractures (Ballantyne, 1994). Then, some tors are in fact masses of densely packed, but otherwise isolated boulders and such heaps of boulders may attain significant height in excess of 10 m. Finally, an

Fig. 3.2 (*opposite*) Tor distribution around the world (selected examples): (1) Sierra Nevada (Matthes, 1950, 1956; Wahrhaftig, 1965), (2) Idaho (Cunningham, 1971), (3) Laramie Range (Cunningham, 1969; Eggler *et al.*, 1969), (4) Mojave Desert (Oberlander, 1972, 1974), (5) Acapulco (Jahn, 1962; author's observations), (6) Cairngorms (Ballantyne, 1994; Hall, 1996), (7) Dartmoor and Bodmin Moor (Linton, 1955; Palmer and Neilson, 1962; Gerrard, 1974, 1978; Ehlen, 1991, 1992), (8) Aurivaara, northern Sweden (André, 2004), (9) southern Finland (Söderman *et al.*, 1983), (10) Harz (Meinecke, 1957; Wilhelmy, 1958), (11) Bohemian Massif (Jahn, 1962; Czudek, 1964; Demek, 1964; Votýpka, 1979), (12) Massif Central (Godard, 1966; Coque-Delhuille, 1978; Godard *et al.*, 2001), (13) Iberian Peninsula (Daveau, 1971; Pedraza *et al.*, 1989), (14) Nigeria (Thomas, 1965; Moyersons, 1977), (15) Uganda (Doornkamp, 1968), (16) Tanzania (Handley, 1952), (17) Zimbabwe (Pye *et al.*, 1984), (18) Swaziland (Gibbons, 1981), (19) Aravalli Range (Sen, 1983), (20) Khangai Mountains (Pękala and Ziętara, 1980), (21) south-east Mongolia (Dzulynski and Kotarba, 1979), (22) Korean Peninsula (Lautensach, 1950; Tanaka and Matsukura, 2001), (23) Japan (Ikeda, 1998), (24) Devil's Marbles, central Australia (Ollier, 1965; Twidale, 1982), (25) Bega Basin (Dixon and Young, 1981; Thomas, 1994b), (26) South Island (Thomas, 1974), (27) Antarctica (Derbyshire, 1972; Selby, 1972)

Plate 3.6 Castellated tor, defined by an orthogonal fracture set, Karkonosze, south-west Poland

individual tor may combine two or more shapes if one jointing pattern gives way to another one. For example, a seemingly chaotic boulder pile can be superimposed on a more coherent cupola-like or blocky pedestal.

Given the range of forms it is not surprising that many attempts have been made to classify tor shapes. Jahn (1962, 1974), working in the Karkonosze Mountains in south-west Poland, distinguished three main types of tors: tabular, castellated, and boulder piles. Thomas (1976: 429) adds hemispherical forms in which 'a domical profile is evident among the constituent boulders', whereas Cunningham (1971) notes the presence of 'fins' delineated by prominent vertical fractures which hardly fulfil morphological requirements for either tors or bornhardts. Gerrard's (1974) classification of Dartmoor tors into summit, spur, and valleyside tors emphasizes location rather than shape, but it appears that in each of these settings specific tors occur. In the Serra da Estrela, Portugal, granite tors assume at least five distinctive forms, namely domes, castellated tors, bouldery tors,

Plate 3.7 Lamellar tor, dominated by closely spaced horizontal partings (pseudobedding), Bellever Tor, Dartmoor, England (Photo courtesy of Karna Lidmar-Bergström)

Plate 3.8 Massive dome-like tor, Serra da Estrela, Portugal

102 Boulders, Tors, and Inselbergs

rock cliffs, and trapezoidal summit outcrops (Migoń and Vieira, in preparation). But it may be doubted if a universal classification, easy to apply in field mapping, ever arises. A feature to note in some tors is the presence of an elongate avenue in their crest parts. The Great Staple Tor in Dartmoor provides a good example (Campbell *et al.*, 1998). A similar phenomenon of twin tors was described by Jahn (1962) from the mountainous region of the Karkonosze in Poland. They occur on the opposite sides of a ridge, very close to the crest, but the crest line itself is actually at a slightly lower elevation. It is hypothesized that the avenue forms in the axial part of a broad ridge or dome, where tensional stresses are the highest. Preferential fracture opening facilitates penetration of weathering within the most elevated part of a hill. Hence after stripping a depression in between less weathered rock compartments emerges.

Origin of Tors

Although tors had been known for centuries, their origin remained mysterious until the mid-twentieth century. The discussion on the origin of tors started with the publication of Linton's 'The problem of tors' (Linton, 1955; Fig. 3.3) although the two-stage concept

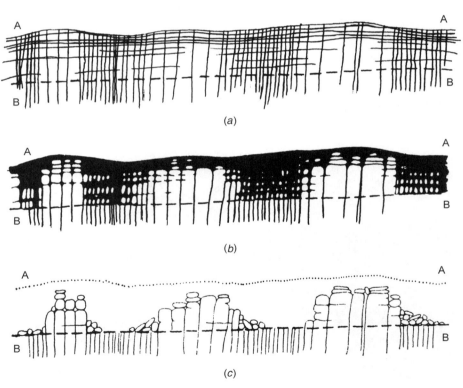

Fig. 3.3 Conceptual model of tor origin (after Linton, 1955, fig. 2)

usually associated with the name of Linton can be traced much further back in time (Thomas, 1968; Twidale, 1978a, 1981a). Furthermore, roughly at the same time, similar ideas were presented independently by German geomorphologists (Meinecke, 1957; Wilhelmy, 1958), but the language barrier has sadly contributed to a general neglect of those parallel publications. The paper by Linton has become a classic (see Gerrard, 1994a). Therefore his own words about the origin of tors are worth quoting: 'A tor is a residual mass of bedrock produced below the surface level by a phase of profound rock rotting effected by groundwater and guided by joint systems, followed by a phase of mechanical stripping of the incoherent products of chemical action' (Linton, 1955: 476). This is a perfectly logical explanation, confirmed by a multitude of studies worldwide, in which instances of selective subsurface weathering and subsequent tor exposure have been documented (e.g. Jahn, 1962; Demek, 1964; Thomas, 1965; Godard, 1966; Eden and Green, 1971). The key evidence validating the theory is the presence of tor-like solid rock masses surrounded by friable products of *in situ* disintegration. These have been encountered in many grus excavations, road cuts, and quarries all over the world (Plate 3.3), and those in the type locality of Dartmoor, at Two Bridges and in Bellever Quarry, have been designated Sites of Special Geological Interest (Campbell *et al.*, 1998).

The process of stripping the saprolite to expose a tor may be more complicated than the simple two-phase scenario suggests. There is no reason to doubt that tors, and the tall ones in particular, may have gone through repeated phases of deep weathering and stripping. The process of tor exposure would have been episodic, as outlined by Twidale and Bourne (1975) for much larger bornhardts. The following field observations are worth citing in this context as supporting circumstantial evidence. First, the examples of yet-to-be-exposed tors show these features as a few to 10 metres high at most, whereas many tors are imposing residuals exceeding 20 m high. Second, the sides of certain tors exhibit features typical of rock face/soil cover junction such as flared slopes (see Chapter 4), now standing high above the surrounding ground surface. Third, elevated parts of tall tors bear traces of protracted subaerial weathering and show a rich assemblage of microforms (weathering pits, karren, tafoni), which are absent or much less developed in their lower parts. This contrast suggests that there was no brief phase of stripping, but a long-term, gradual excavation instead.

Given the current popularity of Linton's model of a two-stage origin of tors, it is difficult to believe that his original paper generated such vehement opposition. In the course of the debate two alternative theories were presented and the two-stage model was either dismissed (Palmer and Neilson, 1962) or considered as an unusual case of very local application (King, 1958).

The first of these is known as the periglacial theory of tor formation, or as a one-stage model (Palmer and Neilson, 1962). The latter denominative refers to the fact that moulding of tors would have occurred entirely at the topographic surface and, in contrast to Linton's model, no preceding (his 'stage one') subsurface weathering is involved. In detail, periglacial tors would have been shaped in three distinct phases. Palmer and Neilson (1962) conceded that the first stage involves stripping of any pre-existing regolith by solifluction and exposure of solid granite. In the next stage granite is subject to frost action and breaks down along partings to form block fields. Finally, downhill movement of released blocks by solifluction would expose a castellated tor. The two-stage origin of Dartmoor tors has been rejected on the grounds that corestones and deep weathering are lacking in Dartmoor, and rounding of constituting blocks can be achieved by atmospheric weathering.

King (1958) accepted that some tors may have originated in the way proposed by Linton and called these 'sub-skyline tors', but vigorously argued that they are rare. Much more common, in his view, are 'skyline tors'. Their significance is summarized as follows: 'Skyline tors are ... the final unconsumed remnants (of once larger rock masses) upon a surface of extremely advanced pediplanation and must carefully be distinguished from the sub-skyline tors which are developing now within the broad valleys as youthful features in a newer cycle of erosion' (King, 1962: 388). This position reflects the general belief by King that deep weathering does not play a significant part in landform evolution and that long-term scarp retreat across unweathered rock is the principal component of subaerial landscape evolution.

At the same time, reconciling views were offered by Pullan (1959), and then by Cunningham (1965), although their studies did not refer specifically to granite areas. Both authors, working in sandstone areas of the Pennines in northern England indicated that isolated rock outcrops (tors) of different origin may occur close to each other and look similar. Thomas (1976) reviewed the evidence and envisaged several possible modes of tor origin:

(1) by regolith stripping, following upon a prolonged period of deep weathering;
(2) by differential denudation during simultaneous chemical weathering and surface erosion;
(3) during slope retreat within variably weathered and irregularly jointed rock, especially in seasonal and semi-arid climates;
(4) during the retreat of frost-riven cliffs under periglacial conditions.

Nowadays tors are regarded as excellent examples of equifinality, that is, they are products of various processes and environments, despite superficial similarities. This is certainly the case of granite tors. Demek (1964) presented the co-existence of tors of different origins in granite and gneiss uplands in the Bohemian Massif, arguing that castellated forms are two-stage tors inherited from the Tertiary, whereas asymmetric rock cliffs and outliers in front of them were produced during the Pleistocene by a complex interaction of periglacial processes. Referring to Dartmoor, Gerrard (1988a) suggests that the present-day tors have their roots in the pre-Quaternary period, when domes and valleys formed and deep weathering operated, but details of their morphology were sculpted in the periglacial environment of the Pleistocene. In the Serra da Estrela the majority of tors are probably two- (or multi-)stage, but rock cliffs and summit skyline tors occur as well, for which there is hardly any evidence of subsurface origin. What is most important in this discussion is that the appearance of a tor provides a very poor and unreliable clue to its geomorphic history.

Tor Evolution at the Surface

The excavation of tors from deep weathering profiles, whether 'instantaneous', gradual, or episodic, is not the final stage on their evolutionary pathway. Surface modelling is equally important and the current appearance of most tors is almost certainly a combined effect of structurally guided subsurface weathering and superficial weathering, not uncommonly in changing environments. Thus, Jahn (1962), while transferring Linton's model to the Karkonosze Mountains in south-west Poland, argued that the geomorphic history of a tor is best resolved into three rather than two separate stages, of which the third involves shaping the tor after its exposure. Twidale (1981b) put forward a model in which bouldery tors ('nubbins' in his paper) are derived from more massive domes, being late-stage products of their mechanical disintegration.

Ehlen (1991, 1994) analysed relationships between the Dartmoor tors and fracture patterns statistically and hypothesized about certain evolutionary pathways of tors which would be as follows: summit tors may develop into spur tors, these in turn into valleyside tors, and these into pinnacles. The former two types may also evolve straight into pinnacles, which then disintegrate into boulder piles (Fig. 3.4). However, it is

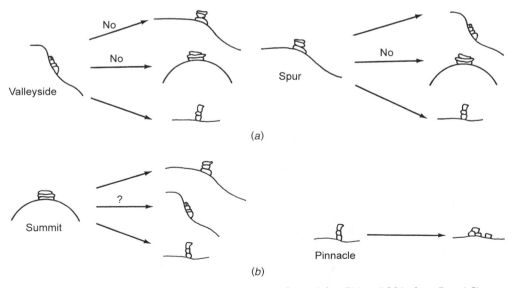

Fig. 3.4 Hypothetical pathways of subaerial evolution of tors (after Ehlen, 1991, figs. 5 and 6)

difficult to visualize how summits can evolve into spurs and valleysides, to make the envisaged scenario possible.

Degradation of tors leads to the origin of such curious granite landforms as perched blocks, otherwise known as logan stones, balanced rocks, or rocking stones (German *Wackelsteine*). These are isolated, massive fragments of rock, which rest on a flattish platform and the area of contact between them is very small. Therefore, the boulders are only conditionally stable and may be forced to move ('rock') if external force is applied. The force needs only to be small, hence some blocks can be moved by a single person pushing or standing on it. Perched blocks are known from a variety of places, including Dartmoor, Sidobre in France, Fichtelgebirge in Germany, Karkonosze in Poland, uplands in the heart of the Bohemian Massif, and Enchanted Rock in Texas.

Another class of minor landforms formed in the course of tor degradation are pedestal rocks, known in German as *Pilzfelsen* and in Spanish as *rocas fungiformas*.

Plate 3.9 Pedestal rock, Jizerské hory, Czech Republic

These are residuals consisting essentially of two parts: a narrow pillar or stem supporting a much larger cap (Plate 3.9). They are most common in differentially bedded sandstone but occur in granites too. Some of the early descriptions come from the American south-west (Bryan, 1925; Leonard, 1927), whereas a more general review has been offered by Twidale and Campbell (1992). The role of wind and sand blasting in their development had already been dismissed by Bryan and Leonard who, as did Crickmay (1935) later, favoured weathering leading to granular disintegration instead. By contrast, Twidale and Campbell (1992) advocate their two-stage origin and equate the pillar/cap boundary with the former land surface. The narrowing of the lower part is the result of more efficient weathering in the moist subsurface position.

On the exposed rock surfaces of tors a variety of minor weathering features may evolve, not uncommonly rising to huge dimensions of many metres across. They are interesting in their own right and will be considered later, but a certain conceptual model of tor destruction due to weathering pits development (Dzulynski and Kotarba, 1979) is worth introducing here. The observational evidence has been furnished by tors in the Ikh Naart Plateau in south-eastern Mongolia, but the model has been found relevant in other areas, such as the Namib Desert (Goudie and Migoń, 1997). Dzulynski and Kotarba note that many tors in their study area are conical, with flattened tops and incipient, shallow weathering pits incised into the top surface. The more flat and extensive the summit surfaces are, the more abundant and larger are the weathering pits. Pit networks and overlapping forms occur. The most extreme case is provided by low shield-like elevations with interconnected pits of various size. These three cases are seen as transitional phases of tor development (Fig. 3.5).

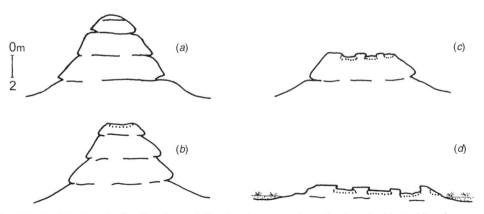

Fig. 3.5 Model of tor destruction through the development of weathering pits (dotted lines) (after Dzulynski and Kotarba, 1979, fig. 10)

Tors and Changing Environments

Few landforms have generated more controversy than tors, and granite tors in particular have been the focus of attention. The model of two-stage tor origin proposed by Linton (1955) in respect of Dartmoor has change in environmental conditions at its very core. He suggested that deep selective weathering, the necessary precursor of tor exposure, took place under warm and humid climatic conditions in the Tertiary. In his own words (p. 478): 'Profound rock rotting to depths equivalent to the heights of our larger tors would appear to require not only a great deal of time but probably also a warmer climate than we now enjoy anywhere in these islands [the British Isles]'.

Later on, tors referred to by Linton have been described as a 'palaeotropical' phenomenon. The rotten granite of Dartmoor, locally with tors ready to be exposed, was seen as the vestige of this vigorous subsurface chemical attack. Climate deterioration by the end of the Tertiary would have brought deep weathering to an effective halt, whereas surface denudation intensified, particularly under ensuing periglacial conditions. Tors started to emerge as the surrounding weathered granite was incorporated into periglacial solifluction deposits and progressively stripped away. Locally, periglacial denudation of the ancient weathering mantle was so efficient that a rock-cut planar surface, a former saprolite/bedrock interface, has got exposed around the tors. The subsequent argument between Linton (1964) and the proponents of the periglacial origin of tors (Palmer and Radley, 1961; Palmer and Neilson, 1962) leaves no doubt that environmental change reflected in the changing intensity of deep weathering was seen as the crucial factor affecting tor evolution. The British 'tor debate' has been frequently referred to in the literature, the most recent treatment having been provided by Ballantyne and Harris (1994).

Independent research on tors in the *Mittelgebirge* uplands of Germany led to similar conclusions (Hövermann, 1953; Meinecke, 1957). Tors in the Harz Mountains, locally still embedded within the weathered granite of the grus type, were interpreted as remnants of a Tertiary landscape which evolved under warmer and more humid conditions than those experienced today. Although they may have been partially exposed in the Tertiary, it was the Pleistocene solifluction and sheetwash that accomplished most work in stripping the grus and revealing hidden tors. Linton's followers in other Central European countries have similarly invoked a Tertiary phase of deep and selective weathering, later replaced by Pleistocene denudation of any residual mantle that was present around the incipient tors (Jahn, 1962, 1974; Demek, 1964).

Further research on tors, evidently inspired by the controversies generated by Linton's paper, has resulted in the recognition of two classes of tor, each with its own specific mode of origin. Two-stage tors were those prepared by selective subsurface weathering and then exposed after surface denudation had for some reasons accelerated. Most of the granite tors in European uplands, in Great Britain, France, Germany, the Czech Republic, and Poland, were considered as members of this family. One-stage tors were those developed through the concurrent action of surface weathering and stripping, basically in the cold periglacial environment (Czudek, 1964; Martini, 1969). Interestingly, granite outcrops have hardly ever been interpreted as one-stage tors, probably because angular blocks are rare products of granite disintegration, except for the very fine-grained variants. The majority of descriptions of one-stage tors concern metamorphic and sedimentary rocks, notably quartzite, greenschist, laminated gneiss, and sandstone. The distinction between two genetically different categories of tors has brought their palaeoenvironmental interpretation even more sharply into focus and made it the ultimate goal of scientific endeavour. Any research on tors inevitably began with the search for either remnants of a thick residual mantle near the tors, or angular, implicitly periglacial, block fields. Each has been considered as the key piece of evidence in favour of one of the competing theories. Attention has been paid to the shape and angularity of tors, in spite of inherent difficulties in how to define the boundary line.

It is beyond dispute that two-stage tors do require an environmental change to be brought to daylight, as there has to be a reason why the balance between deep weathering and surface lowering has changed, the latter increasing in efficiency. Debatable issues are how severe this change ought to be to initiate tor exposure, and whether such changes are regional or local. In theory, there may be a multitude of reasons, including surface uplift and the resultant acceleration of denudation, climatic change towards more arid or cooler conditions, the action of meltwater released from an adjacent snow patch or glacier, the incision of a nearby stream, or even an increase in the erosion rate because of human interference on a local slope, such as overgrazing or vegetation burning. Clearly, there is no need to automatically invoke drastic climatic change and tropical inheritance of tors. To find a site-relevant environmental interpretation a number of approaches can be offered.

First, the distribution of tors can be revealing. Their presence well away from the Tropics casts immediate doubts on the adequacy of the climate-based model. Tors, though not necessarily granitic, have been reported from virtually everywhere on Earth, including high latitude settings such as northern Sweden (Stroeven *et al.*, 2002; André, 2002, 2004), the most frozen Canadian Arctic (Sugden and Watts, 1977), and Antarctica (Derbyshire, 1972; Selby, 1972). These tors have been argued to be preglacial, but no tropical inheritance has ever been implied. Likewise, tor-like features occur widely in present-day arid environments, for instance in the Mojave Desert of North America (Oberlander, 1972), the Namib Desert (Migoń and Goudie, 2003), and the interior of Australia (Twidale, 1982). However, there is a problem of inheritance and lifetimes of tors in these extra-tropical settings. One might argue that tors are durable landforms, have ages in the order of million of years, and significantly predate the establishment of current climatic conditions. In other words, their distribution in respect to present-day climatic zones would be largely irrelevant. In fact, the desert tor landscape in the Mojave Desert referred to above seems indeed, to be inherited from much more humid conditions in the Late Miocene (Oberlander, 1972), but no such recent history of change from humid tropical to arid or cold conditions is applicable for the Namib or for Baffin Island in the Canadian Arctic.

Second, if tors can be shown to be two-stage landforms associated with residual weathering mantles, then the characteristics of these weathering mantles may hold the clue to the environmental conditions at the time of deep weathering. Most granite tors in Europe co-exist with sandy or grus mantles, which are unlikely candidates for tropical residues. In Dartmoor, the properties of *in situ* grus have been investigated by Eden and Green (1971) and then by Doornkamp (1974). Both concluded that the degree of mineralogical change and textural alteration of quartz grains is small and hence incompatible with humid tropical conditions. Grus mantle is better seen as the product of weathering under temperate climatic conditions at the end of the Tertiary, and consequently the tors would be of similar age. Tors in the Harz Mountains, Germany, and the Karkonosze in Poland are similarly associated with grus and their tropical roots are unlikely. Likewise, grus exposures in southern Sweden contain frequent tor-like corestone clusters, and the grus itself is of Late Tertiary age, when the local environment was far from tropical (Lidmar-Bergström *et al.*, 1997). In other places the situation is more complicated because the relationships between tors and weathering mantles are less clear. In the Cairngorms Mountains, Scotland, massive tors exist (Ballantyne, 1994), but their connection with shallow grus on the high plateau (Hall, 1996) is uncertain. Grus is 1–2 m thick, while the tors attain more than 15 m high, and the characteristics of the upper lost horizons of weathering profiles can only be speculated.

Tors in Glaciated Terrains

Although tors impressed Linton by their massive appearance and solitary position in the landscape, he did not believe that they were hard enough to withstand the passage of moving ice. Therefore, he considered tors as indicators of an area which had escaped glaciation and either was located outside its limits, or formed a nunatak (Linton, 1955). In line with this reasoning, tors were used to delimit the extent of local glacier cap on the Aubrac Plateau, French Massif Central (Veyret, 1978), and in the Serra da Estrela, Portugal (Daveau, 1971). However, findings of glacial erratics within and around tors on high-latitude plateaux in the Cairngorms (Sugden, 1968), the Labrador Peninsula (Gangloff, 1983), and the Canadian Arctic (Sugden and Watts, 1977, Watts, 1986) invalidated Linton's claims. A combination of geomorphic and palaeoglaciological research provided an explanation of this apparent contradiction and it is now clear that tors may survive under the cover of ice if this is cold-based and stationary (Kleman and Stroeven, 1997; Stroeven *et al.*, 2002). Thus, the significance of tors in these areas has changed but not been lost, and they are among the tools of a geo-

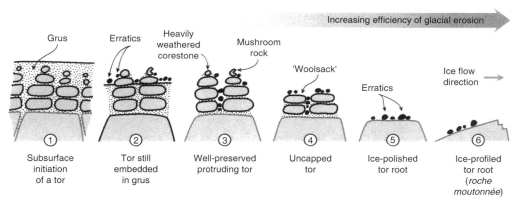

Fig. 3.6 Tors in glaciated terrains, using the example of Aurivaara plateau (after André, 2004, fig. 10)

morphic approach to palaeoglaciology. A recent study of André (2004), focused on syenite tors in the Aurivaara Plateau in northern Sweden, has indicated that tor preservation could be complete or partial, depending on the efficacy of glacial erosion (Fig. 3.6). Indeed, tor geomorphology in formerly glaciated terrains varies enormously, from more than 20 m high massive residuals to subdued, shield-like forms (Ballantyne, 1994).

Inselbergs and Bornhardts

Diversity of Form

Granite inselbergs belong to the most fascinating landforms which occur on Earth. The tallest examples are built of granite, for example Spitzkoppe in the central part of the Namib Desert (Selby, 1982b; Migoń and Goudie, 2003). It is a massive, north-to-south elongated dome built of Jurassic granite that rises steeply above the surrounding plain by more than 600 m. Looking from the south, it appears to point to the sky with a pyramid-like summit, hence it is occasionally advertised in tourist brochures as the 'Matterhorn of Africa'. Spitzkoppe has all the characteristic features of domed inselbergs, such as steep, bare, and upward-convex slopes, a sharp piedmont angle, and a mantle of talus derived from sheet structure degradation around a part of its perimeter. Its slopes show much sympathy to structure and their long-term evolution is primarily controlled by fracture pattern and rock mass strength (Selby, 1982b) (Plate V).

Another fine example from the same area is Vogelfederberg hill, located some 60 km east of Walvis Bay, in the hyperarid part of the Namib Desert. The overall form is one of a dome too, but visually these two hills are unlike each other. Vogelfederberg is much lower than Spitzkoppe, only about 50 m high, and flattened. Its slopes are gentle and can be easily walked on, whereas ascent to the top of Spitzkoppe requires professional climbing. Notably, there is no talus around Vogelfederberg and a sand-filled scarp-foot depression is present around its perimeter (Plate 3.10).

Many other granite inselbergs assume the dome form and these are frequently referred to as bornhardts (Twidale and Bourne, 1978), although it has to be pointed out that bornhardts are not necessarily granitic. Inselbergs of the bornhardt type are distributed worldwide (Fig. 3.7), from equatorial, humid tropical lands in Africa and South America, through the seasonal tropics and savanna environments, hot deserts, and semideserts in North America, Africa, and Australia, temperate regions, both humid and arid/cool, to high-latitude cold environments in northern Europe and North America. Selected examples of detailed studies of granite domed inselbergs are Thomas (1967), Selby (1982a, 1982b), Whitlow and Shakesby (1988), and Twidale and Bourne (1998).

Domed inselbergs show a diversity of form in detail (Plate 3.11). As the examples from the Namib indicate, their height is very variable. Some domes are as low as 20–30 m, whereas others may well exceed 300–400 m high. Thomas (1965, 1967, 1978) reports a similar height range for inselbergs in southern Nigeria, and inselbergs in Zimbabwe vary from a few tens to about 300 m high (Whitlow, 1979). Perhaps more importantly, domes differ in slope angle. Shield inselbergs have a much flatter form, with slope angles as low as 10–15°, and flat summit surfaces. On the other hand, there also occur prominent domes rising steeply above the surrounding plain. The inclination

Plate 3.10 Vogelfederberg—one of bornhardt-type inselbergs in the Central Namib Desert, notable for its lack of talus and the presence of a marginal depression

of their lower slopes may exceed 45–50°, locally it may even approach 90°, and the top surface is of very limited extent. Slopes are typically convex-upward and in many examples apparently adjusted to sheeting planes. However, stepped profiles consisting of alternating convex and vertical or overhanging segments are also present. These evolve in the course of dome degradation, through partial detachment of outer sheets along sheeting and vertical fractures, followed by weathering of an exposed vertical rock face. The slopes of the domes are usually bare, because they are too steep to allow for accumulation of products of rock disintegration. These are subject to gravity-driven transport to the footslope, where they may form chaotic piles of detached boulders and slabs. Around the Spitzkoppe, some talus-forming blocks are truly gigantic, having a length of about 30 m. However, on gently rising domes detached slabs may temporarily remain on slopes and the dome develops a discontinuous blanket of residual blocks. Microrelief of slopes is also diverse. Low-angle dome surfaces typically exhibit an association of weathering pits and pans, connecting channels, and irregular shallow concavities (e.g. Watson and Pye, 1985), whereas steep surfaces are usually smooth, although karren may occur.

The dome slope/plain junction, often called a piedmont angle, is likewise present in many variants (Fig. 3.8). Two of these are particularly interesting. One is the situation in which a steep rock slope of a dome connects with a low-angle surface below, but this surface is also rock-cut. Such footslope surfaces are known as pediments and belong to the most debated landforms, especially in arid geomorphology (Cooke *et al.*, 1993; Dohrenwend, 1994). Rock pediments surround many domed inselbergs in deserts, for

Plate 3.11 Domed inselberg in Kerala, India (Photo courtesy of Yanni Gunnell)

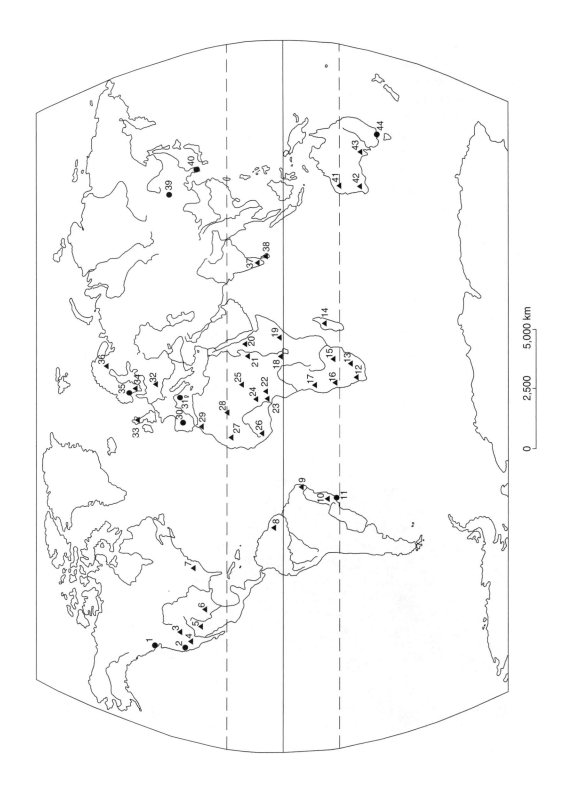

example in the Spitzkoppe group in the Namib, and their possible origin is discussed in chapter 5.

Another variant is the presence of a shallow trough around the perimeter of a dome, so that the plain does not slope away from the hill, but towards it (Plate. 3.10). The troughs, which can be present around the whole or a part of the perimeter, are known as 'scarp-foot depressions' or *Bergfussniederungen* in German (Pugh, 1956; Ruxton and Berry, 1961; Büdel, 1978). On average, the troughs are a few metres deep and a few tens of metres wide. Their origin is related to subsurface weathering, which is enhanced at the slope/plain junction because of increased availability of water, derived from surface runoff. Episodic washing of weathering material excavated an initial trough, which may deepen through both subsequent weathering and episodic fluvial erosion.

There have been attempts to connect different types of domed inselbergs into an evolutionary sequence, but the interpretation of field evidence is ambiguous. One might argue that low domes are end-products of protracted degradation of massive, steep-sided forms, but an opposite scenario can be visualized too. In this one, a low dome protruding from the plain is the top part of a bell-shaped structural dome that awaits complete excavation (Twidale, 1986). Furthermore, construction of such

Fig. 3.7 (*opposite*) Global distribution of inselbergs (triangles) and domed hills, which are not inselbergs because of their position in the landscape (circles) (selected examples): (1) Howe Sound, Vancouver (author's observations), (2) Sierra Nevada (Matthes, 1950, 1956; Huber, 1987), (3) Idaho (Cunningham, 1971), (4) Mojave Desert (Oberlander, 1972, 1974), (5) Sonoran Desert, Arizona (Kesel, 1977), (6) Enchanted Rock, Texas (Twidale, 1982), (7) Georgia (White, 1945), (8) Guyana (Bakker, 1960; Eden, 1971), (9) eastern Brazil (King, 1962), (10) south-east Brazil (Freise, 1938), (11) Serra do Mar (Branner, 1896; Freise, 1938; De Ploey and Cruz, 1979), (12) Namaqualand, South Africa (Mabbutt, 1952; Twidale, 1982), (13) Swaziland (Gibbons, 1981; Watson and Pye, 1985), (14) Madagascar (Godard et al., 2001), (15) Zimbabwe (Whitlow, 1979; Pye et al., 1984; Whitlow and Shakesby, 1988), (16) Namibia (Selby, 1977, 1982b; Ollier, 1978b; Migoń and Goudie, 2003), (17) Angola (do Amaral, 1973, after Twidale, 1982), (18) Uganda (Ollier, 1960; Doornkamp, 1968), (19) Kora, Kenya (Pye et al., 1986), (20) eastern Sudan (Ruxton and Berry, 1961), (21) central Sudan (Mensching, 1978), (22) Adamawa (Pugh, 1956), (23) southern Nigeria (Jeje, 1973), (24) central Nigeria (Thomas, 1965, 1967), (25) Tschad (Buckle, 1978), (26) Sierra Leone (Thomas, 1980, 1994b), (27) Mauritania (referred to in Twidale, 1982), (28) Air (Thorp, 1969), (29) northern Morocco (Riser, 1975), (30) Sierra da Guadarrama (Pedraza et al., 1989), (31) Corsica (Klaer, 1956), (32) Bohemian Massif (Czudek et al. 1964; Migoń, 1996, 1997c), (33) Scotland (Hall, 1991), (34) southern Sweden (Lidmar-Bergström, 1995), (35) Bohuslän (Johansson et al., 2001a, 2001b), (36) Lapland (Schrepfer, 1933; Kaitanen, 1985), (37) southern Deccan (Büdel, 1977), (38) Sri Lanka (Bremer, 1981a, 1981b), (39) Inner Mongolia (Ericson, pers. comm.), (40) Korean Peninsula (Lautensach, 1950; Ikeda, 1998; Tanaka and Matsukura, 2001), (41) Pilbara (Twidale, 1982, 1986), (42) south-western Australia (Twidale and Bourne, 1998), (43) Eyre Peninsula (Twidale, 1962, 1982; Bierman and Turner, 1995; Bierman and Caffee, 2002), (44) Wilson Promontory (Bird, 2000)

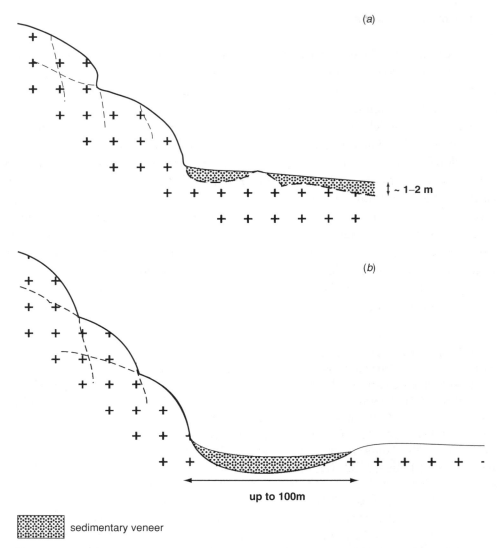

Fig. 3.8 Types of slope/plain junctions around inselbergs: (a) abrupt junction between steep rock slope and rock-cut pediment, covered by a thin sedimentary veneer, (b) wide marginal trough (scarp-foot depression)

evolutionary scenarios is evidently upset by the recognition of a great diversity of slope forms and deposits within a single hill. An excellent example of a detailed morphological analysis of a dome is that of a low hill, less than 50 m high, near Oyo, Nigeria (Thomas, 1967). As many as five different types of slope/plain junction have been mapped and slope inclination varies from 5–10 to more than 35° (Fig. 3.9).

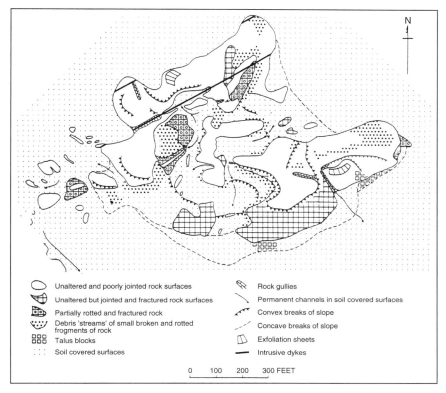

Fig. 3.9 Detailed geomorphology of a bornhardt (after Thomas, 1967, fig. 3)

The dome form is not the only form that granite inselbergs take. Twidale (1981b) identified three principal morphological types, namely domes, nubbins (or block- and boulder-strewn inselbergs), and castle koppies (castellated inselbergs), but the distinction between them in terms of visual appearance is not always easy. Two parallel characteristics appear important, namely the shape of constituent compartments and their isolation from the rock core. In fact, a rigid boundary line is very difficult to draw and domed and castellated forms may grade into a nubbin (Plate 3.12).

Boulder inselbergs, as they will be referred to in this book, are apparently chaotic heaps of detached rock compartments which usually mantle a more massive, bedrock-rooted core or rest on a low-angle rock platform. Individual components of a boulder inselberg can be either angular or rounded, or these two can co-exist. They can also vary in size, from 1–2 m to more than 10 m. The distribution of boulder inselbergs is very wide and they have been described from a wide range of environmental settings. Twidale (1982) argues that they are the most common inselberg type in the seasonally humid tropical regions, but cites examples of their occurrence elsewhere as well. Indeed, they can be found in the arid Namib Desert, in the Mojave Desert of California, but also in humid temperate regions of Central Europe (Migoń, 1996).

Plate 3.12 Group of closely spaced domes, which during degradation gradually acquire the morphological features of a block-strewn inselberg, or nubbin, Mojave Desert, California

Castellated inselbergs tend to have an angular outline and are typically composed of a massive lower part, sculpted into an array of pillars, walls, and clefts in the upper part (Plate 3.13). Visually, they often show a stepped appearance, with vertical faces separated by sub-horizontal benches. The former relate to vertical fractures, the latter to horizontal ones. Castle koppies tend to be smaller than domed inselbergs and their typical height is of the order of tens of metres. As with the other forms, castellated inselbergs are not restricted to particular environmental settings and have been reported from a range of climatic zones, but are probably particularly characteristic of the savannas of south and east Africa.

The mutual relationships between domes, boulder inselbergs and castellated forms are the subject of discussion. Twidale (1982) has argued that nubbins and castle koppies evolve from bornhardts and outlined the process in the following way:

Bornhardts, nubbins and castle koppies are genetically related forms, the two last named being derived from the marginal subsurface weathering of domed forms. Nubbins and castle koppies are the shattered remnants of bornhardts. ... The two derived forms probably reflect both the geometry of the domed structures—small radius, in the case of nubbins, large, in the case of koppies—as well as the degree of exposure of the residuals, for, whereas the nubbins have been weathered over their entire surfaces, the castellated forms appear to have been attacked from the sides, suggesting that the upper part was exposed, or at least close to the surface, during marginal attack. (Twidale, 1982: 175–6).

Plate 3.13 Castellated inselberg defined by an orthogonal fracture pattern, Mojave Desert, California

Further on, he argues that ongoing degradation of these derivative forms leads to the development of a low platform which completes the cycle of inselberg development.

However, there are indicators that some boulder inselbergs are indeed derivatives of the domes, but the evolutionary process took place at the surface rather than beneath it. These include situations such as the outline of an original dome form being discernible in the secondary form of the boulder pile. Moreover, slab-like shapes of constituting boulders suggest they have resulted from the disintegration of a sheet structure. A boulder inselberg appears to be the logically expected product of the degradation of a dome, if the latter had vertical fractures superimposed on dome form. Relict inselberg landscapes in the Bohemian Massif, especially in its northern parts (Migoń, 1997c), and in the southern part of the Mojave Desert (Oberlander, 1972, 1974) offer good examples of boulder inselbergs which have apparently evolved from more massive forms (Plate 3.12), but in neither area is there convincing evidence that this has happened in the subsurface.

The case of castellated forms is even more complicated. Some of them may have evolved from low-radius domes in the subsurface, as Twidale (1981b, 1982) proposed and as their flared sides imply. But in too many examples the visible fracture pattern does not offer any clear indication of the existence of a primary dome (Plate. 3.13). One might rather argue that castellated inselbergs form in orthogonally fractured granite masses, where partings are relatively few in number. Hence, domes (bornhardts) and castle koppies would reflect different fracture controls. Godard (1977) noted that castellated forms tend to be significantly smaller than domical ones. This is understandable given the characteristics of fracture patterns and the dimensions of structural domes, which are hundreds of metres across, versus the dimensions of zones of decreased density of vertical fractures, whose width is measured in tens rather hundreds of metres. Meyer (1967) reports the following two types of granite inselbergs in north-east Transvaal, South Africa: domes with bare rocky slopes and conical forms with boulder-mantled slopes. He emphasizes that there is no form–size relationship present, hence the reasons for different shapes are primarily structural.

However, some authorities see the difference in form as a primarily climate-dependent phenomenon. It is argued that domed inselbergs typify humid tropical environments, whereas less massive castellated landforms are characteristic of seasonally humid tropics. Boulder inselbergs, in turn, are said to be landforms of drier climates (Bremer, 1981*a*) and indicative of the increased efficacy of physical weathering. On the other hand, Thomas (1994*b*) suggests that boulder inselbergs enjoy optimal conditions for development in the humid tropical zone because it is these circumstances that favour selective deep weathering and corestone separation. This issue, important for the development of different concepts in granite geomorphology, will be discussed later, but one needs to emphasize here that the worldwide distribution of the three types, and their co-existence in many areas, severely undermines any attempts to link form and environment, at least in respect of granite residual hills.

The least researched inselbergs are the conical ones, yet they do occur and can be widespread in specific areas, such as central Arizona (Kesel, 1977). They are probably akin to castellated forms, but whereas in the latter a decrease in fracture density occurs at a very short distance, the conical forms reflect a more gradual change in fracture density (Fig. 3.10). It is worth noting that the size of some conical granite inselbergs is not very much different from the highest bornhardts. Klein Spitzkoppe in the Namib

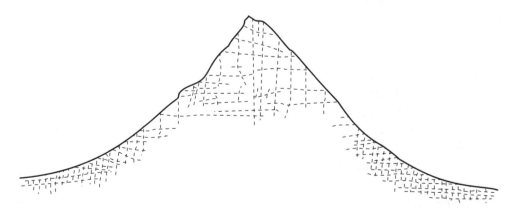

Fig. 3.10 Expected relationships between the conical form of an inselberg and fracture density

Desert is a conical inselberg developed on an orthogonally fractured, but otherwise massive portion of the Jurassic granite and is 3 km long, 2 km wide, and its relative height is 550 m.

Origin of Inselbergs

The discussion about the origin of inselbergs bears many similarities to the discussion about the origin of tors. In most accounts two competing theories are presented. One holds that inselbergs are specific by-products of long-term scarp retreat and pediplanation, and it is usually linked with the name of Lester King and his 'Canons of landscape evolution' (1953). King outlined his position in 'A theory of bornhardts' (1949) and offered a particularly clear statement on the subject some years later, while arguing against another concept, namely that of a subsurface origin of inselbergs. In his discussion note, King (1966) wrote:

Instead these granitoid mountains are distributed very clearly in zones leading down from an upper cyclic planation to a younger landscape which is in process of extending its dominion into the older terrain by multiple scarp retreat. ... It [the two-stage concept] fails utterly to account for the great bornhardt fields of Africa or South America with their many hundreds of feet of relief. (King, 1966: 97–8)

King was reacting to two contemporaneous writers, Ollier (1965) and particularly Thomas (1965), who, using evidence from equatorial Nigeria, developed a model of subsurface initiation and origin of inselbergs, referring particularly to domed inselbergs (Fig. 3.11). In his own words (Thomas, 1965: 65):

The development of rock domes in humid tropical climates can be regarded as a result of the operation of chemical and mechanical processes on a basically cuboid form, under conditions where the speed of ground water weathering is greater than the rate of denudation of the landsurface.

And he then goes on to state that:

The irregular distribution of these landforms over the landscape can be explained only in terms of spatial variations in the characteristics and structures of the underlying rocks. ... It is the primary joint system in any rock mass that governs the development of domed features by the process outlined above.

In this view inselbergs, like tors for Linton (1955), are two-stage landforms, formed through selective deep weathering followed by stripping of the saprolite, and the reasons for selectivity are predominantly structural. A few years later, a key piece of evidence for a subsurface origin of inselbergs was provided by Boye and Fritsch (1973), who described an artificial excavation of a small dome from the overlying saprolitic cover in south Cameroon. Twidale (1982) provided further examples from Australia and South Africa.

122 Boulders, Tors, and Inselbergs

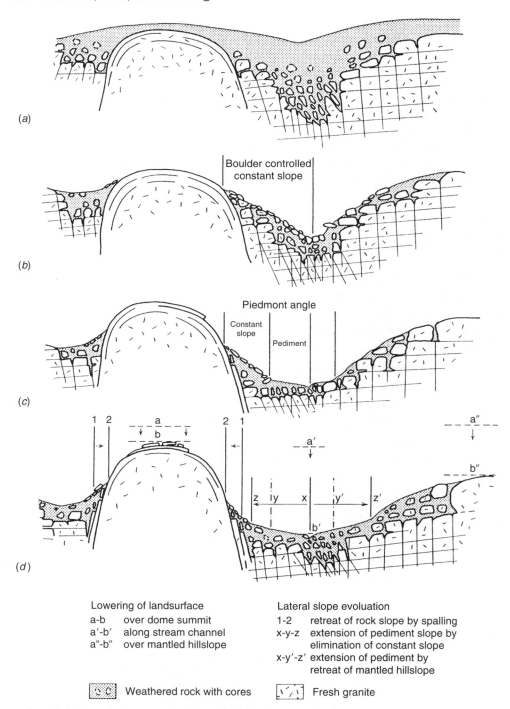

Lowering of landsurface
a-b over dome summit
a'-b' along stream channel
a"-b" over mantled hillslope

Lateral slope evolution
1-2 retreat of rock slope by spalling
x-y-z extension of pediment slope by elimination of constant slope
x-y'-z' extension of pediment by retreat of mantled hillslope

Weathered rock with cores   Fresh granite

Fig. 3.11 Model of dome evolution (after Thomas, 1978, fig. 1)

However, an apparent discrepancy between the height of many inselbergs, measured in hundreds of metres, and the typical depth of deep weathering of less than 50 m, troubled the proponents of the two-stage theory and was used as a corollary argument against its validity (King, 1966, 1975). The solution to the problem was offered by Twidale and Bourne (1975), who introduced a model of episodic exposure of inselbergs. In essence, excessively high hills would have risen through many repeated stages of deep weathering and stripping. Hence their height would have increased through time, and ultimately surpassed the typical thickness of a weathering mantle. The model was applied to domed inselbergs in southern Australia and geomorphic evidence for episodic exposure included the presence of benched slopes, flares, and tafoni across the rock slopes. The latter two are considered as indicators of the former position of the slope/plain junction, where a surplus of moisture facilitates weathering. Arguing for multi-phase, episodic exposure Twidale and Bourne had also to address the issue of the durability of inselbergs, which despite their elevated position survived largely unscathed while the surrounding plain was subject to significant lowering. To explain this apparent geomorphological anomaly, they pointed out the contrasting resistance of granite against weathering in wet and dry settings. An incipient inselberg remains relatively dry as it sheds rainwater off its slopes towards the slope/plain junction. Any water-related weathering is retarded and the hill undergoes very limited degradation. The surrounding plains receive more water and their drainage is much less efficient. Hence *in situ* rock decomposition may proceed uninterrupted for a long time, until an environmental crisis of a local or regional nature induces regolith stripping. In the next formative phase the hill is even higher and there is more runoff to the plain, therefore the contrast in weathering efficacy becomes enhanced, and so on. The growth of an inselberg through time by the mechanism outlined above is cited as one of the prime examples of positive feedback or the reinforcement effect in geomorphology (Twidale and Bourne, 1975).

Today, the two-stage model of inselberg development enjoys widespread support and is applied to both actively forming and apparently relict granite inselberg landscapes (e.g. Hall, 1991; Bremer and Sander, 2000). However, it is worth remembering that it is not a mid-twentieth century discovery and that the key observations were made much earlier, particularly by Falconer in Nigeria (1911: 246):

A plane surface of granite and gneiss subjected to long-continued weathering at base level would be decomposed to unequal depths, mainly according to the composition and texture of the various rocks. When elevation and erosion ensues, the weathered crust would be removed, and an irregular surface would be produced from which the most resistant rocks would project. ... In this way would arise the characteristic domes and turtlebacks which suffer further denudation only through insolation and exfoliation.

The preference given to the two-stage theory does not mean that the concept linking inselberg development and scarp retreat has been completely abandoned. Bremer (1971,

1981*a*, 1981*b*) offered a reconciling view, emphasizing the role of scarp-foot weathering during long-term scarp retreat, using examples from Nigeria and Sri Lanka. In this model, inselbergs may indeed cluster in front of receding scarps, but the key means by which scarps retreat is deep weathering, which is focused within triangular embayments within them but is less intensive on intervening spurs. Non-uniform retreat

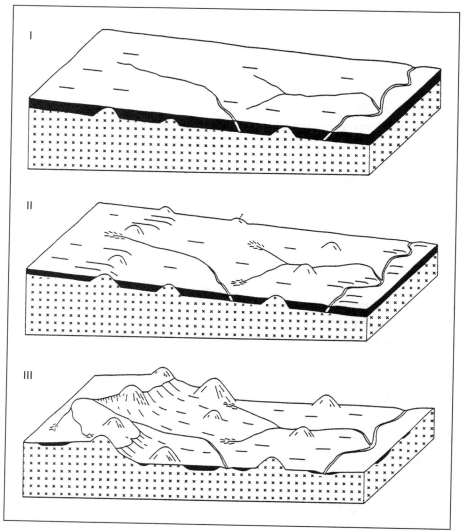

Fig. 3.12 Origin of inselbergs and escarpments during long-term planation lowering on a tilted block (after Bremer, 1993 fig. 3). Surface degradation is more efficient near local base levels (rivers) due to elevated soil moisture, hence isolated inselbergs on the plain. Weathering mantle is shown in black

leaves inselbergs in the place of former spurs, or wherever rock properties dictate increased resistance against weathering (Fig. 3.12).

In other areas, where inselbergs do occur, the evidence of antecedent deep weathering is missing and the two-stage model fails, unless a scenario involving complete stripping applies. One such area, where alternative concepts have been debated, is the Namib Desert. Its central part is dotted with inselbergs, standing either in majestic isolation or in small clusters. Selby (1977: 11) considered the general morphology of the area and concluded that: 'The lack of field evidence for chemical weathering and the clear evidence for physical weathering and erosion around emerging bornhardts, and their isolation on erosion surfaces cut across unweathered bedrock renders inapplicable an origin for the bornhardts through stripping of a chemically weathered regolith'.

His view contrasts with that of Hövermann (1978a) who sees inselbergs as the key evidence that a weathering mantle *must* have existed, saying that 'Granite inselbergs are the cores of rock masses, least affected by deep weathering' (p. 62; present author's translation). Ollier (1978b) agrees with Selby that deep weathering does not occur in the Namib today, but considers the inselbergs as exhumed landforms of probably Mesozoic age, suggesting stripping of saprolite prior to the widespread deposition in the Cainozoic. Finally, Selby (1982b) returned to the subject and presented evidence that the inselbergs of the Namib owe their form and position in the landscape to lithology and structure. In some cases, the domical form was produced at the time of intrusion and nowadays the forms are progressively exposed as the overlying schist is wasted away. Other inselbergs, such as Gross Spitzkoppe, Mirabib, or Amichab have their form strongly adjusted to the fracture patterns. Migoń and Goudie (2003) pointed out that the high inselbergs of the Namib are built of the extremely massive Jurassic granite, whereas the surrounding plain is cut across the much more shattered Salem granite of the Precambrian age. Hence they can be perfectly explained in terms of ongoing differential denudation, without recourse to past deep weathering.

Another area where deep weathering followed by stripping have generally been discarded as significant processes is the inselberg landscape of central Arizona (Kesel, 1977). In addition, in certain specific cases, regional models built on the two-stage theory have to be treated with reservation because direct evidence of deep weathering in the form of surviving saprolite patches is missing. This note of caution applies particularly to arid and high-latitude occurrences. Recognition of the history of inselbergs in each specific area needs to be based on both an analysis of the residual form itself and wider consideration of residual deposits, their sedimentary equivalents, and the history of environmental change in the relevant period of the geological past.

Summarizing the discussion, it appears clear that inselberg landscapes may evolve in different ways and their form is a rather unreliable clue to their origin. One might argue that the two-stage model applies in most instances. Nevertheless, inselbergs, like tors, remain examples of geomorphological equifinality.

Geological Controls

The issue of rock control in inselberg development is strongly related to the problem of its origin. This is because the two-stage model requires a certain structural or lithological predisposition that would account for the selectivity of deep weathering, the isolation of a more resistant core, and its survival during and after exposure.

Perhaps the most common observation about granite inselbergs is the relative paucity of primary fractures within the residual masses (Thomas, 1965; Twidale, 1982; Migoń, 1997c), but it is worth quoting the historic comment of F. P. Mennell who, as Twidale (1981a) reminded us, was aware of this relationship more than 100 years ago:

... the influence of the divisional planes of the rocks must not be overlooked, and it is to the variations in the number and character of the joints that the varied scenic aspects of the Matopos may be traced. Where stretches of comparatively level country occur, it will generally be found that the joints are numerous and irregular in direction, so the rock readily breaks up and presents a large surface to the agencies of disintegration. ... On the other hand, joints are often entirely absent over a considerable area, and the tendency of the rock is to weather into smooth rounded surfaces with a very large radius of curvature. (Mennell, 1904: 74; the passage quoted in Twidale, 1981a)

By inference, the reasons for the separation of inselbergs reside in decreased fracture density and limited access of groundwater to the more massive bedrock compartments at the stage of subsurface weathering. Field examples of relatively fresh, more massive domed or castellated cores within the saprolite confirm the validity of this reasoning, although usually there are only shallow excavations (less than 20–30 m) available for observations. The concept of an intimate relationship between fracture density and efficacy of subsurface weathering applies to both domed and castellated inselbergs, but neither primary fracture patterns nor spatial scales are identical, hence the diversity of the ultimate form, reflected in its size and appearance.

The relationship between an orthogonal pattern and the progress of weathering is relatively uncontroversial. Primary fractures on granite outcrops are generally easy to identify and distinguishable from secondary joints (Ehlen, 1991), and spatial variability in their density is evident. Differential exploitation of orthogonal fractures has been confirmed for tors, and there is no logical reason why a similar differential etching should not act on a larger scale, to produce castellated inselbergs. It remains to be explained what controls the density of fractures, but this problem seldom occupies geomorphologists.

A much more contentious issue is the nature of structural control on the dome form. Since many, though not all, domes show a pattern of fractures parallel to their topographic surface, the problem of the origin of sheeting is difficult to separate from discussion about domes as larger forms or structures. Two general schools of thought seem to be present. One holds that the dome form is largely primary and governed by structure. This is most evident in the writings of Twidale (1973, 1982; Twidale *et al.*, 1996). He argues that domes are due to lateral compression induced by horizontal stresses at deeper crustal levels. Under such stress the rock mass is not only massive and lacking in open partings, but stress trajectories are also arched upwards. Consequently, discontinuities formed in response to compressive stress are also primary, and those near the surface become more pronounced as the locked-in stresses are relieved.

In the Namib, the dome form of certain granite inselbergs reflects the original shape of stock-like intrusions into schist (Selby, 1977, 1982*b*). This may also be the case in the Mojave Desert, where stock-like massive bodies are seen to protrude through densely jointed granite masses, for example in the Granite Mountains within the Mojave Nature Reserve. According to this mode of dome origin, exogenic processes act merely to strip cover rocks and to expose the form that has already existed in the subsurface. Further modifications of the dome after exposure are minor and are guided by pre-existing fractures which now open up and dilate.

An alternative hypothesis states that the domical form results from degradation of a formerly cuboid rock compartment, through the development of sheeting in response to erosional unloading, spalling off the consecutive sheets and rounding due to various subaerial weathering processes. Gilbert (1904) presented this dilatation model, referring to domes in the Sierra Nevada (thus, not really to inselbergs), and his view was reiterated by many subsequent authors, particularly in general textbooks.

It is evident that dilatation does occur on the slopes of inselbergs and that disintegration of upper sheets may lead to an increase in rounding. Selby (1982*b*) demonstrated these processes at work in the Spitzkoppe group in Namibia, whereas Ollier (1978*b*) showed an excellent example from another inselberg in this area, Mirabib (Plate 3.14). However, it is difficult to visualize the entire process of converting a cube to a dome occurring at the surface, and also to explain how such a cube retained angularity before exposure. In addition, nascent domes are not uncommon and dome-like compartments have been encountered in excavations. Therefore, a scenario envisaged by Thomas (1965) seems more likely, that a morphological dome results from subsurface modification of a massive, perhaps compressed rock compartment through deep weathering and sheeting development in the near-surface position. A similar view has been expressed by Twidale *et al.* (1996: 21–2):

Many bornhardts evolve in two main stages, one involving fracture-controlled subsurface weathering, and another the preferential stripping of some or all of the altered rock. As weathering reduced the size and changed the shape of any compact projections, so would strain trajectories accommodate to the changing outlines of the incipient bornhardt: as weathering by groundwaters proceeded, the flanks of the projections would be steepened, because groundwaters would run off them, and new, steep partings would develop in rough parallelism to the

Plate 3.14 Dilatation operating on the slope of the domed inselberg of Mirabib, Central Namib Desert

surface of the projecting mass of fresh rock. In this way, overlapping lenticular patterns would be formed. This is precisely the pattern of sheeting fractures found in many field localities.

Accordingly, sheet fractures and dome forms develop concurrently, influencing each other.

The fate of a dome after exposure would be mainly fracture-dependent. If the dome possesses few partings other than sheeting joints, augmenting of the dome form may be expected, as shown by Selby (1982b). If, however, orthogonal joints are present, they will start to open up and guide weathering, leading to slope failures and the development of ruiniform, castellated relief.

The influence of petrological differentiation on the location of inselbergs is comparatively minor. In only a few cases has the association of inselbergs with a particular rock property been successfully demonstrated. One example is provided by Marmo

(1956) from Sierra Leone. Massive bornhardts of the Gbengbe Hills tend to have developed in porphyroblastic granites, with large microcline megacrysts. In the Jelenia Góra Basin, south-west Poland, some inselbergs are supported by localized intrusions of fine-grained granite amidst the coarse, porphyritic granite (Migoń, 1997c).

Increasing the spatial scale of enquiry, rock control may be suspected in the location of inselberg landscapes. This follows the observations of many authors that one granite area supports inselbergs and another does not, although they are adjacent to each other (Thorp, 1967b; Eggler et al., 1969; Cunningham, 1971). Brook (1978) approached the problem in South Africa and found that inselberg landscapes tend to develop in granites which have undergone potassium metasomatism, or have an elevated content of low temperature quartz and potassium feldspar. A similar relationship has been found in the Kora region of north-eastern Kenya, where inselberg landscapes are clearly associated with granites enriched with potassium feldspars (Pye et al., 1986). In the Namib Desert intrusions of Jurassic granite have been moulded into inselbergs, whereas the older Salem granite tends to underlie the extensive plains. These contrasting landscape effects probably relate to two factors. First, the younger granite postdates the break-up of Gondwana and suffered very little from any crustal deformation. Therefore it is very massive, whereas the history of deformation of the Salem granite is much longer and its fabric much weaker. Second, Jurassic granites have a very high potassium content, exceeding 6 per cent (Goudie and Migoń, 1997), whereas the Salem granites have a more balanced proportion of potassium to sodium. In all these circumstances, lower weathering rates may be inferred for more massive and potassium-rich granites.

Further Development of Inselbergs

Mechanisms involved in the degradation of inselbergs are similar to those already outlined for tors. Exactly how inselbergs are reduced in height and extent depends on their fracture patterns. Massive domed inselbergs are subject to mega-exfoliation due to the opening of sheeting joints. Individual slabs are separated from the underlying rock mass and fall or slide off the slopes, forming big debris cones mantling the lower slopes. In orthogonal patterns, vertical joints open too, topples occur, and the summit assumes ruiniform relief. If the two fracture patterns co-exist, then both mega-exfoliation and vertical joint opening are involved and degradation into a boulder-strewn inselberg may occur at a particularly fast rate.

Minor weathering features play an important part in the development of inselbergs (Watson and Pye, 1985). Selective weathering along fractures and growth of tafoni reduce rock mass strength and facilitate mass movement, whereas horizontal surfaces are destroyed by enlargement of weathering pits. Caves and massive overhangs are frequently reported from granite inselbergs and develop either through mechanical widening of fractures, preferential weathering along sheeting joints, or chaotic accumulation of big boulders on lower slopes. However, although degradation patterns can easily be outlined in a general manner, detailed studies of inselberg degradation have been very few.

The processes of inselberg and tor destruction are thus broadly similar, but because size matters, inselbergs are expected to have much longer lifetimes. It has to be emphasized here that with reference to residual landforms, of which inselbergs are examples, it is more appropriate to speak of lifetimes rather than ages. They are dynamically evolving landforms, even if the rate of their surface change is low in absolute values, and a no-change situation should not be implied. In fact, their boulder-strewn forms and deep clefts speak very clearly about significant form change after exposure.

For a long time, minimum ages of inselbergs could have been inferred using only a morphostratigraphic approach and it is only the recent advance in cosmogenic isotope dating that has helped to constrain the history of certain hills in Africa (Cockburn *et al.*, 1999; Van der Wateren and Dunai, 2001), North America (Bierman, 1993), and Australia (Bierman and Caffee, 2002).

In Central Europe granite inselbergs rise above extensive surfaces of low relief which have locally retained a mantle of early to mid-Cainozoic sediments, which is the best evidence for their inherited nature (Richter, 1963, Czudek *et al.*, 1964; Gellert, 1970; Migoń, 1997*b*, 1997*c*). In addition, clayey weathering mantles have a wide distribution in these areas and their formation dates back mainly to the Palaeogene (see Migoń and Lidmar-Bergström, 2001 for a comprehensive review). A distant Cainozoic age has also been inferred for inselbergs in southern Sweden (Lidmar-Bergström, 1995) and isolated granite elevations in Buchan, north-west Scotland (Hall, 1987, 1991). Boulder inselbergs of the Mojave Desert in southern California are long-lived too, and have existed for at least a few million years (Oberlander, 1972). The Stone Mountain dome in Georgia, USA, appears to erode at a rate of 5 to 10 m/Ma, whereas the adjacent regolith-covered plain erodes at around 25 m/Ma. As the height of Stone Mountain is 240 m, its presence in the landscape dates back to the Mid-Miocene at least, 12 to 16 Ma ago (Bierman, 1993).

The oldest granite inselbergs appear to occur in Australia (Twidale and Bourne, 1975; Twidale, 1997*b*). Some of them are very ancient, but owe their survival to burial under younger sediments and fairly recent exhumation. Such are the inselbergs in Pilbara, Western Australia, which date from at least the Early Cretaceous, whereas those in the Kulgera-Everard range on the border of South Australia and the Northern Territory may have been buried as far back as the Jurassic. By contrast, granite inselbergs in the Eyre Peninsula, South Australia, are epigene landforms of protracted lifetimes, but views about their evolution over time differ (Bierman and Turner, 1995; Twidale, 1997*a*; Bierman and Caffee, 2002). Twidale appears to envisage very little form change in the Cainozoic and suspects that the top surfaces of these inselbergs approximate a Mesozoic surface. Bierman and co-authors, however, argue that none of the samples analysed for cosmogenic isotope content is saturated, which does indicate mass loss from the inselbergs, although its rate is one of the lowest ever encountered on Earth, between 0.3 to 0.6 m/Ma. These rates, if extrapolated back to the Mesozoic, still imply 20–40 m of surface lowering.

4

Minor Landforms

Weathering Pits

Perhaps the most characteristic of all minor landforms on exposed granite surfaces approaching horizontality are flat-bottomed or, less commonly, hemispherical hollows ranging in diameter from 15–20 cm to a few metres. They are known under a variety of local names, such as *Opferkessel* in German, *pias* in Spanish, *vasques* in French, or *gnamma*, which is an Aboriginal word occasionally used in Australia (e.g. Twidale and Corbin, 1963). In English, these superficial features are collectively described as weathering pits. They are not unique to granite, but are also abundant in sandstone and occur in other lithologies too.

A remarkable flatness of floors of many shallow pits is reflected in another name present throughout the literature, namely that of a 'pan' (e.g. Twidale and Corbin, 1963; Fairbridge, 1968; Dzulynski and Kotarba, 1979). However, and despite a more accurate reflection of the form, the term 'pan' for weathering pits has fallen into disfavour, apparently because an identical name is used to describe much larger, closed topographic depressions within low-angle surfaces in arid lands.

The majority of weathering pits are either closed features or there is a narrow outlet in the form of a channel trending away from the pit (Plate 4.1). Another type is an

Plate 4.1 Twinned weathering pit with overhanging rims, Serra da Estrela, Portugal

'armchair pit', which grows into the rock surface from the side of an outcrop. These are hemispherical and wide open. At many localities pits may coalesce to form extensive networks, or else they are joined by channel-like features (Fig. 4.1).

Weathering pits in granite show a wide range of dimensions. Hollows in excess of 10 m long and 3 m deep have been reported, and the largest ever described is probably one in Australia, measuring 18.3 x 4.6 x 1.8 m (Twidale and Corbin, 1963). Unfortunately, there are very few systematic measurements of large populations of pits, and this severely restricts any attempts to generalize about the size of pits. Goudie and Migoń (1997) provided such a data set for two outcrops in the central Namib Desert (Table 4.1). An interesting observation is that weathering pits in this arid area are much larger than their counterparts in humid temperate latitudes.

Weathering pits have a wide geographical distribution (Fig. 4.2) and claims that they are specific to certain environments or climatic conditions are very difficult to sustain.

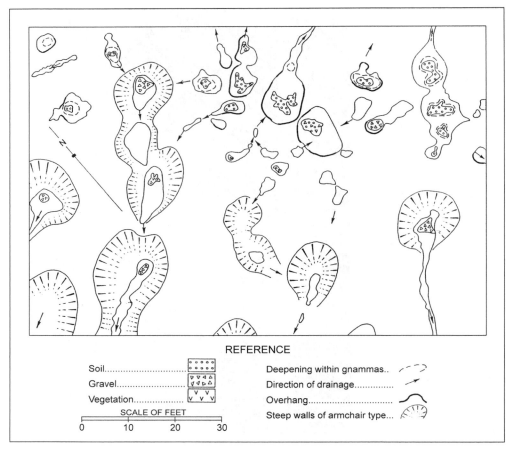

Fig. 4.1 Plan of a weathering pit cluster (after Twidale and Corbin, 1963, fig.4)

They occur on exposed rock surfaces amidst equatorial rainforest in Surinam (Bakker, 1960) and subtropical rainforest in south-east Brazil (Branner, 1913), on tors and inselbergs within the seasonally humid African savannas (Watson and Pye, 1985; Coe, 1986),

Table 4.1 Selected morphometric parameters of weathering pits in the Spitzkoppe area, Namib Desert

	Length (m)		Width (m)		Depth (m)	
	Range	Mean	Range	Mean	Range	Mean
Gross Spitzkoppe site	0.57–8.45	3.39	0.56–5.95	2.0	0.06–0.58	0.24
Klein Spitzkoppe site	0.67–10.70	3.01	0.45–6.50	2.01	0.06–1.00	0.36

Note: Length and width measurements are from rim to rim.
Source: After Goudie and Migoń (1997).

in the semi-arid plains of southern Australia (Twidale and Corbin, 1963) and eastern Mongolia (Dzulynski and Kotarba, 1979), in the semi-arid mountains of the Iberian Peninsula (Centeno, 1989; Pedraza *et al.*, 1989), in the arid Namib Desert (Goudie and Migoń, 1997), in the Mediterranean mountains of Serbia and Macedonia (Gavrilović, 1968), the cool temperate uplands of Central Europe (Czudek *et al.*, 1964), and the high-latitude mountains in Scandinavia (Dahl, 1966; André, 2002). It is true that the high-latitude examples from Scandinavia and North America are much less impressive than weathering pits located in warmer climatic zones. For example, weathering pits researched by Dahl (1966) around Narvik, Norway, are typically up to 15–30 cm in diameter and up to 10 cm deep. Pits reported by André (2002) are somewhat bigger, some well over 1 m long, but they are also very shallow, 6 cm deep at most, and poorly differentiated in respect to the adjacent surface. But one has to be very careful in the interpretation of the difference. Host rock surfaces in Scandinavia were subject to glacial erosion in the Pleistocene and the process of selective weathering and pit development started anew only a few thousand years ago. Hence, one may suspect that the small dimensions of the pits may reflect the limited time available for their growth rather than environmental conditions. This is supported by the observations of Matthes (1950) from the Sierra Nevada. He noted the absence of 'weather pits' (his words) on surfaces overridden during the last mountain glaciation, but they occur in areas covered during the penultimate glaciation as well as in sections of the High Sierra that were never glaciated.

Weathering pits occur on surfaces of different origin. They are present on rock platforms and benches within tors and inselbergs, particularly in their summit parts. They are not uncommon on granite outcrops in coastal settings, for example in Brittany (Guilcher, 1958) or the Malay Peninsula (Tschang Hsi-Lin, 1962). Some of the arid zone examples are located on low outcrops dotting the pediments (Twidale and Corbin, 1963; Dzulynski and Kotarba, 1979; Goudie and Migoń, 1997). Pits are generally rare on loose boulders and this may have three, not necessarily mutually exclusive, explanations. First, weathering pits appear to require a planar surface on which to grow, whereas boulders can be excessively rounded. Second, boulders have rather short lifetimes if compared with tors, therefore there may not be enough time for pits to develop. Third, boulders are usually unstable and may change position, interrupting the process of pit growth at a very early phase.

The origin of weathering pits is a matter of controversy, as is the issue of their growth over time. Most authorities appear to accept the key role of rainwater stagnating in closed depressions on rock surfaces, but individual authors disagree over the actual processes involved. Some indicate solution as the principal chemical process (Fairbridge, 1968; Hedges, 1969) and weathering pits are indeed occasionally described as 'solution pits'. But others see evidence of processes such as hydration (Twidale and Corbin, 1963) or the mechanical action of frost and salt (Fahey, 1986) The uncertainty

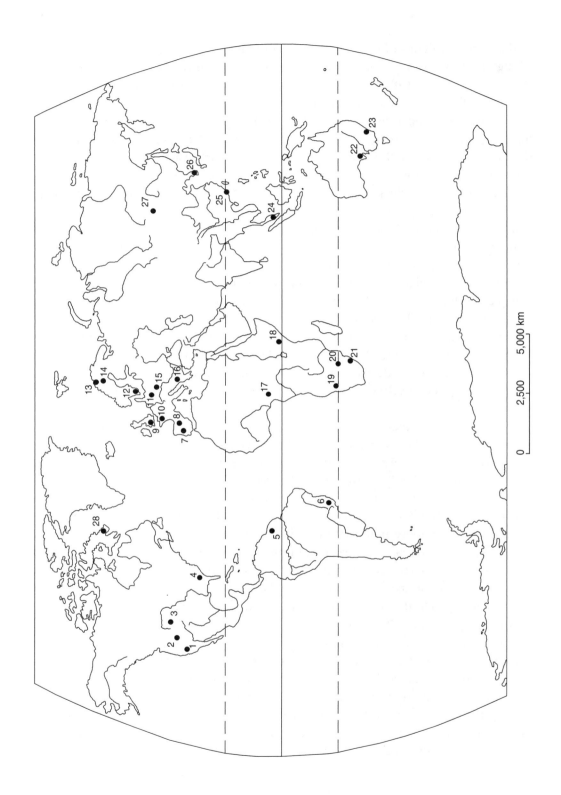

relates to the fact that the origin is usually inferred from the form and known environmental conditions, and not from detailed and systematic measurements of water chemistry or microscopic analyses of the rock. Anyhow, it is likely that at specific sites different processes operate in synergy, or alternate throughout the year as already envisaged by Matthes (1950), and also that weathering pits, as with many other landforms, are examples of equifinality.

The extent to which rock properties control pit initiation and growth is poorly known, but their role must not be dismissed. Clustering of pits on certain outcrops and their absence on adjacent ones is one clue. Coarse granite appears to favour the development of pits, since most reports are from this lithology. It is possible that pits evolve from shallow irregular depressions able to retain water. These, in turn, may relate to minor petrographic differences in the rock, or form due to superficial flaking. That detachment of a flake leaves a shallow rock pool has been observed by a few authors (Twidale and Corbin, 1963; Gavrilović, 1968; Hedges, 1969; Goudie and Migoń, 1997) and evolution from flaking scars may explain why pits smaller than 15–20 cm are practically unknown. Twidale (1982) suggests that pits are initiated under regolith patches, but evidence is very limited. Further growth may follow different pathways, as inferred from the co-existence of different large forms in the Namib Desert (Fig. 4.3).

Growth of a pit requires that the products of rock weathering inside it are constantly or episodically evacuated. This suggests that pit depth is a limiting factor in their evolution because it is more and more difficult to empty the pit if it gets deeper. In fact, the means of debris evacuation are unclear and very deep, closed, and empty pits are very puzzling features (Plate 4.2). Several possible mechanisms have been suggested, namely solutional transport through microfractures, washing out during excessive rainfall and overflow from pits, deflation and flotation. Neither seems to have been actually observed, and each mechanism has its obvious limitations. For example,

Fig. 4.2 (opposite) Global distribution of weathering pits (selected examples): (1) Sierra Nevada (Matthes, 1950, 1956), (2) Idaho (Ikeda, 1998), (3) Laramie Range (Ikeda, 1998), (4) Georgia (Hedges, 1969; Paradise and Yin, 1993), (5) Surinam (Bakker, 1960), (6) south-east Brazil (Branner, 1913; Freise, 1938), (7) Serra da Estrela (author's observations), (8) Sierra da Guadarrama (Pedraza et al., 1989), (9) Dartmoor (referred to in Twidale, 1982), (10) Brittany (Guilcher, 1958; Lageat et al., 1994; Sellier, 1997), (11) Harz (Wilhelmy, 1958), (12) southern Sweden (Lidmar-Bergström, 1989), (13) northern Norway (Dahl, 1966), (14) northern Sweden (André, 2002, 2004), (15) Bohemian Massif (Jahn, 1962; Czudek et al., 1964; Votýpka, 1964), (16) Dinaric Mountains (Gavrilović, 1968), (17) Nigeria (Thomas, 1965), (18) Kora, Kenya (Coe, 1986), (19) Namib Desert (Goudie and Migoń, 1997), (20) Zimbabwe (Żurawek, pers. comm.), (21) Swaziland (Watson and Pye, 1985), (22) Eyre Peninsula (Twidale and Corbin, 1963; Twidale, 1982), (23) south-east Australia, (24) Malay Peninsula (Tschang Hsi-Lin, 1962), (25) Hong Kong (Panzer, 1954), (26) Korean Peninsula (Lautensach, 1950; Ikeda, 1998; Tanaka and Matsukura, 2001), (27) Mongolia (Dzulynski and Kotarba, 1979)

138 Minor Landforms

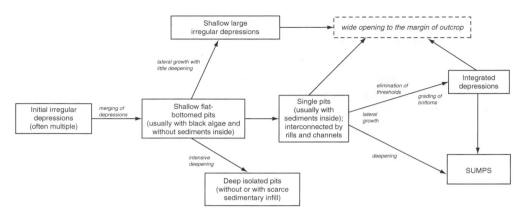

Fig. 4.3 Tentative pathways of weathering pit development, based on observations from the Namib Desert (after Goudie and Migoń, 1997, fig. 13)

Hedges (1969) doubted the efficacy of deflation, except perhaps for the very initial stage of pit development. Washing out of debris does take place from open pits, but is difficult to visualize for very deep and closed pits. In addition, debris observed in pits appears too coarse to be evacuated in suspension, whether by water or wind.

Pits not only deepen, but they also enlarge laterally, perhaps most efficiently if their sidewalls are undercut. At the study site in the Namib Desert, more than two-thirds of the pits investigated show evident undercutting, including deep scallops extending up to 12

Plate 4.2 Giant weathering pit, Erongo, Namibia

cm into the rock, and sidewall collapses have been observed (Goudie and Migoń, 1997). The distinctive elongated shapes of some of them disclose structural predisposition.

How quickly weathering pits grow is little known. There is some scattered information from the Bohemian Massif (Czudek *et al.*, 1964; Votýpka, 1964; Adamek and Kubiček, 1990) which suggests that deepening and widening rates of the order of 0.1–0.5 cm per year are possible. However, there are two possible sources of error. One is the reliability of ancient measurements, another the uncertainty about the identity of profile lines along which pit dimensions have been measured. Most pits do not have clear rims and there is a great deal of liberty in drawing their perimeters. On the other hand, poor development of weathering pits on glacially polished surfaces and moraine boulders suggests that their growth is slow, at least in the initial stage. Megalithic structures also lack well-developed weathering pits. Consequently, pits measuring more than a few tens of centimetres across may be suspected of being older than 10,000 years. Likewise, the lifetimes of weathering pits can only be speculated upon. It is probable that gigantic pits many metres across and deep would have needed a long time to develop, but whether this time period is of the order of tens or hundreds of thousands of years is simply unknown.

Tafoni and Alveoles

Tafoni is a local Corsican name, adopted in geomorphology for the first time by Albrecht Penck (1894) and used hereafter to describe large-scale examples of cavernous weathering incised into vertical and steeply sloping rock surfaces. They are not restricted to granite and occur in a range of rocks, perhaps most commonly in sandstones. (It should be noted that *tafoni* is a plural form, *tafone* being singular.)

Tafoni are difficult to define unambiguously and the literature survey reveals that there has been a great deal of liberty in deciding which features can be described using this term. Specifically, the distinction between two main types of caverns, tafoni and alveolar forms, is problematic. One problem is the size of the hollow. Goudie (2004) suggests that typical tafoni are a few cubic metres in volume. However, many features named as tafoni are evidently smaller, significantly less than 1 metre wide, whereas hollows forming large alveolar networks can be of similar size. Another possible criterion is repetition of form. Alveoles usually occur in groups, not uncommonly along lines of weakness within the rock, and are separated by narrow walls, forming an intricate pattern termed 'honeycomb structures'. Tafoni, by contrast, are singular features and even if two or more tafoni occur side by side, they are usually geometrically different. In addition, many authorities emphasize the occurrence of overhanging margins in tafoni, so the hollows enlarge inside the rock mass (Plate 4.3). Mustoe (1982) suggests that the difference between tafoni and alveoles is genetic and that different processes are involved in their formation, but the evidence is not unequivocal.

Plate 4.3 Tafoni on a boulder inselberg, Mojave Desert, California

Thus, one might agree with the following description of a 'typical' tafone, which states that it has 'spherical or elliptical shape, arched-shaped entrances, concave inner walls, overhanging visors and gently sloping debris-covered floors' (Turkington, 2004: 128), but a wide range of deviations from the standard occurs. In particular, the above characteristics are difficult to apply to an important variant of tafoni, namely to basal tafoni. The latter develop on the undersides of boulders and exfoliation sheets, in the latter case along parting planes, and lack arched entrances and spherical shapes.

The dimensions of tafoni vary. Hollows as small as a few tens of centimetres wide have been reported (Calkin and Cailleux, 1962), but huge caverns many metres wide are also known. Wilhelmy (1977) provided impressive examples from Aruba in the Caribbean, whereas Klaer (1956) noted the presence of large tafoni in Corsica. Tafoni developed in the Spitzkoppe granite in Namibia are up to 8 m wide, 4 m deep, and 3 m high (Migoń and Goudie, 2003; Plate 4.4).

Tafoni are present on granite surfaces of various origin. They are perhaps most common on tor-like outcrops and boulder inselbergs and it is here where the largest dimensions occur. Twidale (1982) noted the presence of side tafoni along distinct horizontal lines, but in other localities their distribution appears random. Large residual boulders may host impressive tafoni, but the smaller ones usually lack the features of cavernous weathering, even if it is present in the surrounding. Likewise, tafoni are very rare on moraine boulder fields, although notable exceptions exist (Calkin and Cailleux, 1962). Coastal outcrops are another setting where tafoni can be found.

Plate 4.4 Big tafone cut into the summit part of a boulder inselberg, Spitzkoppe group, Namib Desert

Tafoni have a very broad distribution worldwide and have been reported from every morphoclimatic zone of the globe (Fig. 4.4). However, their widespread extent must not obscure the fact that they are most common and best developed in arid and semi-arid environments, whereas permanently humid settings appear not to favour tafoni development. For example, they seem absent in the granite landscapes of the British Isles and Scandinavia, and reports about their occurrence in the humid tropics are scarce. On the other hand, Matsukura and Tanaka (2000) analysed an extensive tafoni assemblage from the Mount Doeg-sung in monsoonal South Korea, where annual rainfall is 1,150 mm. One of the classic areas of tafoni occurrence is the Mediterranean region of southern Europe, particularly the island of Corsica (Klaer, 1956). On the other hand, a fine set of tafoni has been described from weathered granite boulders in the Victoria Valley of Antarctica, where a mean annual temperature is $-18°C$ and annual rainfall hardly ever exceeds 100 mm (Calkin and Cailleux, 1962). The tafoni have developed on moraine boulders, the largest being up to 50 cm deep.

The origin of tafoni remains enigmatic. Despite tens of publications focused on tafoni (not all of them granitic), little progress has been made towards elucidation of the problem. Actually, there are two interrelated issues of tafoni initiation, namely the reasons for preferential weathering, and then their development. One hypothesis states that the hollows were originally occupied by a weatherable material, but as Twidale (1982: 291) remarked: 'It is difficult to test this hypothesis because the alleged different

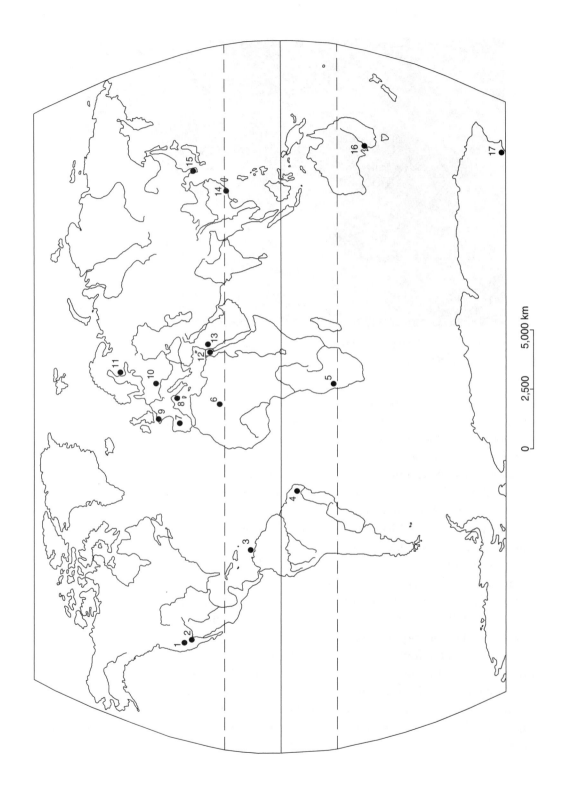

material has been eroded, so that comparisons are impossible. It would be strange, however, if weathering and erosion had everywhere proceeded to a stage where the different material has been completely evacuated'. What might be accepted, is that selective weathering of a schlieren or a pegmatitic vein provided an initial, perhaps small, hollow, which was then enlarged to dimensions exceeding those of the original weak material. It is an unlikely general explanation though, and direct rock control on tafoni localization is generally discarded. However, Matsukura and Tanaka (2000) maintain that initiation of tafoni development is determined by non-uniformity in rock properties and related higher susceptibility to weathering in specific places. They cite the lack of preferential geographical orientation of tafoni openings as corroborative evidence, but the remark quoted above still applies. Thus, the study of tafoni faces the same problem as studies of other weathering features: as soon as weathering processes create a landform, the opportunity to recognize reasons for their selectivity is lost. Twidale's (1982) own favoured explanation is thus the initiation of tafoni beneath the landsurface, due to preferential weathering in the moist scarp-foot zone. Another mechanism is associated with the phenomenon of case hardening (Goudie and Viles, 1979), or surface crusting, widely present on rock surfaces, especially in arid and seasonally dry lands. It is supposed that breaking of the crust exposes the inner, weaker part of the rock mass, which weathers much faster than the rate at which the outer surface is destroyed, hence the inward enlargement of a tafone. Granite tafoni in the Namib Desert are evidently associated with surface crusts (Migoń and Goudie, 2003), but other authors have remarked on the disassociation of tafoni and case hardening (Dragovich, 1969; Conca and Rossman, 1985). Furthermore, crusting may postdate rather than predate the initiation of a cavern.

A range of processes is likely to be involved in tafoni growth. Among these, salt weathering is commonly invoked as the cause of rock disintegration (Bradley *et al.*, 1978), although Dragovich (1969) did not find any clear evidence of salt attack in South Australian tafoni. The dampness of the shaded interior of a tafone is another factor promoting weathering and hollow growth (Blackwelder, 1929). Weathering inside a tafone seems concentrated on the back wall, which typically shows widespread

Fig. 4.4 (*opposite*) Global distribution of tafoni (selected examples): (1) Alabama Hills, California (author's observations), (2) Mojave Desert (Blackwelder, 1929), (3) Aruba (Wilhelmy, 1977), (4) eastern Brazil (Wilhelmy, 1958), (5) Namib Desert (Migoń and Goudie, 2003), (6) Ahaggar (Wilhelmy, 1958), (7) Serra da Guadarrama (Pedraza *et al.*, 1989), (8) Corsica (Klaer, 1956), (9) Brittany (Sellier, 1998), (10) Bohemian Massif (Czudek *et al.*, 1964; Czerwiński and Migoń, 1993), (11) southern Finland (Kejonen *et al.*, 1988), (12) Sinai Peninsula (Wilhelmy, 1958), (13) south-west Jordan (author's observations), (14) Hong Kong (Panzer, 1954), (15) Korean Peninsula (Matsukura and Tanaka, 2000), (16) southern Australia (Dragovich, 1969), (17) Victoria Land, Antarctica (Calkin and Cailleux, 1962, fig. 7)

144 Minor Landforms

flaking and spalling. Flakes and spalls accumulate on the tafone floor, are subject to further disintegration, and are eventually removed by creep, wind, and organisms.

Controls on the growth of granite tafoni have recently been addressed by Matsukura and Tanaka (2000), who systematically analysed rock hardness and moisture content in more than 50 forms hollowed into granite tors on Mount Doeg-sung in South Korea. Their study shows that tafoni interiors are associated with reduced rock strength as indicated by Schmidt hammer readings consistently lower than those for outer vertical rock faces and, at the same time, with elevated rock moisture content (Fig. 4.5). These findings aid our understanding of where and why tafoni enlarge, but the problem of initiation of tafoni development remains unresolved.

Little is known about the rates of tafoni formation but scattered observations point to their rather slow growth. First, it appears that no tafoni-like features have been reported from weathered granite building stones, which is at variance with the situation reported from historical sandstone and limestone constructions, where vertical surfaces are pierced by caverns relatively often. Second, granite boulders forming moraine ridges of late glacial age show few signs of cavernous weathering, which suggests that timescales of $c.10$ ka are insufficient for any more extensive growth of

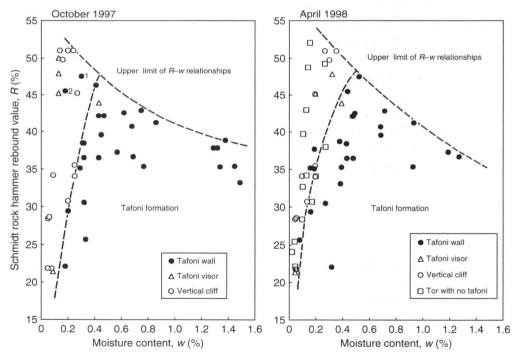

Fig. 4.5 Rock properties within and outside tafoni in Korea (after Matsukura and Tanaka, 2000, fig. 7)

Plate 4.5 Alveolar weathering in granite, Žulova granite massif, Czech Republic

tafoni in granite. Third, very large examples of tafoni tend to occur on boulder inselbergs and collapsing domes, hence host landforms which are likely to have considerable lifetimes, perhaps measurable in hundred thousand or million years.

In distinct contrast to tafoni, alveolar structures, known also as honeycomb structures (or honeycomb weathering), are not very well developed on granite (Plate 4.5). The relative paucity of honeycombs is probably related to grain size, the distance between adjacent points of structural weakness, and the tightness of the fabric, all being much bigger in granite relative to other rocks. One may assume that rough pitted surfaces of granite outcrops are equivalent to honeycombed surfaces developed in fine-grained, loosely cemented, and bedded sandstones.

Polygonal Cracking

Some surfaces of granite outcrops exhibit networks of shallow cracks outlining polygons of variable geometry. Such surface patterns visually resemble tortoise shells and have been described as *Schildkrötenmuster* in German literature (Schülke, 1973). The polygons can be pentagonal or hexagonal in plan, but more often ideal polygonal shapes are deformed. Therefore, the most common pattern appears to be irregular, with polygons of different shape and size occurring side by side (Plate 4.6). Polygonal patterns are known from both isolated boulders and weathered faces of rock outcrops, and occur on flat, sloping, and overhanging surfaces (Twidale, 1982).

Plate 4.6 Irregular polygonal cracking on a boulder, Karkonosze Mountains, south-west Poland

It is difficult to generalize about the dimensions of polygons. It appears that diameters within the range of 10–20 cm prevail, but smaller shapes are not uncommon. Cracks penetrate a few centimetres into the rock, often a mere 1–2 cm. Their width decreases from the surface downwards; at the surface cracks may be more than 2 cm wide. Nested networks may occur, with smaller polygons of less than 5 cm across located within the boundaries of much larger features of more than 20–30 cm in diameter. Typically, polygonal patterns do not have sharp boundaries, but grade into irregular cracks randomly intersecting the rock surface instead. Surface crusting (induration) is a phenomenon associated with many, though not all, polygonal patterns (Plate 4.7).

Polygonal cracking has been reported from virtually all climatic regions of the world (Williams and Robinson, 1989) (Fig. 4.6). Given this widespread distribution, the superficial nature of polygonal networks, and their presence on boulders,

Minor Landforms **147**

Plate 4.7 Extensive polygonal cracking associated with surface crusting, Erongo, Namibia

the latter two both indicating rather young age, it is unlikely that polygonal patterns require any specific macroclimatic conditions to develop. They are azonal landforms.

Polygonal cracking is not restricted to granite outcrops. The phenomenon is known from a variety of rock types, from igneous to metamorphic, and is probably best developed on sandstone (Williams and Robinson, 1989). It has been claimed that coarse granite does not yield to the development of polygonal networks (Twidale, 1982), but some of the networks in Central Europe have in fact developed in coarse rather than even-grained variants (Czerwiński and Migoń, 1993). On the other hand, extremely regular, square shapes resembling a chessboard are known from fine-grained aplite veins cutting granite in the same area.

The origin of polygonal networks is not fully understood and many explanations have been offered, the majority of them being site-specific. Twidale (1982) suggested

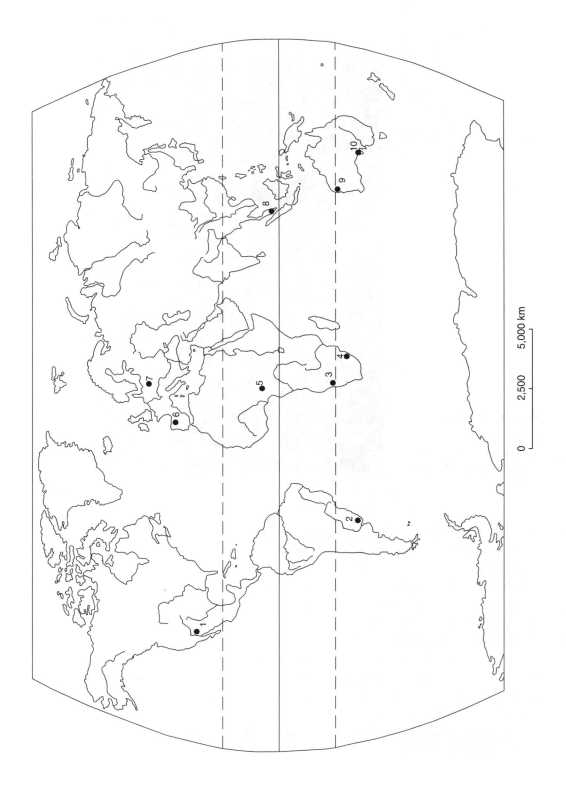

that polygonal patterns develop in the subsurface as a result of precipitation of iron and manganese oxides around corestones, but cracks are also present on boulders which are certainly not exposed corestones (Watson and Pye, 1985; Williams and Robinson, 1989). Williams and Robinson (1989) reviewed most of the proposed theories and concluded that cracking is causally related to surface crusting. It is envisaged that the surface crust is too rigid to accommodate the stresses induced by temperature and moisture changes to which the rock surface is subjected and it therefore cracks. The geometry of networks may be related to rock texture, with more regular patterns developing on more homogeneous bedrock.

Karren

Karren (rillenkarren) or flutes are grooves cut into sloping and vertical rock surfaces. They vary in size, sinuosity, and cross-section, and elaborated classifications have been proposed to account for this variability. Isolated forms do occur, but systems of parallel grooves are much more common. Diverging and converging networks are also known. Karren are usually considered as one of the typical karst surface phenomena associated with limestones, which is correct, but there are surprisingly many examples of similar landforms that have developed on sloping and vertical granite surfaces.

The geographical range of the occurrence of karren is very wide (Fig. 4.7). Most examples have been reported from humid tropical environments (Wilhelmy, 1958) and these are evidently the most impressive, an example being the fluted coastal boulders of the Seychelles, so often portrayed in tourist brochures (Plate 6.7). But they are not restricted to high temperature and high rainfall areas. Karren occur in many European granite uplands, from the Massif Central in the west (Plate 4.8) to the Bohemian Massif in the east, and have been shown to occur in the granite mountain ranges of the Iberian Peninsula. Likewise, they are not absent in drylands, including such very arid areas as the Namib Desert. Few examples, however, have been described from northern latitudes and those present are much less distinctive than their tropical counterparts (e.g. Dahl, 1966). This geographical distribution suggests that the development of fluted surfaces on granite is not climate-dependent, in the sense that there is no narrow range of climatic conditions favouring their origin. But the influence of climatic conditions cannot be fully dismissed, as the restriction of classic examples to the tropics shows. On the other hand, some observers consider karren in silicate rocks an inherently tropical

Fig. 4.6 (*opposite*) Global distribution of polygonal cracking (selected examples): (1) Arizona (Leonard, 1929), (2) Uruguay (Wilhelmy, 1958), (3) central Namibia (Migoń and Goudie, 2003), (4) Swaziland (Watson and Pye, 1985), (5) Cameroon (Mayer, 1992), (6) Serra da Guadarrama (Pedraza *et al.*, 1989), (7) Bohemian Massif (Czerwiński and Migoń, 1993; Chábera and Huber, 1996), (8) Malay Peninsula (Twidale, 1982), (9) Pilbara (Twidale, 1982), (10) Eyre Peninsula (Twidale, 1982)

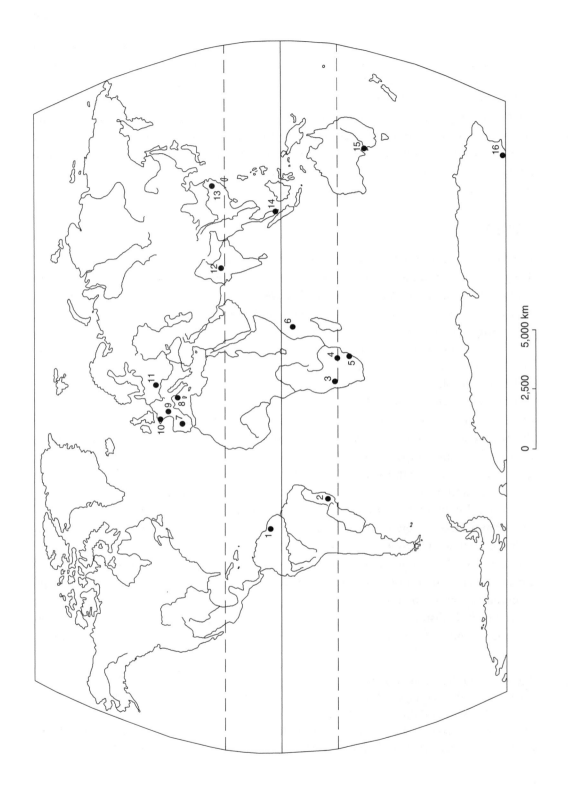

Plate 4.8 Impressive karren networks on granite boulders, Pierre de Jumaitres, Massif Central, France

phenomenon and consequently, they interpret karren occurrences in mid-latitudes as evidence of a tropical or subtropical inheritance (e.g. Wilhelmy, 1958; Czudek *et al.*, 1964). These views have been criticized by Thomas (1976), and the occurrence of fine flutes on megalithic standing stones in Brittany, France, erected in their present position *c.*5,000 years ago, speaks clearly for the possibility of karren development in a cool humid temperate climate (Lageat *et al.*, 1994).

Judging from the published accounts, granite karren are less diversified in their form than similar landforms developed on limestone. Most karren have a wide U-shaped cross-section and are rather shallow. They tend to follow straight lines going downslope, while meandering and deeply incised variants appear unknown. Karren are present on isolated boulders, tors, and sloping surfaces of granite domes in a variety

Fig. 4.7 (*opposite*) Global distribution of karren (selected examples): (1) Surinam (Bakker, 1960), (2) south-east Brazil (Branner, 1913; Wilhelmy, 1958), (3) Namib Desert (Migoń and Goudie, 2003), (4) Zimbabwe (Whitlow and Shakesby, 1988), (5) Swaziland (Watson and Pye, 1985), (6) Seychelles (Wilhelmy, 1958), (7) Serra da Guadarrama (Pedraza *et al.*, 1989), (8) Corsica (Klaer, 1956), (9) Massif Central (author's observations), (10) Brittany (Lageat *et al.*, 1994; Sellier, 1997), (11) Bohemian Massif (Czudek *et al.*, 1964; Czerwiński and Migoń, 1993), (12) Aravalli Range (Sen, 1983), (13) Huangshan (Wissmann, 1954), (14) Malay Peninsula (Tschang Hsi-Lin, 1961), (15) Kangaroo Island, southern Australia (Dragovich, 1968), (16) Victoria Land, Antarctica (French and Guglielmin, 2000)

of environmental settings, from coasts to high-altitude uplands and mountains. They are essentially surface phenomena, although some reach the rock-soil junction and extend beneath the regolith cover (Logan, 1851, cited in Twidale, 1982).

Whether lithology and structure control the localization and development of karren is open to dispute. Watson and Pye (1985) presented examples of karren that had developed preferentially in fine-grained granite on Sibebe Hill in Swaziland, whereas the adjacent coarse granite does not show fluting. However, most authors emphasize the lack of evident control and an apparent independence of karren from fractures.

Much of the discussion of the origin of granite karren has focused on the issue of whether they deserve to be called a 'karst landform', that is, whether they are formed through solution, or if we are dealing with a pseudokarstic phenomenon, that is, the similarity of form does not result from identical formative processes. Thus, Tschang Hsi-Lin (1961) attributes the origin of karren at Pulau Ubin, Singapore, to the erosional action of water flowing off the rock surface, and a similar interpretation has been put forward by Dragovich (1968) in respect of the Remarkable Rocks outcrop in South Australia. The mechanical action of running water was also emphasized in a much earlier work by Branner (1913) on south-east Brazil.

But abrasion as the sole process is difficult to accept on several grounds, and it certainly cannot provide a universally applicable explanation of karren. First, it is a problem of channel initiation. An initial linear depression on an otherwise smooth, often slightly convex rock surface is unlikely to be produced by the mechanical action of water, as there is no plausible mechanism to account for flow concentration along a certain line. Second, it is difficult to visualize abrasion acting efficiently on a vertical rock surface, on which water trickles rather than flows turbulently. Third, the initiation of a concentrated flow in karren starting from the upper edge of an outcrop poses a problem, and in fact catchment areas of karren are invariably very small. Fourth, it is uncertain what the abrasive tools would be. Individual grains loosened by weathering should fall down immediately after detachment, leaving no time and space for abrasion. What needs to be remembered in this discussion, however, is that karren are rather vaguely defined in respect of the slope of their host rock surfaces. Therefore, although abrasion is unlikely on vertical rock walls, it may play a part in the development of karren on gently sloping surfaces, with an incline of 20–30°. Moreover, the problem of flow volume is overcome for rills initiated at the margin of weathering pits and these karren may sporadically act as overflow channels.

Another mechanism has recently been suggested by Mottershead and Lucas (2004), who reassessed the relative significance of mechanical versus solutional processes. They argue that the role of solution has been overestimated and that biotic agencies are important. Organisms act towards loosening rock fabric within karren and create a range of nano-morphologies, which make the outer rock surface susceptible to particle release under the impact of raindrops. Their paper concerns soluble rocks, but this is an even more attractive hypothesis for silicate rocks. The preferential occurrence of granite flutes in subtropical areas subjected to heavy seasonal rainstorms seems compatible with the mechanism proposed.

Difficulties with mechanical action have diverted attention to chemical processes, or mechanical and chemical processes combined. A model advocated by Birot (1958) and Czudek et al. (1964) involves enhanced weathering within karren because of surplus moisture brought about by trickling water, loosening of grains due to various chemical reactions, including solution, and their subsequent evacuation by water flowing down the depression. The presence of lichens may be significant and these may act both directly, through biochemical processes, and indirectly, by retaining water and hence augmenting weathering processes. But lichens are not universally present within karren. Moreover, as Twidale (1982) points out, lichen colonies would preferentially grow in already established depressions, hence the problem of karren initiation remains open.

To overcome these difficulties, Twidale (1982) suggests subsurface initiation of depressions and cites the occurrence of karren in distinct horizontal bands on dome slopes to support the model. However, as he admits himself, direct evidence is not available apart from one report by Logan from Singapore. On the other hand, there is clear evidence for a subaerial origin of karren, such as their occurrence on Neolithic standing stones (Lageat et al., 1994; Williams and Robinson, 1994; Sellier, 1997).

Therefore, despite Twidale's assertion that 'these forms ... have been *merely* enlarged and modified after exposure' (1982: 277; emphasis added), it appears that the karren owe their form principally to surface rather than subsurface processes, although the exact contribution of each process remains poorly understood and is likely to be site-specific. Moreover, it is almost certain that fluted surfaces are the setting for two competing groups of processes. There are those active within karren, acting towards their deepening and enlargement, and those working on the intervening crests, leading to their destruction and hence, attenuation of the karren. It is the balance between these two groups which is crucial for the fate of the karren.

To return to the karstic versus non-karstic nature of karren on silicate rocks, no instances of actual flow within the karren, especially on steep and vertical surfaces, have been reported as far as the present author is aware. Consequently no data about water chemistry are available. Until such data are at hand, the discussion about the development of karren and their karstic or pseudokarstic nature will remain largely speculative and limited to process-from-form inference.

Flared Slopes

A specific feature of many granite outcrops, but again not restricted to granite, is the presence of basal concavities. These concavities essentially occur in two variants, either as a basal zone of rock slope steepening or as an overhang (Plate 4.9). Collectively they

Plate 4.9 Flared side of an outcrop, Serra da Estrela, Portugal

are called 'flared slopes' (Twidale, 1962, 1982) and are typified by the famous Wave Rock, at Hyden Rock in Western Australia and inselbergs in the Eyre Peninsula near Adelaide, in southern Australia (Twidale, 1962; Twidale and Bourne, 1998), but can be found worldwide.

Flared slopes show a wide range of dimensions, even along a single outcrop. They may be as low as 0.5 m, but spectacular overhangs more than 10 m high are also known, such as the 14 m high and about 100 m long Wave Rock. They can be of very localized occurrence, but may also be present virtually all along the perimeter of a hill, for example at the Pildappa Hill, Eyre Peninsula, where they make up 95 per cent of the lower slopes (Twidale, 1962). They accompany large-scale landforms such as inselbergs, but flared sides of individual boulders are not uncommon. Flares are not restricted to the outer margins of outcrops. Fine examples can be found along fracture-guided clefts, which then acquire a distinct, trough-like cross-section. Flared slopes are known to occur at several height levels, one above the other, and these are called multiple flares.

The credit for elucidating the origin of flared slopes goes to Twidale (1962, 1982). After dismissing alternative hypotheses he states: 'The only explanation that takes account of all the evidence is that the flares are a particular form of etch surface or weathering front developed in the scarp foot zone as a result of moisture attack on massive rocks and subsequently exposed' (Twidale, 1982: 250).

The key observational evidence for the subsurface origin of flares includes the occurrence of concavities beneath the topographic surface, at shallow depth below the regolith, and that these tend to follow the local slope, which is also the rock–regolith junction, rather than any horizontal lines. It is assumed that the rock–regolith junction is the boundary between two weathering environments characterized by contrasting intensities of moisture-related decay. Bare rock above sheds rainwater and remains largely dry, therefore the effects of weathering are very limited. On the other hand, water derived from both rainfall and runoff from bare rock surfaces is stored within the regolith, elevating the level of soil moisture, and consequently, the efficacy of weathering increases. With the passage of time, this contrasting behaviour would cause significant steepening of the rock–regolith boundary surface around an outcrop (Fig. 4.8).

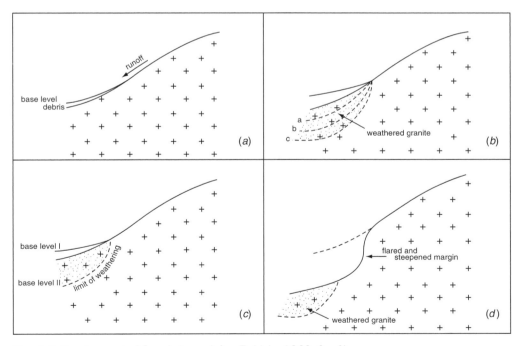

Fig. 4.8 Development of flared slopes (after Twidale, 1962, fig. 8)

The model requires a period of land surface stability, during which flared slopes could form in the subsurface, to be exposed after regolith denudation accelerates. It may be hypothesized that the better developed are the flares that is, the higher and steeper they are, the longer the period of past surface stability. However, other factors may contribute to the development of large flares. Twidale and Bourne (1998) note that reasons for the unusual dimensions of Wave Rock also include higher availability of water from upslope due to its location in a re-entrant and lateral influx of saline water from the nearby river. Extending the argument, alternating periods of stability and instability can be inferred from the occurrence of multiple flares. Indeed, Twidale and Bourne (1975) used flared slopes, among other minor landforms, to propose a model of episodic exposure of inselbergs and tried to constrain the exposure history of inselbergs in the Eyre Peninsula.

Microforms as Climate-Related Features

The wide geographical distribution of microforms might suggest that climatic conditions matter little in their origin and evolution. However, diverging views have been presented in this respect, especially as far as weathering pits, tafoni, and karren are concerned. According to one school of thought, certain environmental conditions favour the development of certain microforms and, at the same time, disfavour or significantly suppress the development of others. This reasoning subscribes to the philosophy of climatic geomorphology and has been presented most elaborately by H. Wilhelmy (1958). In particular, he suggested that karren in crystalline rocks are humid tropical features, tafoni typify Mediterranean climates, whereas the origin of weathering pits is favoured in humid, mainly humid temperate climatic zones. An opposing view holds that no such climate–landforms relationships should be implied, but that it is the rock itself which is decisive for the development of surface microrelief. Another possibility is that whereas regional climatic conditions may be largely irrelevant, local environmental factors such as aspect or moisture availability govern the evolution of minor surface landforms. Twidale (1982), in turn, suggests that microforms are initiated at the weathering front, in the subsurface, and merely become modified after exposure; hence again regional climatic setting does not matter very much. Of course, reconciling views emphasizing both rock and climate are also possible and have indeed been presented.

The key implication of Wilhelmy's concept regards the interpretation of microforms in relation to the present-day climate and climatic history. Since the microforms are claimed to be climate-dependent, then their relationship to contemporaneous climatic regime is one of either equilibrium or disequilibrium. For example, rillenkarren in the humid temperate zone of Central Europe would illustrate a state of disequilibrium because, according to the model, they are surface features characteristic of humid tropical environments. Consequently, their inherited nature is implied and rillenkarren acquire palaeoclimatic significance, becoming a proxy record of past humid tropical conditions. In specific cases, it is maintained, a long-term history of climatic change may be deciphered from surface microrelief. In the opinion of Wilhelmy himself, the granite landscape of Corsica is an excellent example of such a long-term inheritance:

In the spatially restricted area of Corsica there occurs an assemblage of geomorphic phenomena present in massive rocks in most climatic zones of the world: deep arenaceous and ferrallitic weathering, subsurface development of rounded blocks and their subsequent evolution at the surface, tors, pedestal rocks, tafoni, weathering pits and flutings. There are block fields and block streams, convex domes, sharp peaks and jagged ridges. *Tertiary, Pleistocene and recent* forms in different climatic altitudinal belts make Corsica an El Dorado for studies of the diversity of landforms developed on crystalline rocks. (Wilhelmy, 1958: 204; present author's translation; emphasis added)

In various subsequent studies Wilhelmy's model was adopted as a conceptual basis for landform interpretation, not only in basement terrains (e.g. Bremer, 1965, for Ayers Rock, central Australia). In the context of granite, the work by Czudek *et al.* (1964) is a good example. They noted the presence of tafoni, alveoles (dew holes), rillenkarren, and weathering pits on granite outcrops of the Bohemian Massif and interpreted their co-existence as the result of superimposition of relief generations of different ages. Then they compared the geomorphic evidence with the palaeoclimatic record to constrain the timing of microform formation. Rillenkarren would have dated back to the Miocene, tafoni would have been inherited from the seasonally dry environments of the Pliocene, whereas weathering pits are in equilibrium with the present-day climate and continue to enlarge and deepen. The scheme developed by Czudek *et al.* (1964) has been adopted in a number of subsequent papers focused on granite landforms in the Bohemian Massif.

The above model has not gone unchallenged. Thomas (1976: 438) commented that: 'The climatic significance of many of these forms is elusive, although some can be related to prevailing climatic conditions. It must be remembered, however, that it is the microclimate of the rock surface which is important, and that the weathering stage of the exposed rock surface may influence the rate of sub-aerial modification'.

There is also observational evidence of co-existence of different minor surface landforms within a relatively small area, or even within a single outcrop (Watson and Pye, 1985) for which no independent evidence of inheritance is at hand. Moreover, there are no reasons to claim that any of these landforms are currently inactive. For example, Migoń and Goudie (2003) described granite landforms of the Central Namib Desert, which is an area of long-term aridity, established at least in the Middle Miocene. The inventory includes huge weathering pits, big tafoni, rillenkarren, as well as cracked boulders, hence features indicative of all possible earthly environments,

158 Minor Landforms

according to the climatic model (Plate 4.10). But all these landforms are actively evolving at present. Weathering pits are subject to extensive flaking, scalloping, and granular disintegration (Goudie and Migoń, 1997), rear sides of tafoni shed flakes as well, and karren host algae cover and channel occasional runoff as evidenced by grus accumulation at their outlets.

Twidale (1982) took a different approach and emphasized sculpting of many minor features at the weathering front rather than at the surface. Initial concavities are provided by selective weathering attack at the rock/regolith interface and channelization of subsurface flow along this interface. These may be further enlarged after exposure, according to the principle of positive feedback or reinforcement. However, Twidale's hypothesis cannot be universally accepted and many authors have commented that they have not found any evidence for subsurface initiation (e.g. Watson and Pye, 1985; Williams and Robinson, 1994).

In fairness to the proponents of the climatic approach it has to be noted that indeed, in certain areas, some microforms are much more common than others. It has long been observed that weathering pits are abundant in Dartmoor and other granite uplands in south-west England, but tafoni and rillenkarren are extremely rare. Williams and Robinson (1994: 414) even claimed that 'no fluting has been reported on granite outcrops anywhere in the British Isles', but a group of parallel flutes does in fact occur

Plate 4.10 Large weathering pits in the Spitzkoppe group, Namib desert, are actively evolving today, despite the scarcity of rainfall and decaying organic matter. A rock-cut pediment is seen in the background

on Rough Tor on Bodmin Moor. Microforms in the Karkonosze Massif, south-west Poland, occur in similar proportions (Czerwiński and Migoń, 1993; Migoń and Dach, 1995), which suggests that weathering pits are evolving at present, whereas the development of flutes and tafoni is for some reasons severely inhibited.

The controversy concerning the significance of microforms persists because processes involved in their origin and development are inadequately known. For example, enlargement of weathering pits tends to be considered the result of destructive action by standing water rich in organic matter, but the huge weathering pits in the Namib do not hold water nor is there much organic matter there to cause biochemical corrosion. Uncertainty about the processes delays more definite environmental affinities of microforms from being established. In addition, one must not a priori rule out that similar surface features can be produced by the action of different processes. An even more attractive hypothesis is that localized surface conditions on rock outcrops are critical, including microclimatic characteristics as hinted at by Thomas (1976). Rillenkarren may require an abundant water supply to form and this is why they are perfectly developed on sprayed coastal outcrops in the Seychelles, but have also quickly formed on Neolithic standing stones in Brittany, France (Lageat et al., 1994; Sellier, 1997, 1998), where rainfall is abundant and salt is transported from the coast inland. For the development of tafoni and alveoles, high-frequency changes in moisture variability and salt availability appear to play a key role, whereas temperature and annual rainfall matter much less. Therefore, tafoni can and do evolve in such contrasting environments as the arid Namib Desert affected by coastal fog, the Mediterranean island of Corsica, the monsoonal Korean Peninsula, and Antarctic valleys (Fig. 4.4).

An alternative approach to the subject considers the stability of affected surfaces as the central point (Bremer, 1982). Microforms are manifestations of selective weathering and undoubtedly develop from incipient concavities on flat, sloping, or vertical rock surfaces. To attain the present dimensions, up to many metres long or deep, a protracted period of time appears to be required. Unfortunately, the rates of microform growth can only be speculated upon, but it is significant that granite surfaces exposed in historical times have no or very insignificant relief. Whereas stone deterioration literature contains impressive evidence of rapidly developing honeycomb features on sandstone and limestone (e.g. Goudie and Viles, 1997), no comparable examples exist on granite. Even granite surfaces exposed a few thousand years ago, like those in Egypt, have their relief relatively smooth. Flutings on standing stones in Brittany are exceptional in that they have had about 5,000 years to develop (Lageat et al., 1994; Sellier, 1998). Likewise, 10–20,000 years of surface exposure in granite areas covered by late Pleistocene ice sheets and glaciers has not yet resulted in the development of diversified microrelief (e.g. Matthes, 1956; Swantesson, 1989; André, 1995). An additional factor here might be the smoothness of the initial bare rock surfaces, which does not facilitate 'getting the weathering system started'. It seems therefore, that big pits, voluminous tafoni, and deep flutings may require a few tens of thousands of years at least to attain their size. As surface slope and aspect are important factors controlling surface weathering, it may be assumed that if they change, the intensity of weathering attack will change too and the evolution of a microform can be interrupted. Tafoni may require a particularly protracted stability as they often develop inside boulders which have their outer surfaces crusted. Hence, there has to be a time for the crust to develop, and then for a hollow to grow inward, although at a later stage both processes may proceed concurrently.

In this context it might be helpful to consider surface stability in different settings and environments. In polar environments and at high altitudes mechanical rock breakdown, whether into angular fragments or individual crystals, is efficient due to high-frequency changes in temperature, including freezing and thawing, moisture, and radiation, therefore surface stability in the timescale required is hardly possible. Similarly, movement of blocks within block fields and moraines precludes prolonged stability. An apparent absence of microforms in most, though not all, high-latitude settings may thus be not so much an effect of unfavourable climatic conditions, but of insufficient surface stability. In much the same way, the paucity of microforms in arid areas noted by Wilhelmy may be explained by ongoing flaking and granular disintegration induced by thermal shock and salt weathering. It is also significant that some of the best developed microform assemblages occur on domes and platforms built of post-kinematic granites in cratonic settings. This is the situation of granite landforms in Erongo and Spitzkoppe in Namibia (Migoń and Goudie, 2003), in Swaziland (Watson and Pye, 1985), in Matopos, Zimbabwe (Żurawek, pers. comm.), and on Australian inselbergs (Twidale, 1982). All these granites are extremely massive, with wide fracture spacing, and they provide an exceptional surface stability. In fact, surface exposure dating using cosmogenic isotopes performed on some of the Namibian and Australian inselbergs (Van der Wateren and Dunai, 2001; Bierman and Caffee, 2002) confirmed the exceptionally low rates of surface lowering in these settings. With such ample time available, different microforms could have developed and they now co-exist, contradicting claims that climate is the prime control on their development.

Slope Development in Granite Terrains

Rock Slopes

Granite Rock Slopes—Form and Geological Control

Rock slopes developed in granite may take different forms, as reflected in their longitudinal profiles. Field observations and a literature survey (e.g. Dumanowski, 1964; Young, 1972) allow us to distinguish at least five major categories of slopes: straight, convex-upward, concave, stepped, and vertical rock walls (Fig. 5.1). In addition, overhang slopes may occur, but their height is seldom more than 10 m high and their occurrence is very localized. These basic categories may combine to form compound slopes, for example convex-upward in the upper part and vertical towards the footslope. Somewhat different is Young's (1972) attempt to identify most common morphologies of granite slopes.

162 Slope Evolution

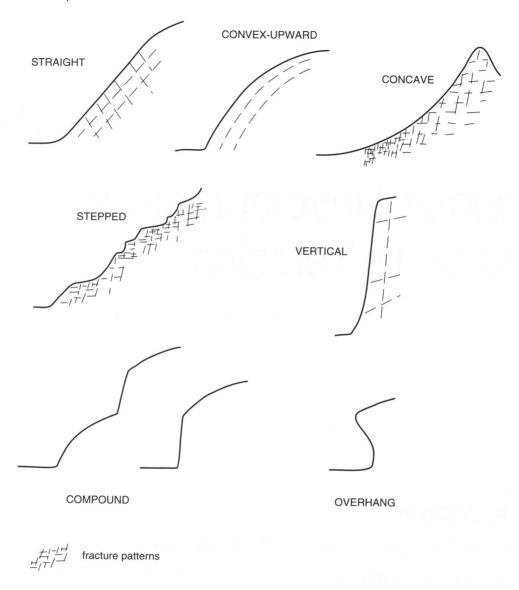

Fig. 5.1 Diversity of rock slope profiles encountered in granite areas and their relationship to fracture density

He lists six major categories:

(1) bare rock domes, smoothly rounded or faceted;
(2) steep and irregular bare rock slopes of castellated residual hills, tending towards rectangular forms;
(3) concave slopes crowned by a free face;

(4) downslope succession of free face, boulder-covered section and pediment;
(5) roughly straight or concave slopes, but having irregular, stepped microrelief;
(6) smooth convex-concave profile with a continuous regolith cover.

The latter, lacking any outcrops of sound bedrock, are not considered as rock slopes for the purposes of this section. Young (1972) appears to seek explanation of this variety in climatic differences between regions, claiming that 'Variations of slope form associated with climatic differences are as great as or greater, on both granite and limestone, than the similarity of form arising from lithology' (Young, 1972: 219). This is a debatable statement and apparently contradicted by numerous field examples of co-existence of different forms in relatively small areas.

Slope forms do not appear specifically subordinate to larger landforms but occur in different local and regional geomorphic settings. For example, the slopes of the Tenaya Creek valley in the Yosemite National Park include, in different sections of the valley, straight, vertical, convex-upward, and concave variants (Plate 5.1). Apparently, multiple glaciation was unable to give the valley a uniform cross-sectional shape. Similarly, inselbergs occur in different shapes and may have all the basic slope forms listed above. One exception is bornhardts, which by definition have convex-upward shape, although their decay may gradually transform these slopes into multi-faceted, step-like ones.

Within each type of granite slope form a variety of surface inclinations occurs. This is seen particularly clearly in the case of convex-upward slopes of domical hills. These

Plate 5.1 Granite rock slopes in the Tenaya Valley, Yosemite National Park, Sierra Nevada, California. Half Dome is seen on the right, the Royal Arches in the foreground left

assume the whole range of geometries, from shield-like low outcrops to prominent steep-sided 'sugarloaf' monoliths typified by Pao de Açucar in Rio de Janeiro (Plate 5.2). Kesel (1977) provided some quantitative data about the inclination of concave slopes of conical granite inselbergs in central Arizona, showing that in their steepest segment they vary from 19° to 44°.

Notwithstanding Young's (1972) assertion, the most plausible explanation for the variety of granite rock slopes, in terms of both their profiles and inclinations, is that they reflect variations in structural control, above all in the density and geometry of fracture patterns. From this point of view, granite masses may be subdivided into two general categories. The first one would include rocks fractured according to orthogonal and rhomboidal patterns, even if spacing of fractures varies widely, from a few centimetres to a few metres. The other category encompasses massive compartments lacking a clearly defined system of vertical and horizontal fractures. These monoliths may be up to 1 km long. However, curved sheeting joints are usually present, spaced from about 1 m to a few tens of metres apart. There are field examples of the two fracture patterns co-existing and intersecting each other, for example at the Amichab inselberg in the Namib Desert (Selby, 1982*a*, 1982*b*).

Straight, concave, and stepped slopes tend to occur within granite masses belonging to the first category. Selby (1982*b*) has shown that the inclination of straight segments of inselbergs in the Namib relates to the spacing of orthogonal joints, hence the overall planar profiles of the slopes reflect a more or less uniform fracture spacing

Plate 5.2 Different geometries of granite domes, Rio de Janeiro, Brazil

along the slope length. Some straight slopes may owe their form to direct control exerted by a steeply dipping planar fracture plane. In the Tatra Mountains, Slovakia, Kalvoda (1994) observed that fracture planes dipping at angles of 50° or more usually support individual rock slope segments. A fine example is provided by the twin granite cone of Kościelec in the Polish part of the High Tatra. The northern slopes of both peaks are adjusted to major discontinuity planes dipping steeply to the north and retain their general inclination of 40° for a few hundred metres. Similar morphology can be observed at the Three Brothers in the Yosemite Valley, whose west faces, inclined by about 45°, steeply follow dipping fracture planes (Huber, 1987).

Details of free face geomorphology are controlled by the presence of discontinuities. Rock slopes in the Tatra are dissected by a dense network of chutes and ravines developed along vertical and sub-vertical fractures. These concavities play a key role in sediment routing, channelizing particle fall, and act as pathways of debris avalanches and debris flows, so that huge scree cones form at their outlets (Plate 5.3).

Plate 5.3 Rock slopes cut by ravines and chutes, through which the material to scree cones is supplied, Tatra Mountains, Slovakia

This type of slope geomorphology can be found in most formerly glaciated granite massifs, for example in the Cairngorms and the Isle of Arran in Scotland, unless the free face is built of an extremely massive rock as in the Yosemite Valley. In the granite part of the Hunza Valley, Karakoram, deep ravines are separated by tall towers and pinnacles, delineated by sub-vertical fractures (Goudie et al., 1984). Zones of tectonic brecciation (mylonitization) are also preferentially eroded into chutes, minor tributary valleys, and passes, and their width may attain tens of metres.

Generally straight slopes are not limited to rock precipices, either in the Tatra or elsewhere. They also occur in less steep variants, being variously inclined between 20 and 40°. Lukniš (1973) emphasized their widespread presence outside the limits of the Pleistocene glaciation in the Tatra and considered them as preglacial landforms. These rectilinear slopes, locally as long as 2 km, carry a cover of boulders set in sandy to grus matrix up to 3 m thick, but the top surface of the solid rock beneath appears planar. Similar straight slopes with a thin veneer of regolith have been reported from other mountainous granite terrains, such as the Chilean Andres (Weischet, 1969) and the ice-free Victoria Land of Antarctica (Selby, 1971). This category of slope fulfils the criteria for the so-called Richter denudation slopes, developing through disintegration and retreat of a rock face and steady downhill movement of resultant talus, which helps to maintain a uniform slope angle near the angle of talus repose.

Following the rock mass strength–slope inclination relationship and the concept of strength-equilibrium slopes (Selby, 1980), one may hypothesize that concave rock slopes form in rock masses which show variability in fracture spacing, with more steep upper segments developed in less jointed rock. This is yet to be tested against field evidence, but testing may become difficult because access to the solid rock is usually obscured by the mantle of boulders. In the south-west USA boulder-covered granite slopes are particularly common, but Melton (1965) failed to find any systematic relationship between the slope gradient and the size of boulders. In his study, for slopes inclined in the range of 16.0–28.5° the median size of blocks remained nearly constant.

Massive granite compartments are typically associated with slopes which are convex-upward. They may occur as distinctive facets within high escarpments (Plate 5.4) or support individual bornhardts. However, the range of inclination may vary considerably and the relationships between the rock mass strength, fracture patterns, and slope gradient are not straightforward. Selby (1982a) found that different sections of a composite profile of a granite bornhardt may fall within or outside the limits of the strength-equilibrium envelope. The decisive factor is the inclination of sheeting planes which inherit their angle from the original dome form. Thus, if a dome was a low-radius one, sheeting planes will have a relatively low angle (20–40°) and the resultant rock slope will be much less inclined than its rock mass strength would suggest (Fig. 5.2). Specific examples cited by Selby are the Mirabib inselberg and the Pondok Mountains in

Plate 5.4 The south-facing escarpment of the Erongo Massif, Namibia. Concave slopes in the foreground are built by effusive rocks, convex slopes in the background have developed on the massive Jurassic granite (Photo courtesy of Andrew S. Goudie)

the Spitzkoppe group, both in the Central Namib Desert, but shield-like, low-angle outcrops built of massive granite can be found in many different environments.

Geological control is also evident in many vertical rock faces which follow fracture planes. However, they are difficult to explain solely in terms of adjustment to structure. Although some rock masses are strong enough to support almost vertical faces, for instance at the Mirabib inselberg (Selby, 1982b), many are too densely jointed to be in strength equilibrium. Basal undercutting usually provides a good answer to why vertical rock faces occur as marine cliffs, rocky sides of river gorges, and bare shoulders of formerly glaciated valleys. The reasons for undercutting vary and include wave action in coastal settings, lateral fluvial erosion, and glacial plucking. Wave action is a particularly powerful means of oversteepening a rock slope and bringing it into disequilibrium conditions. For example, Steers (1948) contrasted vertical granite cliffs on the exposed western coast of the Penwith Peninsula in south-west England with sloping cliffs in the more sheltered southern section. Similar contrasts between windward and leeward coasts occur in Lundy Island (Godard and Coque-Delhuille, 1982). The lack of strength equilibrium implies ongoing mass movement processes acting towards the reduction of slope angle. Indeed, most very steep slopes show evidence of recent failures, and debris piles typically occur at their footslopes (Plate 6.2). In the coastal context rockslide debris

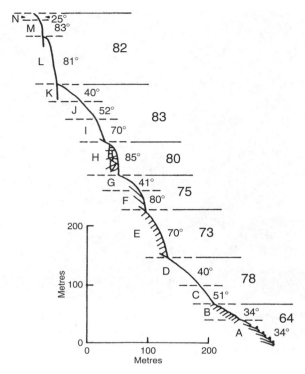

Fig. 5.2 Relationships between slope relief and rock mass strength, on the example of a composite bornhardt profile from the Namib Desert (after Selby, 1982a, fig. 14)

is subject to further disintegration and redistribution by wave attack and if this is accomplished, the rock cliff is exposed to undercutting again. The situation in inherited glacial valleys is different and usually no efficient means of evacuation of rock fall material is available. With time, slope angle is reduced and a strength-equilibrium rock slope grades into a talus apron that has built up in the meantime near the footslope. But there are cases requiring special explanation that would combine the strength-equilibrium concept and glacial undercutting. Although more than 10,000 years have elapsed since the ice melted, the imposing vertical walls of the Yosemite Valley (Plate 5.1), some almost half a kilometre high, still exist and the talus aprons below them are not particularly voluminous. In addition, most catastrophic slope failures in this area have begun on slopes which were rising above the ice surface during the last glaciation. It seems that the rock mass strength of the massive Yosemite granites and granodiorites is high enough to support vertical slopes for a long time anyway and the steepening of valley sides due to glacial undercutting, although considerable, has not moved the slopes outside the limits of the strength-equilibrium envelope.

Plate 5.5 Recent collapse of a basal cave in massive granite, Erongo, Namibia

Overhang slopes are perhaps the only ones not directly controlled by fracture patterns. They are of two main types, of which flared slopes have already been discussed. Another class includes footslope caves developed along sub-horizontal fractures due to preferential weathering and groundwater movement; hence geological control is indirect in this case. Overhang slopes are evidently in disequilibrium. Flared slopes are normally only several metres high and supported by massive rock; hence they are rather stable, but basal caves are frequently affected by roof collapses (Plate 5.5).

Rock Slope Failures

Equilibrium conditions on rock slopes are approached and maintained by constant loss of mass by means of various mass movement phenomena. These may assume a wide range of types, from fall of individual joint-bounded particles to catastrophic failures of entire slopes. Failures leave distinct fingerprints: scars and niches in the upper slope and talus aprons at the footslope, which are used to decipher the nature and magnitude of an event. Very few rock slope failures in granites have been eye-witnessed (e.g. Wieczorek *et al.*, 2000; Wieczorek, 2002).

Particle Fall

If the presence of extensive talus aprons is taken as an indicator of the importance of rock particle fall, then it has to be concluded that it is most significant in high mountains with a recent history of widespread Pleistocene glaciation. Valley sides in such mountain settings typically have two distinct sections: a rock slope in the upper part and a talus slope in the lower part. Representative examples are numerous and include, among others, the Sierra Nevada in California, the Grand Teton range in Wyoming, the Tatra Mountains on the Poland/Slovak border (Plate 5.3), the Karakoram, and the Himalaya. Rock/talus slopes can also be found in once glaciated valleys within dissected plateaux, for instance in the Serra da Estrela in Portugal and the Cairngorms in Scotland. It needs to be emphasized that particle fall is by no means the sole contributor to the talus build-up; occasional rock avalanches play their part too, whereas debris flows actively redistribute material across the talus slope. By contrast, particle fall appears insignificant in the slope development of bornhardts and domed inselbergs, as may be inferred from a rare occurrence of taluses there. Likewise, individual dome-like highs within mountain ranges, such as in the Sierra Nevada, the Sierra de Guadarrama in central Spain (Pedraza *et al.*, 1989), in Corsica (Klaer, 1956), or the Mourne Mountains in Northern Ireland do not shed much fine debris which would accumulate to form talus covers (Plate 5.6). This is understandable given the wide joint spacing in rock masses sculpted into granite domes.

The ratio between the length of the two slope segments involved, described as the H_w/H_s index (H_w being the height of rockwall and H_s being the height of the talus slope below; Kotarba, 1987) varies and may reflect climatic, lithological, and topographic differences and temporal constraints between sites. If analysed within relatively small areas which have been freed of glaciers at roughly the same time, the differences in the index value probably depend on the strength of the rock mass itself. Kotarba (1987) compared rock slopes in glacial cirques in the Isle of Rhum and the Cairngorms, Scotland, and found that the values of the H_w/H_s index are 0.6–0.75 in the former and between 2 and 4 in the latter. His preferred explanation for the differing degrees of postglacial slope transformation is climatic, but bedrock is not entirely identical. Slopes in the Isle of Rhum are built of microgranite, whereas those in the Cairngorms have developed in medium-grained granite. Although details are not given in the quoted paper, microgranite is typically densely fractured and this structural feature might account, perhaps in addition to the climatic factor, for the much faster build-up of talus slopes in the Isle of Rhum.

Very massive granite releases relatively little debris and hence it may support a very steep slope, which would account for much of the total height of the valley side. Talus cover is then proportionally less developed and often discontinuous along the valley. Such is the situation in many of the formerly glaciated valleys of the Sierra Nevada, California, within the boundaries of the Yosemite National Park. Granites and granodiorites are extremely massive and the supply of material to build the talus is insufficient, so that valley sides remain dominated by rockwalls in spite of more than 10,000 years of ice-free conditions. However, some sections of the northern side of the

Plate 5.6 Dome-like residual hill (Hen Mt.) in the Mourne Mountains, Northern Ireland, is not surrounded by talus deposits, apparently because of the massiveness of granite

Yosemite Valley are built of diorite, which is much more densely jointed than adjacent granites and granodiorites. Diorite outcrops have contributed much larger volumes of debris to the talus section, which now covers at least two-thirds of the total slope height. The height of the diorite cliffs has been reduced to less than 100 m and they stand in stark contrast to the 600 m high granodiorite cliff of El Capitan nearby (Fig. 5.3).

Rock Mass Failures

More massive granite, which does occur in mountainous settings too but typically supports more or less isolated domes, tends to fail through rock mass failures, that is, large volumes of material are simultaneously set in motion. These failures may be initiated as slides or topples, either of which can subsequently be transformed into a rock avalanche. Slides are favoured if there are extensive planar fracture surfaces dipping

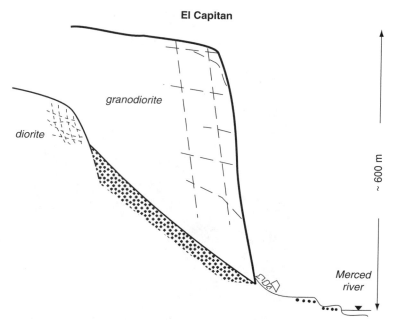

Fig. 5.3 Contrasting development of talus slopes in massive granodiorite and jointed diorite, Yosemite Valley, Sierra Nevada, California

out of the slope and at an angle greater than the angle of internal friction of granite (Hoek and Bray, 1981). For planar and unweathered rock surfaces the latter is 29–35° for fine-grained granite and 31–35° for coarse granite, but the presence of water along the partings reduces the critical angles by a few degrees, whereas an increase in surface roughness results in an elevation of the friction angle (Barton and Choubey, 1977). Hence, steeply dipping smooth sheeting planes developed in response to erosional unloading in deeply incised mountain valleys provide an optimal setting for massive slides.

By contrast to sliding, toppling occurs if the dominant fracture sets are disposed vertically, but it may be associated with sliding if at the same time the inclination of the basal failure plane exceeds the angle of friction and the width/height ratio of the toppling column is less than the tangent of the potential failure plane (i.e. $\alpha > \phi$ and $b/h < \tan \alpha$). Hoek and Bray (1981) reviewed the principal mechanisms of toppling and distinguished its sub-categories (Fig. 5.4). Among them, block toppling and block-flexure toppling appear particularly relevant to granite. The former happens when individual columns of hard rock are separated by widely spaced vertical fractures and shorter columns at the toe of the slope are pushed forward under the load imposed by taller columns behind them. Block-flexure toppling is the sum of minor displacements of individual blocks within a column along flat-lying discontinuities. This variant of

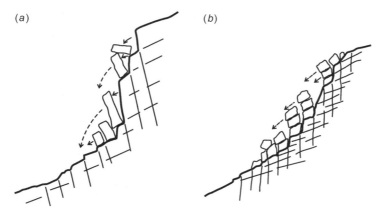

Fig. 5.4 Categories of toppling relevant to granite rock slopes: (*a*) block toppling in vertically fractured granite, (*b*) block flexure toppling in orthogonally fractured granite. Arrows show the direction of movement

toppling has been recognized as a significant component of the development of mid-slope trenches in the Canadian Cordillera (Bovis, 1982).

As a general phenomenon, rock mass failures should be viewed as a rock slope response to local strength disequilibrium. Hence, they are expected wherever slopes are too steep in relation to the inclination allowed by the rock mass strength, and the rock is massive enough for stress build-up over a long period. Therefore they are most common within slopes which are yet to achieve strength-equilibrium conditions, such as those lining recently deglaciated valleys, oversteepened by fluvial incision concurrent with recent and geologically fast uplift, or forming domes recently excavated from saprolitic cover. An area which is well known for catastrophic rock falls and slides is the Yosemite Valley and its tributaries; it has also probably been best researched in this respect (Wieczorek and Jäger, 1996, Wieczorek *et al.*, 2000; Wieczorek, 2002). An inventory of more than 450 historic events from the years 1867–2002 (Guzzetti *et al.*, 2003) reveals that single and multiple rock slides were most common (254 cases, 55%) followed by rock falls (138 cases, 30%). However, falls were much more voluminous, involving up to 600,000 m^3 of material, and 13,975 m^3 on average. At least one prehistoric rockslide, at Mirror Lake, was even larger, involving 1,140,000 m^3 of material. By contrast, the average volume of slides was only 1,404 m^3 and the maximum was 100,000 m^3. Interestingly, most of the massive failures affected the upper slopes of the valley, located above the vertical limit of the ice during the last glaciation. This is interpreted as evidence of the important role of long-term weathering in addition to stress release, and suggests that the return to equilibrium in extremely massive rock as present in the Yosemite occurs at a timescale of tens of thousands of years rather than a few thousands years. The Table 5.1. includes characteristics of a few very large rock slope failures in the Yosemite Valley.

Table 5.1 Characteristics of selected large-scale rock slope failures in the Yosemite Valley

Location	Failure type	Volume of material (m^3)	Geomorphology of deposit	Date
Slide Mountain	slide	1,900,000	blocks > 6 m long landslide dam 12 m high lateral spreading as possible precursor	prehistoric (probably 1739 or 1740)
Mirror Lake	fall	11,400,000	boulders up to 250 m^3 fan-shaped deposit 520 m wide at least 28 m thickness 2 km long dammed Mirror Lake	prehistoric
Old Yosemite Village	fall	20,000	boulders up to 10 m long fan-shaped deposit 300 m wide	1872
Liberty Cap	fall	36,000	boulders up to 10 m long deposition zone 120 m long, 50 m wide and at least 5 m thick	1872
Middle Brother	fall	600,000	large talus produced, roughly 200 × 200 m	1987
Happy Isles	fall	38,000	talus apron against the slope, 250 m wide and up to 70 m long	1996

Source: Compiled after Wieczorek, (2002).

Massive rock slides and avalanches have frequently been reported from the high mountains of Central and Southern Asia, as well as from the Andes. In the granite part of Tian Shan within the borders of Kyrgyzstan some rock avalanches travelled more than 2.5 km and more than 1,000 m downslope (Strom, 1996). A variety of mass movement processes has been recognized to shape mountain slopes in the Hunza Valley, including rock falls, slides, avalanches, and debris falls, so that the rock slopes give way to extensive scree slopes. Some of the scree cones are of gigantic dimensions, up to 500–600 m high and more than 1,000 m long (Goudie *et al.*, 1984).

Perhaps the largest rock slope failures that have ever affected granite slopes were those which occurred at Nevados Huascaran in the Cordillera Blanca in northern Peru. The first rock slide happened on 10 January 1962 and involved a total volume of approximately 13×10^6 m^3 of material, including snow and ice, that descended 4 km downslope with a speed exceeding 100 km/h. Blocks as big as $20 \times 15 \times 12$ m were set in motion. The second slide occurred on 31 May 1970 and was triggered by a powerful earthquake off the coast of Peru. It involved more than 50×10^6 m^3 of material that travelled 14.5 km downslope, attaining a velocity of about 300 km/h. These two events were also most devastating, killing about 5,000 people in 1962 and more than 18,000 in 1970 (Plafker and Ericksen, 1978; Schuster *et al.*, 2002). Evidently, the combination of steep relief and rock fracturing in a tectonically active setting provided the optimal conditions for this environmental catastrophe, and the seismic trigger did the rest.

Massive, catastrophic slope failures are likely to play a most important part in the evolution of domed inselbergs and other bornhardts. However, it appears that no single failure event has ever been reported or witnessed, and the above supposition derives from geomorphic observational evidence. Although there are domes around which no abundant products of disintegration occur, the footslopes of many others are built from the chaotic accumulation of big slabs and rectangular blocks of granite, apparently derived from the upper slope. It has been suggested that in some cases these blocks may in fact be corestone residuals from the eroded saprolitic cover mantling the dome (Thomas, 1978), but this cannot be a universally valid explanation. Blocky aprons around massive domes in arid and semi-arid zones in particular are difficult to explain in this way. The inspection of lower slopes around Spitzkoppe in the Namib Desert shows that both block falls and slab slides must have occurred extensively, involving huge quantities of rock. The south-western rock face of Gross Spitzkoppe plunges beneath a chaotic talus composed of monolithic blocks, some as much as 30 m long. The form of the remaining rock slope clearly indicates the presence of a partially missing sheet, the thickness of which must have been at least 50 m (Plate 5.7). Likewise, irregular piles of slab-like, often angular blocks around domes located along mountain ridges, for example in the Sierra de Guadarrama in Spain (Pedraza *et al.*, 1989) are

Plate 5.7 Dome form and talus resultant from large-scale rock slides, Spitzkoppe group, Namibia

difficult to reconcile with a subsurface origin. On certain low-angle domes, such as the Enchanted Rock, Texas, or in the Erongo Massif, Namibia, one can see disintegrated upper sheets in the process of slowly sliding down towards the footslope. Unfortunately, details of the rates and timing of slab movement are not known.

Similar chaotic blocky accumulation can be observed around domes in Rio de Janeiro. The presence of talus there is not immediately evident because it is quickly invaded by lush subtropical vegetation, as around the most famous dome, Pao de Açucar, but nevertheless it occurs widely. Overhangs and steps across the upper slope indicate that failures have been initiated in the lower slope, possibly above weathering-related basal overhangs or flares, and then propagated upwards.

Disintegration of domes through rock slope failures may, in the long run, result in the loss of an original dome form. The evolutionary sequence has been hypothesized by

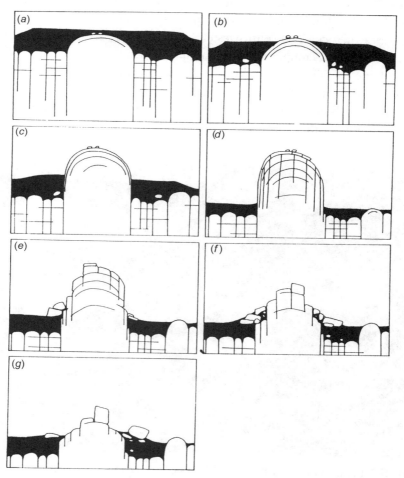

Fig. 5.5 Model of dome growth and disintegration (after Thomas, 1978, fig. 2)

Thomas (1978) in respect of the African inselbergs (Fig. 5.5). Some isolated granite hills in Central Europe have been interpreted as products of advanced dome degradation (Migoń, 1993, 1997c). Witosza hill in the Jelenia Góra Basin, south-west Poland, is a particularly clear example. It is inferred that it used to be an 80 m high, multi-faceted domed inselberg, outlined by steeply dipping (40–60°) sheeting joints and crossed by north-west–south-east trending vertical fractures. The opening of vertical and sheet joints has led to toppling and sliding, the testament to which are extensive boulder fields mantling the lower slopes. Some of these boulders are well in excess of 10 m long. Convex slopes have been gradually replaced by vertical ones in the upper part and have disappeared beneath a talus lower down. Advanced degradation of the uppermost parts has transformed the dome into a ruiniform, castellated relief, with little indication of its primary form (Plate 5.8).

Plate 5.8 Opened fractures superimposed on the dome form, Witosza hill, south-west Poland

Further examples are provided by granite hilly lands in the Czech Republic, in the vicinity of Žulova (Czudek et al., 1964; Migoń, 1996) and in the Nova Bystřice Highland (Votýpka, 1964). It has been suggested that the transformation of domes into block-strewn hills signalled long-term environmental change, from warm and humid conditions prior to the late Miocene to the more recent drier and colder conditions of the Pliocene and Pleistocene, and an increased role of mechanical weathering. However, following the assumption that an exposed steep-sided dome is not necessarily in strength equilibrium with surface conditions, it seems more likely that the degree of form change is a function of time elapsed since the exposure rather than of climate. Given the likely exposure of domes in the Bohemian Massif in the mid-Miocene at the latest (Migoń, 1997c), relaxation processes on their slopes have operated over the last c.15 million years. The frequent co-existence of sheeting and orthogonal fracturing undoubtedly facilitated much more rapid slope degradation than the sheeting alone would have done. Spitzkoppe and other domed inselbergs of the Namib were probably exposed around the same time span (Van der Wateren and Dunai, 2001), but their granites are much more massive than the Central European ones, therefore their dome forms may have survived for much longer.

The association of many major rock slope failures with well-developed sheeting planes implies that the condition of slope unloading plays a critical role in rock slope instability in massive granitoid rocks. However, the triggers of individual failure events may vary. In tectonically active zones seismic tremors may result in the ultimate crossing of stability thresholds and induce slope collapse. Depending on local relief, the results may be catastrophic failures like those in the Yosemite Valley triggered by the 1872 Owens Valley earthquake (Wieczorek, 2002) or the 1970 Nevado Huascarán slide (Plafker and Ericksen, 1970), or minor movements of rock slabs on the flanks of inselbergs (Twidale et al., 1991).

Gravitational Spreading

Another phenomenon observed in high mountains is gravitational spreading of ridges bounded by steep slopes. Its geomorphic expression includes ridge-top trenches and uphill-facing scarps within the upper slope, the former occasionally filled by seasonal ponds. These depressions may attain hundreds of metres long and their bounding scarps can be as much as 30–40 m high. Although spreading phenomena, also known as *Sackung*, appear to be particularly common in sedimentary rocks because bedding planes offer readily available slip zones, they have been reported from granite areas as well. Mountain areas with a recent history of deglaciation, hence significant de-stressing of slopes, provide the best examples.

In the Affliction Creek drainage basin in the Canadian Cordillera, c.100 km north of Vancouver, a series of uphill-facing scarps has developed on recently deglaciated steep (c.60° on average) slope underlain by highly fractured granite in the lower part and

Pleistocene basalt in the upper one (Bovis, 1982). The majority of scarps are in granite and numerous tension cracks are located higher upslope within the basalt cap. Scarp development is best seen as a combination of erosion along tension cracks and antithetic motion in the manner of block-flexure toppling. The latter has been made possible by favourable structural circumstances, that is, main fractures strike parallel to the slope face and dip steeply (35–70°) into the slope. It has also been shown that part of the movement has been very recent, accomplished in the last 140 years, following the retreat of the Little Ice Age glacier. In mountain ranges in Colorado, United States (Sawatch Range, Williams Fork Mountains, Sangre de Cristo Mountains) there occur both double-crested ridges and hill-top grabens, some up to 1 km long, again associated with formerly glaciated valleys and step relief (Varnes *et al.*, 1989). Long axes of the trenches follow major fracture sets and are parallel to the contours, which has led Varnes *et al.* (1989: 1) to comment that 'topography is the primary determinant of the location and direction of the trenches and scarps, but the topographic grain of the terrane is, itself, determined in part by rock structures, such as joints and faults'. The role of structure has also been made clear by Bovis (1982: 804), who remarked that 'deglaciation has provided the potential for deformation, the magnitude of which is controlled by local factors such as structure, lithology and seismicity'. The key role of lithological and local structural predisposition explains why ridge-top trenches and uphill-facing scarps are not as ubiquitous as the recent history of deglaciation of many mountain ranges might suggest. For example, they are very rare in the granite part of the Tatra Mountains in Poland and Slovakia.

The origin of the long and deep clefts cutting across granite domes in Bohuslän in south-west Sweden is uncertain (Olvmo *et al.*, 1999; Johansson *et al.*, 2001b). They also occur in abundance in the adjacent gneissic area, and are 40–300 m long and 0.5–9 m wide. Their depth varies from a few metres to as much as 25 m at Träleberg. Preferential deep weathering along vertical fractures is the mode of origin favoured by the authors, and the presence of locked-in corestones and relicts of clayey weathering residuals gives some support to this hypothesis. On the other hand, the location of at least some clefts in the axial parts of steep-sided domes, hence a zone of maximum tensile stress, suggests the role of slow gravitational opening, possibly acting along earlier etched fractures. At Hunnebo also, where the dome has steep slopes and a flattened summit surface (Olvmo *et al.*, 1999), the pattern of clefts is consistent with their opening due to slow gravity-driven movement (Fig. 5.6).

Caves Associated with Rock Slope Development

Cave-like openings arising from differential weathering and saprolite removal are generally small and form a minority of granite caves. Most granite features to which the term 'cave' is applied are causally associated with gravitational mass movement phenomena on rock slopes.

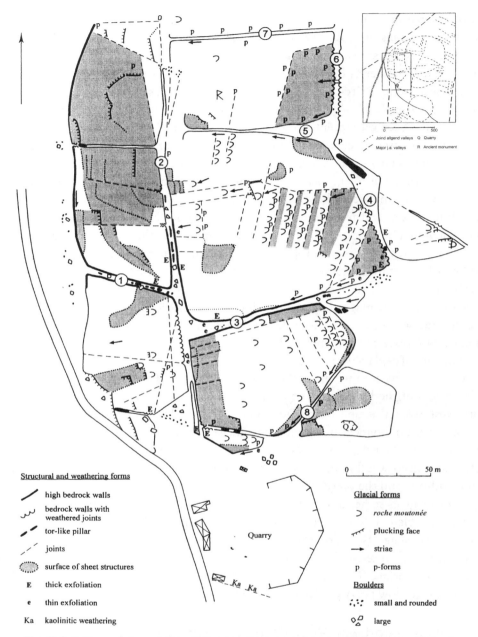

Fig. 5.6 Pattern of deep clefts within a granite dome, Hunnebo, Bohuslän, south-west Sweden (after Olvmo et al., 1999, fig. 3)

In relation to the slope, caves originate in two settings and this is reflected in their genetic classification and morphological characteristics. Within upper slopes subjected to tensile stress caves develop along vertical or sheet fractures which, while opening, become sufficiently wide to be penetrated by humans. These fracture-aligned caves are usually single passage features and rarely exceed 20 m in length. However, a few larger-scale examples have been reported in the literature. At the Enchanted Rock dome, Texas, a two-storey fracture cave exists which is more than 300 m long and 2–3 m high (Twidale, 1982). In the Karkonosze Mountains, Czech Republic, an extensive system of underground passages has developed within the cirque wall of Labská jama (Pilous, 1993). Its total length is 102 m and approximate depth is 10 m. A smaller cave, 9.5 m long, occurs nearby. The opening of fractures on such a scale implies quite significant locked-in stress, released during dome decompression, deglaciation, or rapid fluvial incision.

The opening of fractures within steep slopes may ultimately lead to rock mass failure by fall, slide, or topple. Instantaneous transfer of large volumes of material will result in massive but chaotic talus accumulation at the footslope. The size of talus blocks involved in the failure and the toughness of granite, which does not disintegrate easily upon impact, dictate that talus is made of big lumps of rock, with little fine material available to fill the voids between the big blocks. In a number of examples, interconnecting voids and openings under and between individual blocks are big and extensive enough to resemble small caves.

Witosza dome in south-west Poland is known for the co-existence of cave-like features of different origin (Migoń, 2000). In the hill-top part there is a fracture cave 18.5 m long and up to 4 m high, developed along a vertical discontinuity parallel to the rock cliff. In the middle slope there are a few passages developed along opened vertical and steeply dipping sheet fractures, for which the roofs are provided by huge boulders fallen from above. The longest of these caves is 8.5 m. Finally, extensive accumulation of detached boulders on the lower slope has resulted in the origin of the Rock Chamber cave, which is 8.5 × 6 m in area and up to 3 m high. Its roof is formed by a single boulder more than 15 m long (Fig. 5.7). A similar genetic variety of caves has been reported from Fichtelgebirge in Bavaria, Germany, where fracture caves, talus caves, and forms of combined origin occur (Striebel, 1999). The largest of these, the Ochsenkopf-Nivationshöhle, is more than 100 m long.

A specific origin is inferred for many granite caves in Sweden. They are located within *roche moutonnée* hills and within extensive talus accumulation at the base of rock slopes (Sjöberg, 1987). The largest of these caves is the labyrinth of Bodagrottorna on the coast of the Bothnian Gulf, more than 2,800 m long in total. De Geer (1940: 203; quoted in Sjöberg, 1987) observed that in this area 'the bedrocks are split, often with open fissures, sometimes forming caves', and hypothesized that 'such masses of angular boulders ... seem likely to be caused by considerable earthquakes'. Recent

Fig. 5.7 Geomorphic setting of granite caves on the slopes of Witosza hill, south-west Poland: (1) fracture caves, (2) roofed open clefts, (3) boulder caves

research points in the same direction, and it is argued that catastrophic slope collapses occurred shortly after ice wastage in Fennoscandia and were triggered by powerful earthquakes, with a magnitude of up to 7 on the Richter scale (Sjöberg, 1987; Mörner, 2003). The Boda caves would have formed in 9663 BP, as suggested by strong disturbance of a varve dated to this particular year.

Pediments

Pediments have been defined as terrestrial erosional footslope surfaces inclined at a low angle and lacking significant relief in all three dimensions (Whitaker, 1979). They occur widely in arid and semi-arid granite areas in Australia (Mabbutt, 1966; Twidale, 1978b), northern Africa (Dresch, 1959; Mensching, 1978), South Africa and Namibia (Mabbutt, 1955; Migoń and Goudie, 2003), and particularly in south-west USA (Dohrenwend, 1987). In most instances they are backed by the steep rock slope of a mountain range or residual inselberg (Plate 4.10). A distinct piedmont angle separates the low-angle surface of the pediment from the much steeper slope above, but both are essentially rock-cut. However, pediments may also terminate in pediment passes which connect them with similar surfaces on the other side of a local divide, gently sloping in the opposite direction. In the Mojave Desert in particular, but elsewhere too (Mabbutt, 1955), there occur pediments, named pediment domes, sloping radially in all directions, forming together a shield-like elevation, not crowned by any residual hill. Pediments can be found in different rocks, although for rather weak lithologies a term *glacis d'erosion* tends to be used (Cooke *et al.*, 1993). None the less, claims have been repeatedly made that they develop preferentially in granite because of its disintegration into bimodal debris, that is, boulders and gravel/sand fraction, with little intermediate size material. Twidale (1982) offers further suggestions as to why pediments and granites are so intimately associated, emphasizing rock vulnerability to moisture attack, easy redistribution of weathering-derived sand and grus across the surface, the typical sharpness of the weathering front because of low permeability of granite, and fracture density contrasts instrumental for the development of piedmont angle.

The origin of pediments is controversial and still widely disputed. The key issue is the correct separation of pediment-forming and pediment-modifying processes, but these are, as pointed out by Oberlander (1989) and Dohrenwend (1994), too often not properly differentiated. It used to be maintained that pediments are formed mainly by sheetfloods, but these are better considered as secondary processes which occur precisely because a planar sloping surface is already in existence. Fluvial processes are responsible for most of the deposition that takes place on pediments, but the origin of pediments as rock-cut landforms is likely to be different. Mabbutt (1966: 78) advanced a model of mantle-controlled planation to explain pediments, saying: 'Such [granite] pediments are extended by subsoil notching at the level of the mantle and are broadly levelled by subsoil weathering and eluviation beneath the mantle, where there is a thin layer of weathered rock. ... In these ways the levelness of depositional profiles is transmitted to bedrock floors'.

His concept found support in the work carried out in the American south-west, where a mantle of disintegrated rock has been found beneath planar pediment surfaces (Cooke and Mason, 1973). Moss (1977) expanded the argument and showed that the thickness of the weathering mantle beneath the pediments in the Pass Mountain and the Union Hills, near Phoenix, Arizona, is up to 20 m, and considered deep weathering as a crucial preparatory process. A closer inspection of some of the best-developed pediments, such as those around the Granite Mountains south-east of Baker, California, reveals that their topography is rather rough, with numerous boulders and boulder tors separated by pockets and patches of grus (Norris, 1996) (Plate 5.9). Moreover, the capping of some pediment surfaces by lava flows dating back to 8–9 Ma ago indicates that they are, at least in part, inherited landforms which formed under warmer and more humid climatic conditions than today (Oberlander, 1974; Dohrenwend *et al.*, 1984). This finding undermines claims that pediments are specific arid zone landforms, pointing to the possibility of their formation in different environments. On the other hand, extensive pediment slopes in the Namib Desert can hardly be explained in terms of inheritance and develop today by a combination of weathering, fluvial erosion and deposition, and aeolian processes. For the Mojave Desert pediments Dohrenwend (1987) suggests that they may have evolved from gently dipping detachment fault surfaces, typical of this extensional

Plate 5.9 Many granite pediments in the Mojave Desert, California, have very rough relief, with numerous boulders and low rock platforms

geotectonic setting. If true, this model cannot be applied to most of the African pediments because recent crustal extension does not occur in areas where they are to be found.

Slope Development in Weathered Granite Terrain

Weathered granite terrains are topographically highly variable and range from extensive plains underlain by, or cut across, a continuous saprolitic cover such as those in east Africa and Australia to mountainous areas with steep slopes and deeply incised valleys, so characteristic of Hong Kong or coastal ranges in south-east China. Likewise, they may differ in the thickness and properties of the weathering mantle. Whereas deeply weathered plains are landscapes of general stability, because of negligible slope gradient, more hilly terrains are prone to various types of mass movement processes. In specific areas, where the thickness of saprolite is of the order of tens of metres, slopes are steep, and extreme precipitation events are frequent, landslides may become the dominant geomorphic process of mass transfer. They may also have disastrous consequences for local populations. Areas such as Hong Kong (So, 1971; Au, 1998; Chau *et al.*, 2004), the mountains of Japan (Oyagi, 1989; Chigira, 2001; Onda, 2004) and south-eastern coastal

Plate 5.10 Subtropical landscape near Rio de Janeiro, Brazil, is frequently affected by rainfall-triggered landslides, mainly shallow slides and debris flows (Photo courtesy of Nelson Fernandes)

Brazil (Jones, 1973; Fernandes *et al.*, 2004) are notorious for rainfall-triggered massive, and deadly, landslides affecting slopes developed in weathered granite (Plate 5.10).

Apart from landslides, these areas may be subject to severe gully erosion which propagates through both the weathered granite itself and colluvial covers developed at the expense of the primary weathering mantle. Vegetation clearance, nowadays almost ubiquitous in the Tropics and Subtropics, usually exacerbates the problem of gullying, which can quickly turn large areas into unproductive wasteland. For example, gully development in weathered granite has become a major conservation issue in south-east China (Xu, 1996; Sheng and Liao, 1997) and Swaziland (Morgan *et al.*, 1997), whereas many hilly landscapes in the foothills of the Serra do Mar range in south-east Brazil show an array of landforms produced by shallow landslides and gullying, with the first usually paving the way for the second (Coelho Netto *et al.*, 1988; Coelho Netto, 1999; Plate 5.11).

Plate 5.11 Shallow landslides and subsequent gullying in the weathered terrain near Bananal, south-east Brazil

Types of Movement and Association with Weathering Zones

The variety of slope instability phenomena encountered in granite terrains does show some dependence upon the characteristics of the weathering mantle involved. Durgin (1977) proposed the following relationships (Table 5.2). He contends that most sliding occurs in decomposed granite rather than in saprolite and cites the following reasons: a significant decrease in bulk density, practically no cohesion of either saturated or dry material, efficient seepage, and usually an abrupt weathering front. The latter feature appears to be spread wide in the Japanese Alps and its role as a significant barrier to

Table 5.2 Types of mass movement in different types of granite-derived weathering mantle

Stage of weathering	Mass movements
Saprolite	Rotational slides and slumps
Decomposed granite (< 15% of fresh rock left)	Debris flows, debris avalanches, debris slides
Corestone stage (15–85% of fresh rock left)	Rock fall avalanches, rolling rocks
Fresh rock (< 15% of weathered material along joints)	Rock falls, rock slides, block glides; debris avalanches and slides over sheeting surfaces

Source: After Durgin (1977).

groundwater movement is emphasized. Saturation of the thin superficial grus layer is claimed to be the principal reason for frequent shallows slides on grus-covered slopes (Chigira, 2001; Onda, 2004).

Unfortunately, many landslide reports remain vague about the characteristics of the material involved in the failure, although it is true that these characteristics are difficult to assess *post-factum*. In addition, the abruptness of the weathering front under decomposed granite is highly variable. It has been shown before that in many locations the zone of grusification passes into solid rock gradually rather than along a sharp boundary. On the other hand, the grus mantle may retain its primary fracture surfaces throughout, especially if these are indurated by iron oxides or have developed a tectonic polish (Ehlen, 1999), and these may act as suitable sliding planes.

Another reason why landslide phenomena may be more frequent in slightly weathered granite is the relationship between weathering mantle characteristics and local slope. In general, a grus-like mantle, by virtue of its strength (Table 5.3), can support slopes which are steeper than those developed in fine-grained saprolites. These steep slopes may be more prone to abrupt yielding induced by an extreme precipitation event, whereas gentler slopes underlain by heavily weathered granites may release stress through creep and shallow soil flow, which occur in a more steady manner. Chigira (2001) has identified two types of slope failures affecting weathered granite terrain in western Honsiu island, Japan, confined to the upper saprolitic zone and the lower zone of micro-sheeting, respectively. He maintains that the latter are more common and demonstrates it by referring to the 29 June 1999 event in the Hiroshima area. An extreme precipitation episode, when daily total rain amounted to 250 mm and hourly intensities were in excess of 50 mm, induced widespread landsliding on the valley sides, where micro-sheeting was present.

Table 5.3 Strength of weathered granite approximated by *in situ* Schmidt hammer values and uniaxial compressive strength

Weathering grade	Schmidt hammer rebound values			Uniaxial compressive strength (MPa)
	Coarse granite (Hong Kong)	Granodiorite (Hong Kong)	Granite (Britain)	
Completely weathered	10–18	< 18	< 20	5–10
Highly weathered	14–32	15–30	20–45	15–60
Moderately weathered	30–58	25–50	45–58	40–120
Slightly weathered	> 55	45–68	53–58	100–180
Fresh rock	57–60	59–68	58–66	260–280

Source: Data from Irfan (1996). Schmidt hammer data from Irfan (1996: 31), compressive strength values from Goodman (1993: 223).

Corestone Movement

Corestone movement downslope is a type of slope instability very specific to granite, because few other rocks disintegrate and decompose into large more or less rounded boulders set in a fine-grained matrix. Depending on local slope topography, corestone movement may be accomplished by single falls or topples, rolling down the slope, or multiple bouncing. Likewise, travel distances may be variable, from a few to several hundred metres.

Corestone movement occurs in two variants. One involves boulders that have already been isolated from the weathering mantle and rest in a perched position in exposed settings. Further weathering at their base may lead to the displacement of the centre of gravity outside the area of contact and ultimately to fall. Rocking or logan stones are particularly vulnerable to movement and seismic shocks are known to trigger boulder falls. During the 1995 Kobe earthquake in Japan boulders of up to 3 m long moved several tens of metres downhill, causing property damage.

Selective erosion of the saprolite or colluvial deposit may also result in corestone movement. On a steep slope, after fine-grained material from around corestones is washed away, exposed boulders may become unstable and begin moving. This type of movement is probably very rare in the natural environment because the inclination of slopes cut in the saprolitic cover is normally not sufficient for boulders to cross the threshold of stability, but its probability of occurrence increases on slopes subject to human intervention.

In many low latitude areas rapidly growing cities encroach on steep slopes developed at least partially in saprolite, too often without proper planning, engineering advice, and hazard assessment. Examples from granite areas include Rio de Janeiro and its suburbs, Hong Kong, and Acapulco in Mexico. Excavations for access roads and housing not uncommonly produce steep to vertical cuts many metres high, in which poorly consolidated weathered material of grus and sandy type is exposed. This saprolite is prone to water erosion during heavy rainfall events, which would be much restricted under forest and shrub vegetation. Evacuation of grus leaves any corestones present in the saprolite unsupported and these may in turn begin to move down the slope, gaining momentum and inflicting much damage on structures. The hazards posed by such boulders are not fully appreciated, although in certain susceptible places like Hong Kong remedial structures of rock trap ditches are placed at the base of hazardous slopes (Grigg and Wong, 1987).

The devastating role of boulder movement can be exemplified using the natural disaster which affected Acapulco in October 1997, triggered by Hurricane Pauline and associated heavy rainfall of 411 mm during a 24-hour period (Lugo et al., 2002). Acapulco is a seaside resort which has grown rapidly in the last 40 years, increasing its population from less than 50,000 in the early 1960s to over 900,000 today. To accom-

Plate 5.12 Lag deposit of massive boulders, the legacy of catastrophic corestone movement down the channels during Hurricane Pauline, 1997

modate this population expansion, new neighbourhoods have been built on steeper and steeper slopes of the Veladero mountain massif that overlooks the Acapulco Bay. Many of those located highest uphill consist of primitive houses built on cut-and-fill benches backed by unprotected slopes in corestone-rich saprolite. Channels in turn have been partially incised into ancient debris flow deposits, containing rounded boulders of up to 10 m long. Under high discharges during Hurricane Pauline selective erosion of these deposits took place, followed by boulder movement down the Camaron River. Residual boulder piles choked the channels and the streets of Acapulco, and the diameter of individual stones was up to 9 m (Plate 5.12). What proportion of the boulder load came directly from unprotected slope cuts cannot be ascertained, but should not be neglected. Similar phenomena have been mentioned by Durgin (1977) regarding Rio de Janeiro and DeGraff *et al.* (1989) in respect of the Upland province in Puerto Rico.

Landslides

Landslides are rare in many granite terrains, especially in mid-latitude uplands, but they can be both frequent and widespread if a combination of steep relief, deep weathering, and high rainfall occurs. South-East Asia is one such area where landslides in weathered granite abound; south-east Brazil is another. Other areas that suffer from landslides include the islands of the Caribbean, the northern Andes, the dissected uplands of east Africa, and weathered granite massifs of southern Italy. That the role of landslides in shaping granite terrains is rarely addressed in general accounts reflects a biased view from plains and mid-latitude uplands, whereas the perspective from mountainous Japan is very different as illustrated by the following statement by Chigira (2001: 219): 'Granitic rocks are known to be very sensitive to weathering and are very vulnerable to landsliding; an exception are the granites of North America and Europe, which are not generally associated with landslides, because most of the weathered granites on these continents have been eroded by glaciation'. One may doubt the validity of the final remark, but the general view is expressed clearly.

Landslides frequently affect the steep mountainous terrains of Japan (Fig. 5.8), not uncommonly turning into disasters (Table 5.4). The mountains of Rokko near Kobe are but one example of an area fashioned by frequent landslides caused by extreme precipitation events. Tens of landslides per square kilometre were observed on 9 July 1967, when daily rainfall amounted to 319 mm, including a large slide from the slope of Mount Yotsugi that involved 5,200 m^3 of weathered granite. Some slides are triggered by seismic events, such as the great Kobe earthquake from 17 January 1995. One of the landslides induced by this shock was the Nojima Ohkawa slide in the Awaji island, measuring 160 × 370 m and extending about 15 m down to the weathering mantle/solid rock boundary. More deadly was the Nigawa debris slide near Kobe which occurred in decomposed granite and then travelled for 250 m as a debris flow, destroying a residential district and killing 34 people.

Landslides have played a key part in the topographic evolution of Hong Kong and today, because of the extremely rapid progress of urbanization encroaching on steeper and steeper slopes, they are a subject of much concern (Au, 1998; Dai and Lee, 2001). In fact, inappropriate land use and neglect of existing drainage systems and protective works are cited as significant factors behind many failures (Lumb, 1975; Chen and Lee, 2005). Hong Kong is a mountainous area, with a highest elevation of 957 m a.s.l., frequently affected by heavy rainfall. Mean annual rainfall is in excess of 2,000 mm and mean monthly values during the rainy season of May to September are 350–400 mm. In reality, these values have not uncommonly been recorded as 24-hour totals.

Almost every sizeable storm passing over Hong Kong triggers widespread slope instability, landslides, and earth flows (Fig. 5.9), causing significant economic loss and, not uncommonly, deaths. One particularly devastating typhoon hit Hong Kong on 12 June 1966, when an hourly rainfall maximum of 157 mm was recorded in Aberdeen (So, 1971). More than 550 individual landslides occurred on granite slopes, often located so close to each other that the post-event slopes acquired a distinctively scalloped nature. In a certain area of the territory, more than 17 landslide scars per square kilometre were recorded. Interestingly, the survey revealed that most of the

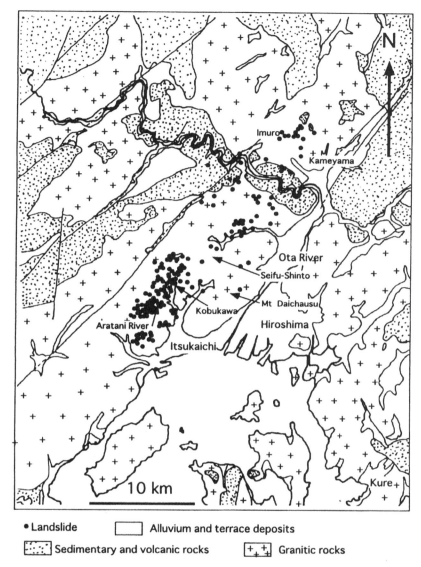

Fig. 5.8 Landslide distribution in weathered granite in the Hiroshima area, Japan (after Chigira, 2001, fig. 2)

slides occurred in forested areas (34.8%), which may seem to contradict the common perception of the stabilizing role of vegetation. However, in areas such as Hong Kong, where rainfall is very high and saprolite production is very efficient, the role of tree cover on steep slopes is more complicated. Dense vegetation growth prevents rain wash, thus allowing for an increase in regolith thickness, which in turn leads to slope instability on a larger scale. So (1971: 64) specifically comments that: 'In Hong Kong,

Table 5.4 Death toll associated with landslides and associated debris flows in weathered granite terrains in the mountains of Japan

Place of event	Year of occurrence	Number of fatalities
Rokko	1938	600
Minami Yamashiro	1953	221
Shimane	1964	108
Uetsu	1967	131
Rokko	1967	91
Kure	1967	88
Obara	1972	64
Kure	1999	8
Hiroshima	1999	20

Source: Adapted from Chigira (2001).

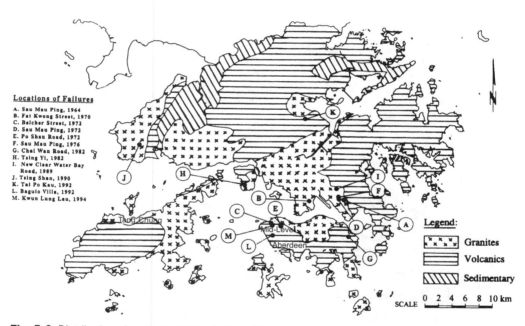

Fig. 5.9 Distribution of major landslides in Hong Kong in relation to bedrock geology (after Au, 1998, fig. 3)

where some of the steep, vegetation-covered slopes provide sites for an important residential zone and a back stage to the congested city, over-reliance on the vegetation cover for stabilization of slopes may turn a mere accident in nature into a great man-made disaster'.

Au (1998) provided a comprehensive analysis of rainfall-induced mass movement phenomena in Hong Kong and offered the following conclusions concerning the patterns and mechanisms of slope failures in weathered granites.

(1) Debris avalanches and debris flows are the most common mass movement phenomena, whereas debris slides are much less frequent and deep-seated slumps are extremely rare.
(2) The majority of slope failures are small and superficial, with the thickness of the failed mass seldom exceeding 3 m.
(3) Significant variations in the physical properties of granite saprolites cause differential saturation and pore-water build-up, therefore most failures are local.
(4) The presence of solid rock compartments within saprolites, or less weathered veins, decreases hydraulic conductivity and leads to the rise of pore-water pressure and, eventually, failure.
(5) Shallow landslides, especially if located on valley sides, turn into long run-out debris flows after reaching drainage lines.

An event in the Tsing Shan area from September 1990 is a good example of the latter phenomenon, known as the bulking process. A small debris avalanche involving some 400 m^3 of material reached a colluvial bench and imposed loading that triggered a much larger debris avalanche with an approximate volume of 2000 m^3. This in turn travelled to a drainage line and turned into a voluminous debris flow (about 19,000 m^3 was involved) that caused bedrock incision and levee formation. The fan-shaped toe of the flow was 120 m wide (Au, 1998). Further examples of landslides have been reviewed by Zhou et al. (2002).

Another region where the importance of landslides cannot be underestimated are granite and granite-gneiss mountain massifs in the more humid parts of South America. Extensive tracts of the Serra do Mar in coastal south-east Brazil are deeply weathered, up to 60–80 m in depth, due to a combination of a steep hydraulic gradient, fracturing of rock, and a subtropical climate (Thomas, 1995). Annual rainfall is 1,000–2,300 mm, with a distinctive rainy season from January to April, when storms exceeding an intensity of 100 mm per day are not uncommon. One such event centred upon Serra das Araras west of Rio de Janeiro, on 22 January 1967, resulted in precipitation of between 218 mm and 275 mm in 3.5 hours, with hourly intensities in excess of 100 mm. It caused tens of thousands of individual landslides of different types, many turning into debris- and mudflows after reaching valley floors. The Rio de Janeiro–Sao Paulo highway was much damaged and so were some hydroelectric power stations, whereas the death toll was probably near 1,700 and included the wiping out of entire small villages (Jones, 1973). De Ploey and Cruz (1979) analysed catastrophic planar slides in the Caraguatatuba area near Sao Paulo induced in March 1967 by massive rainfall exceeding 650 mm in just two days. The bedrock is actually gneiss, but their conclusions about the role of vegetation are relevant for adjacent granite and granite-gneiss areas too. On steep slopes (>20°) root systems of trees appear to actually promote landsliding because the root matrix increases infiltration and throughflow, which then contributes to the reduction of shear strength in the footslope area. Minor slides are initiated in the lower slope, often within the colluvium mantle from past slides, and these propagate upslope to cause massive slides capable of stripping the entire weathering mantle down to the bare rock slope, as was the case on the Morro do Tinga near Caraguatatuba (De Ploey and Cruz, 1979, photo 1). Another such event occurred in Rio de Janeiro itself on 13 February 1996, when two major storms hit the Tijuca Massif (1022 m a.s.l.) in the morning and then in the late afternoon, with total precipitation in excess of 350 mm. Hundreds of landslides were triggered on slopes, mostly shallow translational slides turning into devastating debris flows after converging in major valleys (Fernandes et al., 2004).

Slides leave concavities of variable depth in the upper slope zone of depletion. These topographic hollows are invaded by tropical forest rather quickly, suggesting a re-establishment of stability, but their existence actually sets the stage for future mass movement. It has been shown that the presence of slope hollows significantly affects both surface and subsurface hydrology, with most runoff being directed to the hollows. This results in elevated pore-water pressures in regolith and in consequence to failures, especially if an external trigger comes into play (Montgomery and Dietrich, 1994; Fernandes et al., 2004; Vieira and Fernandes, 2004). It seems that unless a large catastrophic failure causes complete stripping of weathered material, hillside hollows and landslides will develop concurrently, providing an example of positive feedback in granite landscapes.

In terms of short-term landscape effects, the significance of landslides appears most clearly when the sliding plane coincides with the regolith/solid rock boundary. In such circumstances, the convex rock slopes of domes are exposed and thick colluvium accumulates on the lower slope. Whereas the latter can be quickly re-colonized by subtropical forest and its existence is no longer obvious, the former become prominent, and probably become long-lived landscape facets. Thus slides within a saprolitic cover are important contributors to episodic dome excavation by the weathering mantle. This model of dome evolution resembles one advocated by Hurault (1963) for domes and inselbergs in Guyana. The presence of thick colluvium at the footslope is no less important, as this becomes subject to secondary slope instability and extensive gullying, especially if land clearance has occurred. The suburbs of Petropolis north of Rio de Janeiro, characterized by steep relief, offer many examples of severe land degradation (Plate 5.13).

The importance of landslides is also evident in the Upland province of Puerto Rico in the Caribbean. The highland is built, among other things, of intrusive rocks, mainly granodiorite and diorite. Copious precipitation and rock fracturing provide optimal

Plate 5.13 Deforested hills built of weathered granite subject to shallow slides and gullying. Slope hollow in the upper right is probably an old landslide scar, Petropolis, Serra do Mar, south-east Brazil

conditions for deep penetration of weathering and slopes as steep as 45° are underlain by saprolitic cover. Tropical storms passing over the island typically trigger debris slides and flows, some as long as 750 m and up to 25 m thick (DeGraff *et al.*, 1989).

Characteristic multiconvex hilly areas in subtropical locations provide another example of a terrain which is moulded by landsliding, although its consequences are less disastrous because the general relief is less steep. A look at these landscapes reveals two major types of hills (Plate 5.14). Some resemble symmetric cupolas, but there are also hills with distinctive hemispherical hollows cut into their slopes. These amphitheatre-like features narrow down into funnel-like shapes and widen again to form footslope fans. In addition, an erosional incision may be present along the axis of the form. This association of landforms is so inseparable that locally it is described using a single name. In south-east China a term *benggang* is in use and literally means a 'collapsing hill' (Xu, 1996), whereas in south-east Brazil such hollow-and-fan pairs are referred to as *rampa complexes* (De Meis and Monteiro, 1979). Similar topographic features have also been recorded on the deeply weathered and moderately dissected Nyika Plateau in northern Malawi in Africa (Shroder, 1976). The significance of mass movement in the latter area is shown by the following figures: 235 landslides have been identified in a 277 km^2 area using aerial photographs, which means almost one landslide per square kilometre, and many of them are more than 500 m long.

Plate 5.14 Landscape around Bananal, south-east Brazil, with abundant signs of past and recent slope instability in weathered rock

As the Chinese name discloses, the hollows result from mass movement which causes a part of the slope to collapse as a slump. Landslides are favoured by the widespread presence of a thick saprolitic cover, in which multiconvex hilly landscapes are cut. Observations from locations such as south-east Brazil (Coelho Netto, 1999), equatorial western Africa (Thomas, 1994b), or south-east China (Xu, 1996) show that many less elevated hills are built entirely of weathered rock, whereas in the higher ones a solid rock core is mantled by a saprolite many metres thick. In addition, in south-east Brazil it has been demonstrated that ascent of groundwater preferentially occurs at fracture intersections, which induces seepage erosion and tunnelling in the near-surface material. This leads to further reduction of strength of the saprolite. These lithological and hydrological circumstances are conducive to the development of curved slip planes within the saprolite. Hence the resultant landslides are predominantly rotational in the upper zone, but typically turn into earthflows in the lower zone. Subsequently, the resultant topographic depressions within the hillslopes act as channels for runoff and may become the loci of gully erosion, responsible for sediment transfer further downstream. On the other hand, non-gullied hillside depressions collect runoff and support vegetation, the focus of further deep weathering, which in turn leads to renewed instability (Shroder, 1976).

Visual similarities between landscapes of 'collapsing hills' do not mean that their ages and controlling factors are identical. In China, extensive *benggang* development

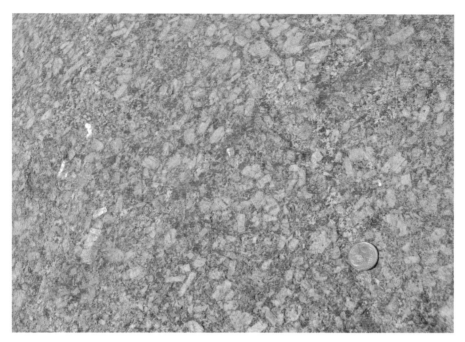

I. Coarse Revsund granite from Sweden, with large potassium feldspar phenocrysts

II. Deeply weathered granite, with abundant joint-bound corestones, at Lake Tahoe, Sierra Nevada, USA

III. Boulders within a hilly relief of the Erongo Massif, Namibia

IV. Haytor in Dartmoor, south-west England

V. Spitzkoppe—one of the highest inselbergs in the world, Namib Desert, Namibia

VI. Granite landscape of the Serra da Estrela, Portugal

VII. Plain with residual hills in the Mojave Desert (Joshua Tree National Park), California

VIII. High-mountain granite landscape, Yosemite National Park, Sierra Nevada

appears related to the human disturbance of natural landscapes. In the light of historical reports and oral traditions from the Guangdong province, this started around 1830 and coincided with the onset of massive tree felling (Xu and Zeng, 1992; Xu, 1996). Individual landslide events are caused by excessive precipitation. For example, more than 10,000 slumps within a total area of 2,800 ha have been recorded in the Wuhua County after the passing of Typhoon No. 7 in 1986 (Xu, 1996).

The history of landslide activity in south-east Brazil has been traced much further back in time. Near Bananal six colluvial layers representing different phases of slope instability have been recognized, the oldest dated as $17,980 \pm 1,720$ BP. The third, fourth, and fifth layers come from the Pleistocene/Holocene transition, pointing to widespread landscape instability at that time, whereas the uppermost layer is associated with recent deforestation and coffee plantations (Coelho Netto, 1999). Widespread landsliding and associated gullying caused overloading of local streams and induced massive redeposition of weathered material along drainage lines. The thickness of the resultant valley fill is up to 20 m and the flat surfaces of alluviated valley floors stand in stark contrast to steep slopes. At present, most rampa complexes around Bananal appear to be fossilized despite widespread deforestation and thick alluvial fills are subject to dissection. On the high elevation plateau of Campos do Jordão in Serra da Mantiqueira, Modenesi (1988) has distinguished three generations of landslide-related slope hollows and suggested that they may record the history of environmental change during the entire Pleistocene. These examples show that the geomorphic role of landsliding in the development of granite landscapes extends far beyond the spatial scale of individual slope units affected by movement.

Debris Flows

Debris flows are material-laden (25–75% by volume) slurries capable of downslope movement with a speed within the range 0.5–20 m/s, and occasionally more than that, depending on slope characteristics and sediment size and concentration. Their sediment load is typically bimodal and consists of very large material, including big boulders a few metres long, suspended in a fine-grained, clayey-silty matrix. Debris flows belong to the most hazardous mass movement phenomena, but also play a considerable role in slope and valley floor modelling. Debris torrents are phenomena similar to debris flows, but are evidently depleted in fine material and travel within a steep, pre-existing channel. The predominantly granitic Coast Mountains in British Columbia, Canada, are the type locality for debris torrents (Slaymaker, 1988).

Material availability and topographic requirements dictate that in granite landscapes debris flows occur in two distinct settings. One includes Alpine-type mountain topography and glacial troughs cut into plateau-like relief, but debris flow occurrences in these settings will not be reviewed here. Although certain lithological controls are not denied, debris flows do not appear to create a specifically granitic high mountain morphology. Selected references to debris flows in granite mountains include Innes (1983), Ballantyne (1986), Kotarba et al. (1987), and Bovis and Jakob (1999).

The other setting includes tropical and subtropical deeply weathered mountainous terrains, in which topographic gradients are high enough to allow regolith slides to reach watercourses in valley floors. Then the content of water in the sediment increases, the pore-fluid pressure builds up, and the sliding mass liquefies (Iverson et al., 1997). Mobilization of landslide material into a debris flow may also occur if the former reaches a slope hollow. These hollows are usually poorly drained places into which groundwater from adjacent slopes flows. Hence conditions for pore-water pressure and ultimately liquefaction increase. In yet another scenario, slides involve debris mass which is nearly saturated already. Therefore it transforms easily into a debris flow if topographic circumstances are favourable.

Examples of debris slides coalescing and turning into debris flows have been reported from Hong Kong (Lumb, 1975; Au, 1998), Japan (Marutani et al., 2000), Serra do Mar in Brazil (Fernandes et al., 2004), Puerto Rico (Jibson, 1989), and other areas. Some of these events were catastrophic in terms of life and property loss. For example, debris flows in Rio de Janeiro triggered by extreme precipitation in February 1996 destroyed 222 houses and caused more than 40 deaths (Amaral, 1997). Debris flows of this type are not confined to low-latitude weathered mountains. They may also be present in mid-latitudes if a combination of steep relief due to erosional dissection, sparse vegetation, and relatively thick saprolitic cover occurs. Calcaterra et al. (1987) prepared an inventory of mass movement phenomena in selected drainage basins of the Serre Massif in Calabria, southern Italy, and found that more than 350 of them, c.60 per cent of the total, were debris flows. The most impressive examples were more than 500 m long, up to 15 m thick, and involved transport of more than 100×10^3 m^3 of debris.

Tropical debris flows transporting corestones are important agents of change in valley floor morphology. Their geomorphic effects include widespread deposition of boulders, often as a few metres thick blankets, and formation of levees a few metres high. Subsequently the fine-grained matrix may be eroded away, leading to excavation of individual boulders as a lag deposit. The valley floor of Rio Camaron in the Veladero Massif near Acapulco, Mexico, is filled by a thick deposit consisting of large granite boulders, most of them rounded or semi-rounded, and fine, non-stratified, and non-sorted grus to sandy material. Boulders are not uncommonly 4–5 m long and occur all across the width of the valley floor (Plate 5.12), causing channel branching and diversions. In places they form steps up to 3 m high. These deposits are highly unstable, therefore episodes of high energy flow are capable of undermining and re-mobilizing the boulders. It appears that boulders which choked the river bed and caused widespread destruction to the city during Hurricane Pauline in October 1997 were derived largely from old debris flow deposits filling high-gradient mountain valleys above the coastal plain of Acapulco.

Gully Erosion

The presence of saprolitic and colluvial mantles makes hillslopes susceptible to gullying, particularly if the slopes have been cleared of vegetation, as is indeed the case in many subtropical and tropical regions. Gully networks, not uncommonly turning into extensive badlands (Plate 5.15), have been reported from many locations and are often given local names. In South Africa and Swaziland they are called *dongas* (Price Williams *et al.*, 1982), in Madagascar they are known as *lavaka*, but these are preferentially associated with gneiss rather than granite areas (Wells *et al.*, 1991). Some of the Chinese *benggang* described in the previous section are actually extensive gully networks rather than landslide features (Xu, 1996). Table 5.5 provides references to selected gully erosion studies carried out in granite areas.

There is a great variety of gully erosion landscapes in granite terrains, depending on local geomorphic setting, the properties of the underlying material, the magnitude of human disturbance, and the evolutionary stage of gully development. Likewise, there exists great diversity of shape and dimensions of individual gullies. Gullies occupy various geomorphic settings and there have been attempts to classify them accordingly.

Plate 5.15 Deep gullies (donga) in colluvium derived from weathered granite, Swaziland (Photo courtesy of Andrew S. Goudie)

Table 5.5 Gully erosion studies in granite areas (a selection)

Geographical location	Reference
Massif Central, France	Cosandey et al., 1987; Muxart et al., 1990
Hong Kong	Ruxton and Berry, 1957; Berry and Ruxton, 1960
South-east China	Xu, 1996; Luk et al., 1997; Sheng and Liao, 1997
Swaziland	Price Williams et al., 1982; Morgan et al., 1997
Madagascar	Wells et al., 1991
Serra do Mar, Brazil	De Oliveira, 1990; Coelho Netto, 1999
New South Wales, Australia	Crouch and Blong, 1989

Thus, Berry and Ruxton (1960) distinguish between hillslope and spur crest gullies, which corresponds to their former division into lateral and longitudinal gullies (Ruxton and Berry, 1957). Wells et al. (1991) categorize the *lavaka* of Madagascar into toe-slope, mid-slope, and valley-forming forms. In the Bananal area of south-east Brazil there are gullies incised into slopes and those dissecting alluvial valley floors (Plate 5.16), as well as more complex forms involving both types (De Oliveira, 1990). There exists a multitude of cross-sections of gullies, including shallow box-like forms, deep rectangular troughs, V-shaped gashes, and wide shallow slope hollows. Wells et al. (1991) argue that this diversity of form reflects an evolutionary sequence which may be subdivided into five stages: initiation, incision, inactivation, initial filling, and final filling. However, in other cases gully shape may depend on the erodibility of the

Plate 5.16 Expanding gully network within a valley floor, near Bananal, south-east Brazil

underlying material. Some of the hillslope gullies around Petropolis, south-east Brazil, are cut into colluvium in the upper part of their cross-section and saprolite in the lower part. The saprolite/colluvium boundary corresponds with significant widening of a gully upward and a bench-like feature may develop on top of the saprolite.

The areal extent of gully networks is also highly variable. At the extreme stage of development they may cover more area than the remaining original slope surface, as is the case with deeply weathered hilly relief in central Madagascar. In the Guangdong province of south-east China, where deeply weathered granite constitutes bedrock, there may be more than 100 individual gullies within small catchments, less than 1 km^2 (Luk et al., 1997). The depth of individual gullies may exceed 20 m, whereas their length may be measured in hundreds of metres.

Gullies originate in many ways, as dictated by local surface and subsurface conditions. Ruxton and Berry (1957) emphasize the connection between eluviation within deep weathering profiles and gully evolution. Removal of fine particles lessens the cohesion of the saprolite and may lead to crack development during the dry (winter) season. These cracks in turn become rapidly enlarged by flowing water during storm events in the next wet season, and initial gullies form. The thickness of the weathering mantle is often greatest under secondary ridges. Hence the effects of eluviation and resultant cracking are also more pronounced there. This explains the presence of longitudinal gullies along spur crests and common right angle bends in gully systems. The latter form if a lateral gully propagates upward and eventually intersects with a crack running along the spur. Another mechanism is envisaged for gully growth in the Bananal area (De Oliveira, 1990; Coelho Netto, 1999). The majority of gullies connect with river channels and enlarge upward along axes of tributary dry valleys by a combination of spring sapping and tunnelling. In the first stage of their development colluvial material is eroded and mobilized, but then incision progresses into the saprolitic cover. A model emphasizing throughflow in the development of gullies has found support in the study of Beavis (2000), who found a good correlation between gully orientation and bedrock fractures in the granite terrains of New South Wales in Australia. However, in Swaziland gullies are claimed to be initiated by surface rather than subsurface processes (Morgan and Mngomezulu, 2003). Whatever the exact mechanism of gully propagation, it is usually slope deforestation and overgrazing which have promoted the progress of slope erosion. Vegetation removal in turn may be related to various human activities, in both direct and indirect ways. In south-east Brazil forest clearance was associated with the expansion of coffee plantations in the eighteenth century, whereas in south-east China and in South Africa forests have been cut to increase agricultural land. Many Swaziland gullies have developed in response to excessive soil compaction along trackways. On the summit plateau of Serra da Estrela in Portugal shallow gullies in grus-type saprolite extend after shrub communities have been destroyed by fires lit by shepherds.

Gullying is closely associated with landsliding, both spatially and temporarily. Usually, funnel-shaped transitional zones of landslides and loose aprons of landslide colluvium favour the subsequent development of gullies, but in other locations it is slope undercutting by gully extension that precedes landsliding (Xu and Zeng, 1992; Coelho Netto, 1999). Thus, gullies are geomorphic indicators of landscape instability, usually human-induced, and trigger further instability themselves. These mutual relationships determine the significance of gully development for geomorphic evolution in weathered granite areas. Gullies act as sediment transfer routes from upper slopes to main valley floors and provide an otherwise non-existent link between weathered upper hillslopes and river channels. The quantity of material carried down the gully network is not uncommonly so large that lower gully sections and valley floors cannot cope with the sediment delivered to them and aggradation occurs. In Hong Kong recent valley fills have attained a few metres and thicknesses as much as 12 m have been recorded (Berry and Ruxton, 1960), whereas in the Guangdong, China, the alluviation rate since widespread deforestation in the 1950s has been of the order of 10 cm per year. In Serra do Mar, Brazil, the notion of a 'drowned landscape' has been introduced to describe convex hills rising about flat alluviated valley floors (Plate 5.17).

Plate 5.17 'Drowned landscape' near Bananal, south-east Brazil

Slope Sediments

It has already been shown that granite, if not particularly massive, is susceptible to mechanical breakdown and chemical decomposition. In effect, loose material rather than solid rock underlies topographic surfaces. Because of generally low cohesion, little of this material remains in place and the majority begins to move downslope under the influence of gravity and water. Depending on its size, slope angle, and environmental conditions this transport may be accomplished in a variety of ways. Each means of mass movement, from creep to rock fall, has been reported from granite areas. Movement may proceed at a very fast rate, even catastrophically in the case of mud and debris flows, or very slowly, with a speed of less than a few millimetres per year. Similarly, it may involve a large thickness of near-surface material (up to tens of metres) or be restricted to a thin superficial layer. The consequence of this variety is the diversity of slope sediments present in granite landscapes. They include block fields composed of large angular boulders, stratified or poorly sorted sand and gravel, diamictic deposits with occasional larger clasts, and homogeneous clayey-silty masses, among others. The term 'colluvium' is used to describe slope deposits formed by mass movement and surface wash, but not by channelized flow.

Slope deposits are very important for the geomorphology of granite areas. In most cases, they increase the smoothness of slope surfaces, hiding any minor steps and concavities which may have resulted from differential weathering. In the footslope zone they cover topographic breaks which would otherwise occur between the rock-cut slopes above and alluvial valley floors below. However, some of these deposits, such as block fields and block streams, are very distinctive in themselves and may extend over kilometres.

In addition, certain granite-derived deposits have long been suspected to contain clues to palaeoenvironmental reconstructions. Following developments in analytical techniques, dating methods, and modelling, more and more palaeoenvironmental information can be extracted from these, unappealing at first sight, products of granite weathering, erosion, and deposition within the slope system. In this section, we present slope deposits produced by slow-acting processes, in contrast to the catastrophic landslides already discussed.

Mixed Slope Deposits in Temperate and Cold Uplands

Most of the uplands and lowlands underlain by granite in the present-day humid temperate zone are mantled by an almost continuous cover of slope deposits derived from the weathering of granite itself. These deposits are commonly considered a legacy of the cold climate conditions of the Pleistocene, hence the discussion about their lithology and distribution patterns is inseparable from a discussion about the

Table 5.6 Reference list for studies of granite-derived slope deposits in temperate zone uplands

Study area	References
Dartmoor, England	Waters, 1964; Green and Eden, 1973; Gerrard, 1989
Land's End and Scilly Islands, England	Scourse, 1987
Massif Central, France	Coque-Delhuille, 1979, Van Vliet-Lanoë et al., 1981
Harz Mountains, Germany	Hövermann, 1953
Bavarian Forest, Germany	Völkel, 1995
Karkonosze, Poland	Flohr, 1934; Büdel, 1937; Jahn, 1968; Traczyk, 1996
Serra da Estrela, Portugal	Vieira, 1998

environmental conditions of their formation and their palaeoenvironmental significance.

Superficial slope deposits have been extensively researched in the granite uplands of western and central Europe, including Dartmoor, Land's End, the Scilly Islands, the Massif Central, Harz Mountains, Bavarian Forest, and Karkonosze Mountains. Table 5.6 provides a list of selected references.

The common denominator of all these studies is the recognition of different lithological units (or facies) within slope deposits, forming stacked sequences. Scourse (1987), working in south-west England, distinguished as many as five distinct facies and linked them to specific modes of downslope transport under cold climate conditions (Table 5.7).

Table 5.7 Facies of slope deposits in south-west England

Facies symbol	Facies description	Facies interpretation
E	Large granite blocks in granular matrix; occasional erosional contacts between blocks and finer material beneath	Solifluction sediment, with possible 'ploughing boulder' component; coeval with block streams and stripes
D	Matrix-supported diamicton, with angular clasts showing clear downslope orientation	Most common periglacial, solifluction deposit
C	Matrix-dominated sandy to gravelly deposit; larger clasts few; signs of stratification	Soliflucted topsoil from the early stage of periglaciation
B	Mainly large boulders, clast-supported, matrix occasionally absent	Frost-weathered and heaved blocks; limited transport downslope
A	Basal facies ('deformation breccia'), with signs of bedrock overturning and incorporation of underlying sediment	Basal bedrock deformation during movement

Note: The reverse alphabetical order of facies description approximates facies stratigraphy, with 'A' being the lowermost horizon and 'E' being the surface deposit.
Source: After Scourse (1987).

However, most of the literature predates the proposal by Scourse and different names for individual sediment types are employed. The two used most often are 'bedded grus' and 'head'. Broadly speaking, the former is an equivalent of the facies C, whereas the latter fulfils the criteria for the facies D or E. In French the relevant terms are *arènes fauchées* and *limoneuses à blocs*. Sections in slope deposits often reveal a tripartite mantle consisting, from the bottom up, of granite weathered into grus, bedded grus, and a heterogeneous unit with a significant blocky component (Plate 5.18). Such sequences have been reported across Europe, from uplands in south-west England (Waters, 1964; Gerrard, 1989), through the French Massif Central (Coque-Delhuille, 1979), to the Karkonosze Mountains in the east (Büdel, 1937; Jahn, 1968), which suggests that their origin has been governed by at least regional rather than site specific conditions. However, in Germany, slope deposits are traditionally divided into *Basislage* (basal layer), *Mittellage* (middle layer), and *Hauptlage* (main layer) (Semmel, 1985; Völkel, 1995; Völkel and Leopold, 2001).

The thickness of mixed slope deposits is variable and mainly dependent on slope angle. Green and Eden (1973) report from Dartmoor that bedded grus is up to 0.7 m thick, whereas head deposits may reach 2 m in thickness. In the adjacent Bodmin Moor upland, sections revealing as many as 3.5 m of slope deposits have been described (Gerrard, 1989). In the Bavarian Forest in Germany the thickness of slope deposits on

Plate 5.18 Mixed slope deposits in Dartmoor, south-west England. Grus *in situ* in the lower part of the section grades into bedded growan, overlain by blocky horizon (head), itself consisiting of a few separate horizons.

granite varies from a few tens of centimetres to 1.5–2 m (Völkel, 1995). In Harz mixed deposits are only several tens of centimetres thick on steep slopes, but may be up to 1.5 m on gently inclined surfaces (Hövermann, 1953).

The characteristic coarsening of the deposits upward has been interpreted in terms of an inverted weathering profile (Waters, 1964; Fig. 5.10). Accordingly, an early phase of denudation affected the upper, finer horizon within the pre-existing weathering profile and resulted in the formation of a matrix-dominated lower deposit, whereas block-rich top units could only have formed after the deeper parts of the saprolite had been exposed and corestones had been made available for transport. However, the simple grus—bedded grus—head sequences are not present everywhere, nor is large block concentration dominant in the near-surface layer as the model implies, the latter because of derivation from local highs within the weathering front (Green and Eden, 1973; Gerrard, 1989). Therefore the inverted weathering profile model is best considered as a special, simple case, whereas the reality is much more complicated.

The origin of basal bedded grus is particularly controversial. In fact, two separate categories of basal grus can be distinguished, namely cambered grus and overlying bedded grus *sensu stricto* (Coque-Delhuille, 1979). In the former, the observed lamination can be traced down to discontinuities in weathered bedrock, hence it evidently has structural roots. Aplite veins, biotite-enriched zones, and major fracture planes in particular may be subject to downslope bending. Deformation occurs principally due to shearing at the bedrock/slope cover interface, if shear stresses imposed by the overlying mass exceed the shear strength of loosened weathered bedrock. The origin

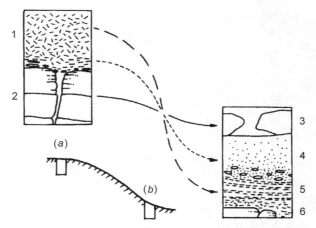

Fig. 5.10 Inverted weathering profile and resultant profile of slope deposits (after Waters, 1964, fig. 5): (a) weathering profile ((1) grus, (2) fractured bedrock), (b) slope deposit ((3) block cover, (4) structureless sand and silt, (5) bedded sand and grus, (6) fractured bedrock)

of lamination in bedded grus used to be ascribed to surface wash (Waters, 1964). This hypothesis has, however, been discounted and bedded grus is now considered a product of frost creep and the associated internal restructuring of the deposit. Laminated structure results from the development of ice lenses in the sediment, which are roughly parallel to the topographic surface as dictated by the pattern of isotherms within the slope sediment, and stresses imposed by ice bodies onto adjacent material (Van Vliet-Lanoë, 1988). Mixed deposits are less problematic and their origin as a product of solifluction over frozen ground is generally accepted (Büdel, 1937; Jahn, 1968; Van Vliet-Lanoë, 1988; Gerrard, 1989; Ballantyne and Harris, 1994).

The key theme in the discussion about the origin of slope sediment sequences has been the recognition of environmental conditions during their formation. Facies changes within the profiles have been interpreted as a response to changing climatic conditions and attempts have been made to associate lithological transitions with the climatic history of the last glacial period (e.g. Jahn, 1968; Van Vliet-Lanoë, 1988; Völkel, 1995). That geomorphic conditions on slopes were changing is beyond doubt, as shown most clearly through the occurrence of gullies cut within head deposits on Dartmoor and subsequently infilled (Gerrard, 1989). However, the paucity of datable material makes correlations very tentative.

Lithological and structural factors have been emphasized less but they may explain local differences in slope deposit inventories. For example, Van Vliet-Lanoë *et al.* (1981) observed in the Margeride Plateau (Massif Central) that slope sediment mantles are common if bedrock is granodioritic but become very thin or are even missing on monzogranites. They link this variation to the efficacy of fine material production from each lithology involved. Granodiorite is relatively depleted in potassium feldspar, enriched in biotite (up to 18%) and enclaves, and shows mineral alignment, hence sand liberation is much easier than in quartz and potassium feldspar dominated monzogranite.

The ubiquity of slope sedimentary sequences and their similarity along north-south and west-east transects across Europe account for similarities in the appearance of upland granite slopes. Long planar, poorly dissected slopes are the repetitive feature and their apparent monotony is only locally interrupted by granite outcrops of tor or rock cliff type (Plate 5.19). The thickness of several metres typical of the less inclined slopes appears sufficient to smooth minor bedrock irregularities. Larger isolated boulders may occur and these are usually residual products of recent selective denudation of upper mixed deposits. On moderately steep and sufficiently wet slopes these boulders may slowly wander down, at the rate of a few millimetres per year, leaving the elongated furrows of a U-shaped cross-section behind and piling soil material in the front; hence, their name of ploughing blocks. In the Cairngorms, ploughing granite blocks as long as 3.5 m occur.

The properties of the regolith are important for the hydrology of granite slopes. The generally coarse texture of bedded grus and mixed deposits allows for relatively efficient infiltration of rainwater, hence slope surfaces remain generally dry and little

Plate 5.19 Long planar slopes underlain by a mantle of mixed slope deposits, with occasional tors, typical of granite uplands in western Europe (Dartmoor, south-west England)

dissected. Such surface conditions are good for grazing, hence many upland slopes have long been deforested and used as pasture ground. This, however, may have an adverse effect on vegetation and slope cover stability, leading to gully erosion in the specific circumstances of overgrazing, such as in the Cévennes in southern France (Cosandey et al., 1987). Except for extreme meteorological events, runoff is mainly accomplished by subsurface flow, which may even occur as a pipe flow. A decrease in slope gradient, especially in the proximity of broad valley floors, may slow down runoff, induce saturation-excess overland flow, and lead to widespread waterlogging. In the granite uplands of south-west England wet valley floors with characteristic tussock vegetation typically give rise to ascending dry grassy surfaces which terminate against tor groups and block fields near the interfluves.

Block Fields and Block Streams

Block fields and other forms of block clustering are much more distinctive and visually impressive than otherwise significantly more widespread mixed slope deposits. Block fields (German *Felsenmeer*) are defined as extensive covers of coarse, usually angular material on flat or gently sloping ground, which lack fine material filling the voids between the blocks (Plate 5.20). In south-west England a local name 'clitter' is in use to

Plate 5.20 Summit block field in fine-grained granite, Les Monts de la Marche, north-west Massif Central, France

describe 'veritable streams, aprons or garlands of blocks and boulders' (Waters, 1964: 91). They usually occur on interfluves and on upper slopes, not uncommonly forming continuous mantles many hundreds of metres long and wide. Rock fall-derived talus aprons below high rock slopes are not considered as a category of block fields.

Block fields may be associated with tors and rock cliffs or occur independently, for example as mountain-top detritus (Ballantyne and Harris, 1994). For the latter, an origin through *in situ* disintegration of the underlying bedrock is envisaged and these are termed autochthonous block fields. In the former case, an additional component of a block field is debris liberated by weathering of tor surfaces and moved downslope over low gradient (Plate 2.3). Block fields are generally considered as cold climate phenomena, hence their apparently relict nature in the present-day temperate conditions of Europe, Asia, and North America. However, the exact origin of block fields remains a subject of debate. A view prevailing in the past, that block rubble is a product of highly intense frost shattering of bedrock, is too simplistic in the light of current knowledge. It appears that frost weathering is but one of the many processes involved, the others being chemical weathering, frost creep and sorting, selective erosion from surface wash, and locally even glacial deposition (e.g. Ballantyne and Harris, 1994).

Block accumulation may assume different forms, depending on local geomorphic setting, the means of downslope transport involved, and subsequent surface processes.

Flat interfluves tend to host block fields which are very irregular in shape and do not display any visible spatial patterns in block distribution. These have been reported, among others, from the Karkonosze Mountains (Dumanowski, 1961), parts of the Scottish Highlands (Ballantyne and Harris, 1994), and the Serra da Estrela in Portugal (Vieira, 1999). Low gradient slopes allowed for some movement, during which sorting could have occurred in response to repeating frost heave. In this way distinctive stone stripes could have formed which are such a notable feature of block-covered surfaces in Dartmoor (Fig. 5.11). Some block accumulation forms rise above the surrounding slope surfaces by a few metres and are more or less distinctly elongated downslope. Particularly long features are called block streams and may be as much as a few hundred metres long. The upper surfaces of block streams, however, may show a number of small enclosed depressions, arcuate ridges, and steps. Such characteristics

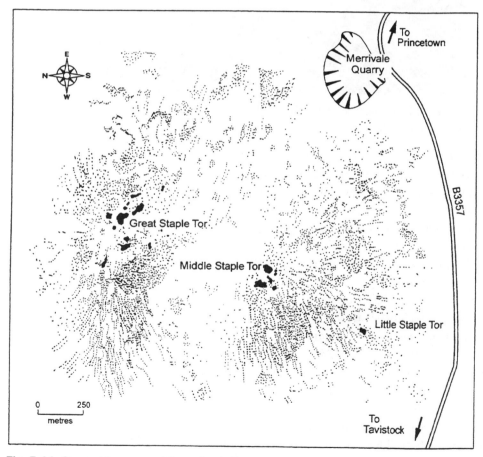

Fig. 5.11 Stone stripes around Great Staple Tor, Dartmoor, south-west England (after Bennett et al., 1996, fig. 31)

are considered indicative of former cementation by ground ice and slow downslope creep. Melting of ice has left the block mass at rest and contributed to the diversification of its upper surface. Block streams may have more than one origin. Some authors prefer to see them as residuals of solifluction tongues and aprons, from which finer material has been washed away (Wilhelmy, 1958), while others consider convex block streams and aprons as relict rock glaciers. The latter have been suggested to occur in Dartmoor (Harrison *et al.*, 1996), the Margeride Plateau in the Massif Central (Coque-Delhuille, 1979), the Karkonosze Mountains (Chmal and Traczyk, 1993; Żurawek, 1999). Steep sides of deeply incised valleys may also support block fields and block streams, particularly if rock outcrops occur near the crest. In the Serra da Estrela, Portugal, a spectacular block field has developed on the western slope of Alto da Pedrice, built of fine-grained variant of granite. It covers 500 m of vertical distance from the mountain top to the valley floor and is a few hundred metres wide across the slope (Plate 5.21). Similar landforms typify many river gorges in the Bohemian Massif, for example that of the River Dyje in south Moravia, where both extensive block fields and clearly defined block streams more than 100 m long occur.

A separate class of block accumulation features is the occurrence of big boulders stacked one upon another within well-defined valleys, called *chaos de blocs* in France. They may extend for a few hundreds metres, such as in the Sidobre granite massif near Castres, southern France, or in Huelgoat, Brittany. Two characteristics are distinctive. One is clear topographic control, another one is rounding of constituting blocks,

Plate 5.21 Block slope on a fine-grained granite, Serra da Estrela, Portugal

whereas all the other forms reviewed here are composed of angular blocks, possibly slightly rounded on the edges. *Chaos de blocs* are products of selective denudation of a corestone-rich weathering mantle and have nothing in common with mechanical disintegration of rock outcrops.

Although most block fields on crystalline rock are regarded as components of a periglacial environment, either present or past, their occurrence is not limited to areas which remained outside the extent of continental or local glaciations during the Pleistocene. Extensive block fields are known from granite uplands which were repeatedly under the ice cover, including the summit plateau of the Cairngorms (Ballantyne, 1984), the northern Scandinavian Mountains (Rea et al., 1996; Whalley et al., 1997), and various parts of the Canadian Arctic (Sugden and Watts, 1977; Watts, 1986; Dredge, 2000). The presence of erratics among the material of block fields shows that these areas were not unglaciated enclaves, as formerly presumed. The prevailing opinion nowadays holds that the survival of block fields under ice indicates a cold-based ice sheet and very limited glacial erosion (Kleman, 1994; Whalley et al., 1997).

A closer look at the distribution of block fields and related features shows that lithology and fracture patterns play an important part in their development. In general, block fields are more common in fine- to medium-grained granites, which break down into angular blocks more easily and do not produce an excessive amount of grus due to their tight fabric. Thus, block streams and rock glaciers in the Margeride Plateau appear restricted to outcrops of fine-grained leucogranite (Coque-Delhuille, 1979). In the Karkonosze, block fields built predominantly of fine-grained granite are widespread on the summit plateau (Dumanowski, 1961), but disappear in the northern part of the massif, where coarse granite occurs. In the hilly area of Les Monts de la Marche in the north-west Massif Central, France, two adjacent elevations bear different cover deposits, apparently depending on lithology and structure. The northern elevation, the Pierres Jaumâtres, is crowned by a cluster of huge rounded boulders, the length of which approaches 10 m. The granite here is coarse and rich in potassium feldspar, and primary fractures are at least a few metres apart. The southern elevation, in the village of Toulx Ste Croix, is covered by a block field composed of angular blocks 0.5–1 m long on average. The granite here is finer and much more densely fractured (Plate 5.22). In the Serra da Estrela continuous block covers are associated with fine variants of granite, whereas with coarse varieties scattered rounded and semi-rounded boulders prevail.

Granite Colluvium in Humid and Seasonally Humid Low Latitudes

Low-latitude colluvial deposits have long been a rather neglected subject in spite of their potential as a proxy record of slope development and as palaeoenvironmental archives. They are also important contributors to tropical and subtropical landscapes

Plate 5.22 Hilltop boulder pile in coarse porphyritic granite, Les Monts de la Marche, north-west Massif Central, France, contrasts with a nearby block field developed in fine-grained granite (Plate 5.20)

and may account for a significant proportion (in excess of 50%) of land surface, as reported from Swaziland (Price Williams *et al.*, 1982) or Cameroon (Hori, 1977; cited in Thomas, 1994*b*).

Colluvium may result from various processes, including landslides, mud- and earth-flows, sheet wash, and soil creep. In addition, soil fauna plays an important role in reworking colluvium and bioturbations may be frequent (Plate 5.23). Their internal texture is usually the key to the formative process. Landslide-derived colluvium is a poorly sorted, heterogeneous sediment, whereas colluvium produced by sheet wash shows lamination related to individual wash events unless destroyed by post-sedimentary processes of bioturbation. Of course, at any one site, different processes may be active at different times, producing complex colluvial stratigraphy. The source of material for colluvium is weathered bedrock, and this is why colluvium becomes widespread in low-latitude granite areas. The colluvium itself may prove unstable and become re-mobilized to form secondary colluvial mantles on even lower slopes.

Landscape facets dominated by colluvium are long planar, usually low-angle (less than 5°) slopes linking steeper slopes supported by weathered bedrock with alluvial valley floors. Their length may be of the order of hundreds of metres and the thickness of colluvium deposits may well exceed 10 m. In the Jos Plateau, Nigeria, the thickness

Plate 5.23 Colluvium deposit near Bananal, south-east Brazil. Complex origin may be inferred from the presence of a stone line (below the hammer) and voids caused by burrowing organisms

of the main colluvial unit is locally as much as 20 m (Hill and Rackham, 1978). Deep gullying may affect colluvial aprons and examples are known from Swaziland (Price Williams *et al.*, 1982), south-east Brazil (Coelho Netto, 1999), and Hong Kong (Berry and Ruxton, 1960). Interestingly, there seems to be no agreement about terminology to describe slope surfaces developed upon colluvium. Price Williams *et al.* (1982) refer to colluvial deposition on pre-existing rock-cut pediments, whereas De Dapper (1989) introduced the term 'pedisediment' to emphasize the presence of a sedimentary layer. Bigarella and Becker (1975) in turn prefer the term detrital pediment. In addition, local terminology is occasionally followed. For example, in the Brazilian work on colluvium the term *rampa* is widely used to describe low angle depositional surfaces (Plate 5.24), whereas a *rampa complex* denotes a pair consisting of an amphitheatre-like hollow in the hillslope caused by landslide and the aggradation surface in front of it (De Meis and Monteiro, 1979; Coelho Netto, 1999).

The stratigraphy of colluvial deposits may contain a few distinct horizons stacked upon one another, separated by stone lines or erosional surfaces which are indicative of the changing balance between deposition and erosion. Thus, the Middle Veld in central Swaziland is dominated by colluvium mantles which can be resolved into two principal components (Price Williams *et al.*, 1982). The lower colluvium, the *Bovu* Colluvium, had been transformed by pedogenic processes and locally incised before the deposition

Plate 5.24 Rampa complex consisting of an amphitheatre-like hollow cut into the slope and a planar depositional surface below, near Bananal, south-east Brazil

of the upper *Mphunga* Colluvium started. Goudie and Bull (1984) envisage colluvium deposition through successive stripping of saprolite from the upper slopes by surface wash and link colluvium deposition with periods of aridity during the late Pleistocene. A two-layered colluvium has also been reported by Berry and Ruxton (1960) from Hong Kong. Its lower layer contains completely weathered small boulders in red silty-clay matrix, while in the upper layer large boulders are more numerous and the matrix is brown-grey. The authors do not explain this succession but it seems compatible with the model of inverted slope profile and, if true, colluvium would record deeper and deeper stripping of the saprolite. Successive stripping of weathering mantle has also been recognized by Modenesi (1988) in colluvial aprons in eastern Brazil

The presence of colluvium is not restricted to gentle footslope surfaces. In fact, they frequently cover much steeper slopes, locally as steep as 30°. In these settings colluvium is rather thin (1–2 m) and in metastable equilibrium, therefore deforestation usually triggers shallow landslides which may affect large areas of hillslopes. Cuts into steep slopes of residual hills around Acapulco, Mexico, reveal a mantle of colluvium 1 to 3 m thick, composed of boulders of various size and a degree of rounding, set in a non-sorted sandy-silty matrix. This colluvium grades into weathered bedrock without any clear boundary, whereas on the surface the continuity of the mantle is locally interrupted by tor-like outcrops and exposed corestones. Downslope the colluvium thickens and forms a continuous apron (Plate 5.25).

Plate 5.25 Bouldery colluvial deposit below a steep slope, Acapulco, Mexico

It is beyond the scope of this book to review the palaeoenvironmental significance of colluvial deposits in tropical granite areas. A review of the subject is available in Thomas (1994*b*).

Pediment Mantles

Pediments are essentially erosional features, but bedrock is typically covered by a thin sedimentary veneer, whose thickness may be as little as a few centimetres, but is more commonly around 1–2 m. The pediment mantles may be derived from *in situ* bedrock weathering, fluvial transport across the pediment surface, and there may be an aeolian input.

Twidale (1978*b*), referring to pediments in the Eyre Peninsula, South Australia, maintained that the mantle is of very local provenance, derived from bedrock weathering, and the efficacy of transport across low-angle (0.5–1.5°) pediments is very limited. However, granite pediments in south-west USA are steeper and both their sedimentary covers and fluvial landforms indicate the role of overland flow. The flow may take the form of sheetflow, but more commonly it occurs within systems of anastomosing washes which migrate laterally across the pediment. Cooke and Mason (1973) have found evidence of buried channels 1–3 m deep on the Desert

Knoll pediment in the Mojave Desert and Cooke *et al.* (1993) claim that they are common elsewhere too. Indeed, semi-permanent shallow channels are a typical feature of granite pediments in the marginal, wetter part of the Namib Desert (Plate 4.10). Rahn (1967) showed that sheetflow is more likely if a cloudburst takes place over the pediment itself, whereas streamflow usually happens after rainfall has affected the mountains behind.

The occurrence of buried channels and patches of relict alluvial sediments in the upper parts of pediments shows that the boundary between predominantly rock pediment above and mantled pediment below may have migrated in the past. This migration would indicate changing relationships between rates of debris supply and removal, likely to be caused by climatic oscillations during the Quaternary (Cooke and Mason, 1973).

6

Granite Coasts

Granite Coasts—Are They Specific?

Although no estimate of the aggregate length of granite rock coasts around the world is available, they surely make up quite a significant proportion of the total, especially around the Fennoscandian and Canadian Shield (Bird and Schwartz, 1985). However, in contrast to the vast amount of literature about inland granite landforms, granite coastal scenery has attracted significantly less attention, in spite of the fact that some of the most spectacular coastal landscapes are supported by granite (Plate 6.1). Detailed studies of granite coastal geomorphology are surprisingly few, although the structural adjustment of the coastline in plan at the regional scale is a recurrent observation (Bird and Schwartz, 1985).

One probable reason for this discrepancy between the length of granite coasts, their scenic values, and scientific knowledge are the low rates of geomorphic change expected along them. Therefore they are poor candidates for any process-oriented studies, which dominate contemporary coastal geomorphology. It is probably because of this scarcity of information that contrasting opinions have been expressed about the specifics of granite coasts. Whereas Twidale (1982: 2) asserts that: 'In coastal contexts, too, the gross assemblage of forms is due to the processes operating there and not to

Plate 6.1 Granite cliffs, stacks, and arches, Land's End, south-west England

properties peculiar to granites. . . . Orthogonal fracture sets also find marked expression but, with few exceptions, granite coasts are much the same as most others'; Trenhaile (1987: 173) goes on to say: 'Igneous coasts are usually quite different from other rock coasts'.

On the one hand, many granite coasts consist of an all-too-familiar assemblage of cliffs, coves, joint-aligned inlets, stacks, and sea arches. From this point of view, no components of coastal morphology are likely to be demonstrated to be unique to granite. But this is also true for granite landforms in general, as was indicated in the introduction to this book. On the other hand, there seems to be enough observational material to claim that certain granite coastal landforms have developed specific characteristics, different from those supported by other rocks, as well as that there exist certain very specific sections of granite coasts which hardly have parallels in other lithologies.

First, granites are highly resistant to erosion and abrasion, therefore they will tend to support relatively stable coastal landforms even along high-energy coasts, such as the Atlantic seaboard of western Europe or in South-East Asia, with its frequent typhoons.

Second, these low rates of change brought about by marine processes imply that the present-day coastal geomorphology may be partially inherited and processes other than coastal ones have been instrumental in shaping the morphology, which now happens to be located in a coastal setting. Given the strength of granite and the relatively recent establishment of sea level at the present elevation, it is not reasonable to try to explain the current coastal granite morphology solely in terms of contemporary processes and environments.

Third, shore platforms are rather uncommon along granite coasts, which again is believed to be due to the much higher resistance of granite against shearing. In this respect, granites stand in contrast to the typical platform-supporting lithologies such as sandstone, limestone, mudstone, or shale (Trenhaile, 1987).

Fourth, there is little alternation between rock-cut sections and sandy beaches along granite coasts, even in low-energy sheltered environments. This is because the sand supply from granite itself is insufficient, as granite tends to break down into big blocks under wave attack. Further wear of products of mechanical disintegration produces abundant cobbles and gravel, but very little sand. For example, beaches are rare along long coastal tracts of Scandinavia if these have developed in solid granite. However, if granite is deeply weathered, then the availability of fine material is much greater and wide beaches form. Godard (1977) cites relevant examples from the low-latitudes coasts of the Ivory Coast, Liberia, and Brazil, but pocket sandy beaches are also common in mid-latitudes, such as in Brittany and the Isles of Scilly in western Europe.

Cliffs and Platforms

Granite cliffs occur in abundance, varying in height, steepness, and lateral extension, which is likely to reflect structural predisposition and differences in energy supply to the shore. For lithologically homogeneous material, and in many settings granite can be classified as such, Emery and Kuhn (1982) distinguish two basic forms of cliffs. Steep cliffs develop if marine erosional processes are dominant, whereas the domination of subaerial slope processes leads towards convexo-concave cliff profiles. The former typify headlands and other exposed coastal sections where material supplied to the cliff base from above can be effectively crushed and then removed by wave attack. The latter occurs within bays and sheltered coasts where dispersion of wave energy results in less efficient redistribution of slope sediments along the coast. Hence net deposition of debris occurs at the cliff base changing its longitudinal profile into a convexo-concave. An example of the relationships between cliff shape and the amount of energy delivered to the shoreline is provided by the coast of the Penwith peninsula in Cornwall, south-west England. Vertical cliffs alternating with coves occur along the section exposed to the west, to the most vigorous wave attack, whereas sloping cliffs prevail in the adjacent, somewhat sheltered, south- and south-east-facing section between Treen and Mousehole (Steers, 1948). In the same area, the alternation of steep and gentler cliffs can be observed on a smaller scale between adjacent headlands and bays around Land's End (Plate 6.2).

Straight vertical cliffs tend to develop in massive, orthogonally fractured granite, with joint spacing of at least 1 m. Their height varies, from a mere few metres to 20–30 m. They are particularly common at exposed headlands of high-energy coasts, for example

Plate 6.2 Headlands and bays, Land's End coast, south-west England. Note numerous entrances to marine caves under the cliff

in Cornwall (Bird, 1998) and Lundy Island (Godard and Coque-Delhuille, 1982) in south-west England, the Shetlands, western Brittany, France, Galicia, north-west Spain, and Portugal. More dense jointing results in the development of steep, but not vertical cliffs, as predicted by the rock mass strength–slope angle relationship. The cliffs of Point-du-Raz in western Brittany belong to this category, as do the Quebrada cliffs in Acapulco, famous for diving into the sea from a height of 30–40 m (Plate 6.3). Domical compartments, in turn, do not support vertical cliffs, but sloping structural surfaces of various inclination, dictated by the primary structure of a dome, whether low-radius or parabolic (Hills, 1971; Twidale and Vidal-Romani, 1998).

Mass movements are an important component of cliff development and granite cliffs are no exception, even if they show significantly more long-term stability than their counterparts in other lithologies. The exact nature of failures depends on the structural characteristics of the granite, particularly on its fracture pattern. Well-jointed granite, which gives rise to castellated cliffs, is subject to toppling that develops from pseudo-continuous seaward flexure of granite columns delimited by orthogonal joints. This category of toppling has been termed a block flexure type (Hoek and Bray, 1981) and has been reported, from Land's End, among other sites. Cliffed coasts in sheeted granite develop through sliding of consecutive sheets, or their fragments, on the underlying rock surfaces. The development of shelters and overhangs in the lower slope due to basal wave attack greatly assists in the process of degradation.

Plate 6.3 Steep cliffs in densely jointed granite at La Quebrada, Acapulco, Mexico

Cliffs retreat in the long term, but little is known about the rates of retreat of granite coasts. Figures as low as 0.001 m yr^{-1} appear in the literature, which would translate into a negligible few metres of retreat in the last 6,000 years if a simple extrapolation is performed (Masselink and Hughes, 2003). This is one order of magnitude less than in limestone, two orders less than in shale, and four orders less than in poorly consolidated Quaternary deposits. However, the reliability of such an extrapolation is uncertain.

The spatial pattern of retreat, even if this is indeed slow, is fracture-dependent in that more jointed compartments retreat faster, making the process non-uniform. Rocky promontories are left in front of the main cliff line and may be completely cut off from the retreating slope, evolving into steep-sided coastal residual landforms called stacks. The Enys Dodnan stack off the tip of Land's End, south-west England, pierced by a magnificent tunnel, is an example (Plate 6.1). Stacks may also be isolated through preferential wave attack, weathering, and erosion along dominant fracture lines (Plate 6.4).

Plate 6.4 Stacks off the cliff line at Point Lobos, California. Their elongation perpendicular to the shore suggests that they have been isolated through preferential erosion of parallel lines of structural weakness

Although stacks along granite coasts do not appear to be as common as they are in bedded sedimentary rocks, their occurrence raises an interesting issue of the timescales necessary for their development. Given the relatively recent establishment of the current sea level one might infer quite high rates of cliff retreat and the efficacy of wave attack. However, care is required here. Cliff sections in south-west England and Brittany show the presence of deeply weathered granite occurring alongside massive compartments, and the efficacy of coastal processes needs to be viewed in this context. Pre-weathering of structural lines of weakness and isolation of tower-like residuals at the weathering front would help to isolate a stack in a much shorter time than if cliff retreat proceeds in unweathered, solid granite (Fig. 6.1). The numerous stacks in front of the cliff line in Pays de Léon, Brittany, are interpreted in this way (Guilcher, 1985).

Two specific types of cliffs have been reported from granite coasts, although they occur in other lithologies as well. These are slope-over-wall and plunging cliffs (Bird, 2000). The slope-over-wall cliffs occur in mid-latitudes and consist of a very steep rocky section near the base which turns into a less inclined, regolith-mantled slope higher up. Sequential development of these forms, causally linked to global climate and sea level change, is envisaged. During a glacial period and sea level drop a former marine cliff is removed from direct wave influence and subject to degradation in periglacial conditions. The resultant form is a convexo-concave, moderately inclined rock slope covered by periglacial slope deposits, locally of substantial thickness of 5–10 m.

224 Granite Coasts

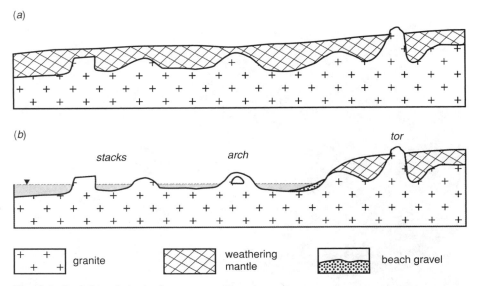

Fig. 6.1 Evolution of stacks from pre-weathered granite

Subsequent sea level rise brings the former cliff back to its littoral setting. Renewed undercutting by waves results in the steepening of the lower slope, which again assumes verticality, but the upper slope retains its periglacial morphology and gently convex shape for a significantly longer time (Fig. 6.2). Slope-over-wall cliffs are common along the south-western coast of Britain, including the granite Penwith Peninsula

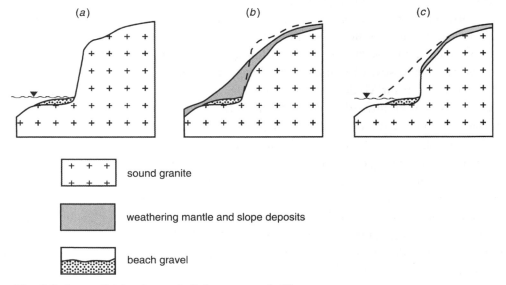

Fig. 6.2 Sequential development of slope-over-wall cliffs

in Cornwall. Bird notes that they are better preserved in sheltered locations, whereas they have almost disappeared from the more exposed coastal sections, and the previously quoted observations by Steers (1948) may document just such a case.

Plunging cliffs are those which are not bordered by any wave-cut platform on the seaward side. Instead, the steep terrestrial slope plunges deeply into the sea and terminates at much greater depth. Steep slopes of massive granite domes located in coastal settings are examples of plunging cliffs, such as those present in the Wilson Promontory, Victoria, Australia (Bird, 1968), around the Guanabara Bay in Rio de Janeiro, or at the Tasman Point of the South Island, New Zealand (Thomas, 1974). Vertical slopes of Pleistocene glacial troughs, now partially submerged, are another sub-class of plunging cliffs, common along the high-latitude coasts of the Kola Peninsula in north-eastern Russia, western Greenland, Maine in the USA, and southern Chile. High cliffs on the coast of Peru, called megacliffs in recognition of their height, which exceeds 300 m (Bird, 2000), plunge deeply into the sea because the landward side has been up-faulted. Altogether, plunging cliffs demonstrate that granite is too resistant for a shore platform to develop within the 10,000 years of postglacial sea level rise. In cases of profound structural influence, as with granite domes, plunging cliffs may have been in existence for a much longer time.

Platforms, as noted earlier, do not belong to the assemblage of landforms typically present along granite coasts, but nevertheless they do occur locally. However, their rough, fracture-controlled morphology contrasts in detail and limited width with the great extent of marine platforms developed on sedimentary rocks. For example, along

Plate 6.5 Marine caves and tunnels cut along fracture lines, Point Lobos, California

the shore of Monterey Bay, central California, almost planar raised platforms cut in soft Tertiary sediments attain hundreds of metres in width and can be traced for many kilometres along the coast, whereas granite headlands are typified by numerous tor-like stacks, joint-guided inlets, plunging cliffs, and only sporadic platforms (Plates 6.4, 6.5) (Griggs and Johnson, 1979). The reasons for the relative under-development of platforms along granite coasts are at least twofold. First, as noted earlier, it is the high resistance of granite to shearing which makes the work of planing off the rock more difficult to accomplish. Second, platforms develop at the expense of retreating cliffs, but these retreat extremely slowly if built of massive granite. Furthermore, the density of fracturing of granite is usually highly variable over very short distances. Abrasion and weathering will tend to focus on the densely jointed zones, so that over time local relief may actually increase and perfect platforms may not develop.

Inlets, Sea Caves, and Arches

Much in the same way as less jointed masses of granite become cliffed headlands and sea stacks, their densely fractured counterparts are preferentially exploited by wave attack and subaerial processes, giving rise to a range of negative forms, including inlets and sea caves. Different local names are used to describe these landforms. Fracture-aligned clefts are called geos, but the term 'zawn' is used in Cornwall, and similar landforms in Scotland are called yawns (Bird, 2000).

An example of how structure determines the occurrence of these negative landforms is provided by the headlands in the Point Lobos peninsula near Monterey, California. They are built of coarse Santa Lucia granite, emplaced $c.80$ Ma ago. Numerous fractures running north-west–south-east and south-west–north-east provide readily available lines of weakness along which a wave attack may be concentrated. Exploitation of fractures has led to the origin of narrow inlets, a few tens of metres long and a few metres wide, which account for the very irregular outline of granite headlands, jutting out into the sea such as at the western end of the peninsula (Plate 6.5). In a number of places fracture-guided straits separate rocky islands, particularly in the north-east and south-east sections. Locally, higher density of fractures has facilitated the development of sea caves and tunnels joining two adjacent inlets. It may be hypothesized that some of the straits may have evolved through inward inlet development from opposite sides, tunnel development, and ultimately collapse.

Along the granite coast near Land's End wide entrances to sea caves are numerous (Plate 6.2). The caves are located at the base of cliffs within bays and have formed along the zones of closely spaced sub-vertical joints. They extend for more than 10 m into the cliff face and their openings are as much as 5 m high. Fracture-guided coastal caves are known from other areas too. Sjöberg (1981) described pear-shaped tunnel caves in Sweden and attributed this shape to the abrasive action of cobbles.

In other cases the exploitation of fractures has probably been a two-stage process, as illustrated by the marine cave at Bullers of Buchan, north-west Scotland. The presence of a 30 m deep cauldron-like hollow c.30 m inland, joined to the open sea through a tunnel (Fig. 6.3), suggests that fracture zones had first suffered from deep weathering into grus. Then, grus was washed away from the fracture zone to form a tunnel that allowed the waves to access the circular pocket of weathered rock. Disintegrated rock has been sucked out and evacuated to the open sea, leaving behind the closed hollow a few tens of metres away from the present-day coastline. It may be speculated that a further retreat of the granite cliffs will eventually consume the tract of land between the hollow and the sea, and the present-day cauldron will become a half-circular cove. In popular accounts the cauldron is referred to as having been formed due to sea cave roof collapse, but given its huge dimensions, $c.100 \times 30$ m, this seems a less likely scenario. The weathering-and-washing out hypothesis, on the other hand, is supported by the common presence of deep pockets of weathered granite on the Buchan plateau further inland (Hall, 1986).

Sea arches belong to the most spectacular and scenic landforms associated with rock coasts, but they tend to occur in abundance in rocks weaker than granite such as limestone and sandstone. Nevertheless, there are arches and tunnels in granite too. One example is the arch pierced through the Enys Dodnan stack off the Land's End promontory, developed along a zone of more densely fractured rock. Similar landforms occur along the coast of Brittany in France, in Chile, and in other places.

Granite Weathering in Coastal Settings

Little is known about the weathering of granite in coastal environments, and particularly about the rates at which it proceeds. As with granite coasts in general, this paucity of information is probably related to the general resistance of the rock, which does not favour weathering process studies. The poor development of coastal platforms in granite, which are the settings where most weathering studies are carried out, is another factor responsible for our limited knowledge. Biological weathering and erosion are probably much reduced as granite shores are not known to support a rich microflora, nor is boring effective in usually hard granite (Trenhaile, 1987).

The most common weathering phenomena on granite outcrops in the littoral zone are flaking and granular disintegration, reported from both high and low latitudes. They are attributed to hydration induced by frequent wetting and drying cycles and to salt crystallization from sea spray. Indeed, enhanced effects of salt weathering and hydration are logically expected in shoreline settings and Haslett and Curr (1998) infer as much as 1.5 m of surface lowering on the seaward end of a probably Eemian wave-cut platform in southern Brittany, France, due to salt weathering. However, Bird (1970), working in the Trinity Bay coast in Queensland, NE Australia, observed similar

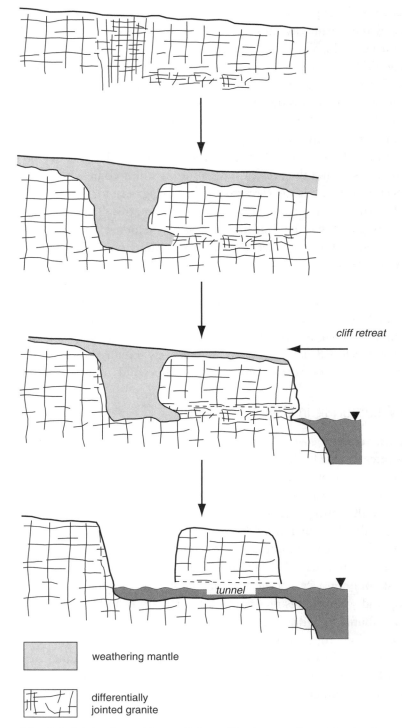

Fig. 6.3 Origin of a cauldron at Bullers of Buchan, north-east Scotland

features further inland and doubted if contact with sea spray is truly essential. Likewise, Guilcher (1958) noted the presence of various features of selective granite weathering along the north-western coast of France (Plate 6.6) but was hesitant to relate them solely to the input of salt aerosols from the sea. Given the general resistance of bare granite surfaces to weathering, and the short time span of a few thousand years available for weathering in the coastal setting as we see them today, it is perhaps not surprising that the effects of littoral weathering are difficult to separate quantitatively.

In recent years, some insights into the mechanisms and efficacy of granite weathering in coastal settings have been offered by stone conservation studies. Signs of deterioration, apparently resultant from their long exposure to sea spray and salt weathering attack, have become a matter of concern and detailed studies focused on stone durability have followed. Rivas *et al.* (2003) have shown in an experiment that the presence of different ions in solution lowers the solubility of NaCl, which then crystallizes at depth rather than at the surface, causing the development of minute scales and granular disaggregation. These micro-weathering features are replicas of the weathering morphology observed on coastal outcrops and granite monuments exposed to sea water spray.

A characteristic feature of granite boulders in low-latitude coastal settings is the presence of well-developed vertical flutings (karren). They typically occur in groups

Plate 6.6 Diversified microrelief on coastal outcrops, Trégastel, Brittany, north-west France

Plate 6.7 Fluted boulders on the coast of the Seychelles (Photo courtesy of Andrew S. Goudie)

and run roughly parallel to each other. The classic site of their occurrence is probably the Seychelles Islands in the Indian Ocean (Wilhelmy, 1958; Plate 6.7), but they are also known from the Malaysian Peninsula (Logan, 1851, cited in Twidale, 1982; Tschang Hsi-Lin, 1961; Swan, 1971) and the Kangaroo Island on the southern coast of Australia (Dragovich, 1968). Flutings are rare along mid- and high-latitude coasts but they are not totally absent. Some cliff faces at Point Lobos, California, do show a pattern of parallel grooves although they are much shallower than their tropical counterparts. The mode of origin usually inferred for these picturesque features involves a combination of mechanical erosion and chemical attack from salt-laden sea spray running down the boulder faces.

Non-littoral Inheritance

Some granite coasts, especially in mid- to high latitudes, have developed a very unusual appearance, so that the spatial configuration of their landforms is difficult to reconcile with the rules of relief development due to wave attack and abrasion. These coasts are examples of an inherited coastal geomorphology which has been created in the past, under the influence of non-marine agents such as deep weathering, fluvial erosion or glacial scour. Subsequent sea level and land level changes have brought these landform assemblages into direct contact with sea water, but because of a combined effect of insufficient time and rock resistance the effects of previous, non-coastal morphogenesis have not yet been obliterated.

Glacial inheritance is probably the best known. In Maine, eastern USA, steep granite cliffs owe their form to glacial erosion in the Pleistocene, while subsequent modification in the Holocene has been minor. The presence of glacial striations on cliff faces, down to the water level, provides the evidence that postglacial change of cliff form has been minimal (Shepard and Wanless, 1971). Commenting about the sheer cliffs on Mt Desert Island, whose height approaches 250 m, they issue a warning that 'These walls might give the impression of being great wave-cut cliffs, but actually they have been very little modified by wave erosion' (p. 34). These are examples of plunging cliffs as discussed before.

The coastal morphology of Scandinavia, particularly of Sweden and Finland, is dominated by the presence of skerries, which is a term used to describe low rocky islets, typically with a gently convex profile. Undercuts may be present on one side and if the range of tides is sufficiently high, they may be alternatively drowned and re-exposed. Skerries occur in thousands in areas such as the Stockholm Archipelago or Åland Islands in the Baltic Sea (Plate 6.8). They may carry evidence of a recent wave attack, but their general form is inherited from the glacial period as testified by the presence of strations, crescentic gouges, plucked lee sides, and, not uncommonly, the

Plate 6.8 Granite skerries along the coast of Bohuslän, south-west Sweden

overall *roche moutonnée* shape. The survival of glacial sculpting indicates that very little change has been brought about by marine agencies in the last few thousand years. In other sections of the Swedish coast the situation is even more complicated. The rounded shapes of bare granite skerries dotting the coastal waters north of Göteborg, along the coast of Bohuslän, are similar to those of classic *roches moutonnées* and undoubtedly, many of the offshore islands must have been moulded by passing ice, perhaps on more than one occasion. However, the history of the hilly, partially drowned landscape along the western Swedish coast goes much further back in time, to the Mesozoic (Johansson et al., 2001a, 2001b). Skerry coasts are not limited to Fennoscandia and can be found in Scotland, Ireland (Galway Bay), and in formerly glaciated parts of North America.

Other coasts show a clear legacy of deep weathering, as emphasized by Godard (1977). The Isles of Scilly, off Cornwall in south-west England, have a complex coastal morphology, in which rock-cut and depositional features are intertwined (Steers, 1948). The archipelago consists of 140 islands, including 12 of sizeable area. Some of the larger islands, such as St Mary's, St Martin's, and Tresco, have two or three granite cores, connected by sand bars and tombolos. Relicts of wave-cut platforms can be seen at a height of 3–8 m above sea level, mainly in weathered granite, but the gross features of the archipelago, including the spatial pattern of granite islands, mainly reflect the differential weathering of granite, probably in Neogene times. Straits separating adjacent islands, such as Tresco/Bryher and Gugh/St Agnes, follow structural lines of weakness, exploited first by deep weathering and only subsequently by marine agencies (Fig. 6.4).

A coastal landscape similar to that of the Isles of Scilly has developed along various sections of the coast of Brittany, France, notably between the towns of Trégastel and Ploumanac'h in the north, which is known as the 'Pink Granite Coast'. The high degree of indentation of the coastline is due to the occurrence of numerous low-radius domes, tors, basins, and defiles (Plate 6.9). Dome and tor morphology typifies the headlands and continues further inland, although the relative relief decreases and concave features are less pronounced. The scenery of the 'Pink Granite Coast' is thus the product of selective wave attacks acting upon pre-weathered granite morphology, with marine rather than subaerial processes having been responsible for stripping the saprolite, which is a rarely reported case. Outside the extent of wave attack, the grus cover remains in place and its continuing presence explains a less varied morphology inland. A similar situation exists in the Pays of Léon in north-west Brittany, where numerous stacks rising above a wide submerged platform are interpreted as residuals left after the weathering mantle and later periglacial deposits were stripped by waves (Guilcher, 1985). Coude (1983) offers further examples from Ireland, arguing for long-term differential weathering and erosion within the Galway Batholith to explain the origin of the Bertraghboy Bay and Kilkieran Bay in Connemara (Fig. 6.5). In the Mediterranean

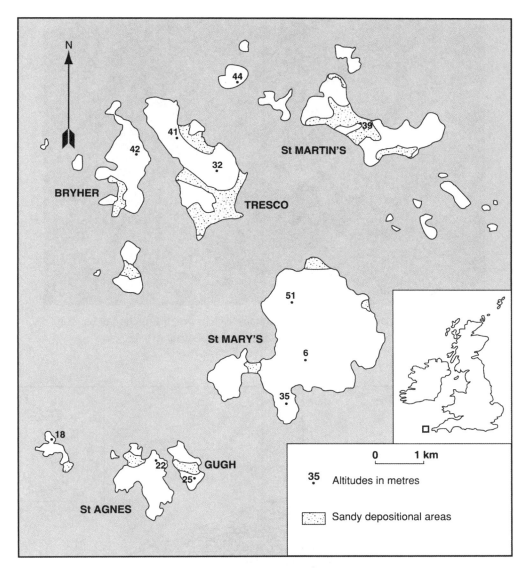

Fig. 6.4 Isles of Scilly—an example of inherited coastal geomorphology

Basin, the pattern of deep weathering is reflected in the present-day configuration of bays and headlands in the granite coasts of Corsica and Sardinia (Godard, 1977).

In the low-latitude context, the profound impact of antecedent deep weathering explains the coastal geomorphology at the southern tip of the Malay Peninsula (Swan, 1971). The present-day assemblage of low hilly islands used to be a hill-and-ridge landscape during the Pleistocene low sea-level stands and was subject to intense weathering under humid tropical climate. Today, the former regolith is being stripped away

234 Granite Coasts

Plate 6.9 Coastal landscape of northern Brittany, with its irregular pattern of bays and headlands, owes its form mainly to the evacuation of weathering mantle

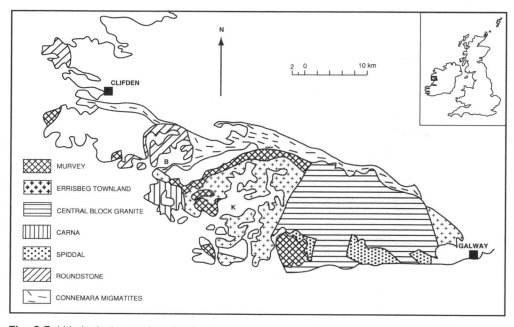

Fig. 6.5 Lithological control on the development of sheltered bays, Connemara, Ireland, based on Coude (1983): (B) Bertraghboy Bay, (K) Kilkieran Bay

by waves and unweathered granite cores are becoming exposed in the form of islands or residual boulders, depending on the size of more massive compartments. These inherited features are incorporated into the present-day coastal geomorphology by the development of a tombolo, which joins them to each other and to the mainland (Wong, 2005). The coastal granite geomorphology of Hong Kong, despite being exposed to frequent typhoons and associated vigorous wave attacks, owes its gross form to structure-guided long-term weathering and erosion. Specifically, the nature of the coastline, whether it is a rock cliff or a bouldery beach, depends upon the characteristics of the weathering mantle along a given section of the shoreline (So, 1987).

The wide range of inherited granite coastal landscapes provokes the question as to what is the norm and what is the anomaly. Indeed, Trenhaile (1997) predicts that most steep coastal slopes in high latitudes are inherited glacial features and few true marine cliffs exist because of generally weak wave action and coastal configuration. In low latitudes it is the structure and patterns of antecedent deep weathering which dictate the shape of many granite coasts. Even in such classic coastal settings as the cliffed coast of Land's End, former deep weathering of granite into grus appears as an important contributor to the observed coastal morphology. Thus, the most distinctive aspect of coastal granite geomorphology is probably the fact that marine processes operating in the last 6,000 years have accomplished so little and the inherited component shows up much more clearly than in any other rock type.

7

Cold-Climate Granite Landscapes

Inselbergs, tors, boulder fields, and pediments are repetitive landforms of many low- to mid-latitude granite landscapes, whether in humid or in arid environments. Although there have been attempts to link these landforms to certain specific climatic environments, their actual distribution, as shown in the preceding chapters, speaks clearly for minor climatic control in their development. Therefore, identification of a 'typical' granite rainforest, or savanna, or desert landscape does not seem possible. Each of these environments is known to host a variety of distinctive landscapes supported by granite, which will be explored in the next chapter.

Likewise, cold environments in high latitudes have long been considered as having a very distinctive geomorphology, in which the factor of rock control matters little, but repeated freezing and thawing is critical. This view is difficult to maintain any longer, especially in the light of recent progress in periglacial geomorphology. The effects of glaciation are more evident, but even there the role of bedrock must not be neglected and formerly glaciated granite terrains do show certain specific features.

Granite Periglacial Landscapes

Many granite terrains are located in cold environments, or have experienced cold-climate conditions in the relatively recent past of the Pleistocene (Fig. 7.1). Therefore, it is reasonable to expect that their geomorphic evolution has been influenced by a suite of surface processes characteristic of such settings, collectively termed as 'periglacial'. Present-day periglacial conditions typify such granite areas as the uplands of Alaska, Yukon, and the northern Rocky Mountains, much of the Canadian Shield, coastal strips of Greenland, northern Scandinavia, extensive tracts of Siberia, and the Tibetan Plateau. Granite areas located further south, in the British Isles, the Iberian Peninsula, the Massif Central, the Harz Mountains, and the Bohemian Massif, were affected by periglacial conditions for most of the Pleistocene. In fact, the most elevated parts of these mountains and uplands experience a mild periglacial environment even today and winter temperatures may remain below 0°C for weeks. The efficacy of present-day frost action is however limited by the insulating snow cover. Some of the granite areas of the southern hemisphere are, or were, within the periglacial realm too. In addition, high granite mountains evolve under the prevailing influence of cold climate conditions, but weathering products are rapidly transported down their steep slopes and 'periglacial' landform assemblages hardly have time and space to develop.

Many present-day high-latitude granite areas were repeatedly glaciated during the Pleistocene and their periglacial history is rather short. They tend to be dominated by inherited glacial landforms, whereas geomorphic features related to frost action are poorly developed. Hence, the best insights into the effects of cold climatic conditions on granite landscapes can be obtained through the examination of upland areas which are now in the temperate climatic zone but have retained their Pleistocene periglacial legacy. Areas such as Dartmoor in England have already become a classic location to look at the efficacy of frost-driven processes.

Before proceeding further, one needs to ask about a typical periglacial landform assemblage, what it actually is, and what components it has. In fact, various definitions of the term 'periglacial' have been offered in the literature, some of them highlighting the proximity of a glacier, but they all tend to converge towards an emphasis on the key geomorphic role of frost. Its action can be resolved into frost wedging, frost heave, frost cracking, and frost sorting (Washburn, 1979). In addition, alternating freezing and thawing is important for mass movement phenomena and the downslope transport of near-surface layers, called solifluction, proceeds at rates of up to 1 m yr^{-1} (Matsuoka, 2001b). Such high rates of non-catastrophic mass movement are unique to periglacial environments.

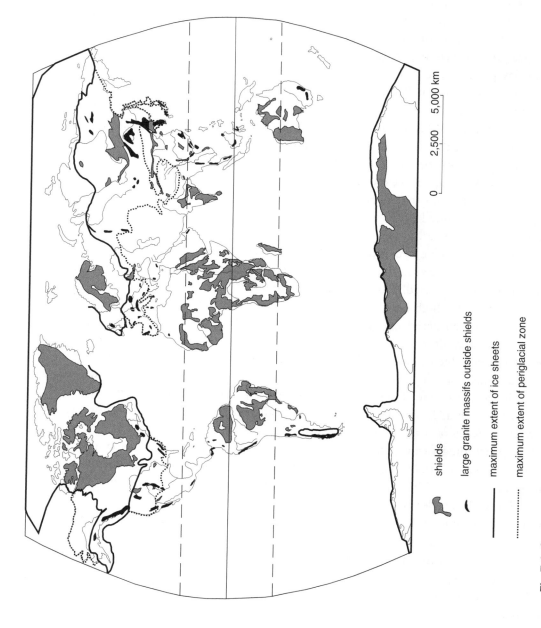

Fig 7.1 Granite terrains in relation to recent and Pleistocene periglacial environments

The typical ingredients of a conventional view of periglacial landscapes developed in solid rock are the following (Jahn, 1975; Washburn, 1979; Ballantyne and Harris, 1994; French, 1996):

- angular products of rock disintegration due to frost wedging and other mechanical processes, not uncommonly forming extensive block fields and block slopes;
- angular outcrops of solid rock supplying debris, in the form of either rock cliffs (frost cliffs) or castellated tors;
- stepped slope profiles with alternating cliffs and benches known as cryoplanation terraces;
- a thin near-surface mantle of disintegration products subject to frost creep and solifluction;
- small-scale patterned ground features attributable to frost sorting.

A closer examination of 'periglacial' granite uplands reveals that some of these uplands have a morphology markedly different than the picture presented above.

One of the best-researched granite landscapes in the context of its cold climate inheritance is the Dartmoor upland in south-west England (Palmer and Neilson, 1962; Waters, 1964; Green and Eden, 1973; Coque-Delhuille, 1987; Gerrard, 1988a, 1989; Harrison *et al.*, 1996). It is a particularly suitable area for assessing the long-term effects of periglacial morphogenesis because it has never been overridden by inland ice and remained an unglaciated cold upland throughout the entire Pleistocene. Hence, the singular effects of frost action may have cumulated towards the present state without interruption. Waters (1964) assembled the evidence for periglacial conditions, emphasizing the occurrence of solifluction covers (head deposits) and inverted slope profiles, block fields, and stone stripes in particular (Plate 7.1). Subsequent researchers challenged Waters' rather simplistic view of slope profile inversion (Green and Eden, 1973; Gerrard, 1989), but did confirm the Pleistocene age of slope deposits. More recently, similarities between the arrangement of granite boulders within some block slope deposits and rock glaciers have been brought to attention (Harrison *et al.*, 1996).

The interpretation of classic Dartmoor tors is more controversial, as has already been seen in Chapter 3. Linton (1955) considered them as two-stage landforms excavated from deep weathering profiles during the Pleistocene, but assumed pre-Pleistocene age of weathering itself. A fundamentally different view has been offered by Palmer and Neilson (1962). In their opinion, tors are truly periglacial landforms formed through frost shattering of bedrock and solifluction transport of detached blocks. Had this view stood the test of time, most of the Dartmoor landscape might be regarded as periglacial inheritance. The current opinion about the tors of Dartmoor is however much in favour of Linton's two-stage theory, although modified to some extent (Gerrard, 1974). Tors are long-lived, pre-Pleistocene features, which have been developing concurrently with the evolution of the Dartmoor landscape as a whole and have merely been modified by periglacial processes. Widely open clefts within the tors

240 Cold-Climate Granite Landscapes

Plate 7.1 Block streams and angular tors, Dartmoor, south-west England

and common pseudobedding structures indicate that mechanical unloading of the rock mass has been one of the most significant factors controlling their geomorphic evolution (Fig. 3.6; Plate IV). Dartmoor tors exhibit a variety of shapes, but angular rock cliffs, usually described as frost-riven cliffs in periglacial geomorphology (e.g. Demek, 1964), and indeed common in many European uplands, are extremely rare. Grus, too, has little in common with either hydrothermal alteration or cold climate bedrock shattering and its formation proceeds nowadays as well (Eden and Green, 1971; Williams *et al.*, 1986). Sections of weathered granite with massive compartments embedded into it, such as those at Two Bridges or at Bellever Tor, provide clear evidence of ongoing tor formation. One of the key arguments for an entirely periglacial Dartmoor has lost its strength.

Another notable absence, besides angular cliff lines, are mid-slope benches known as cryoplanation terraces. Once widely identified all around the world and claimed as diagnostic periglacial features (e.g. Demek, 1969*b*; Washburn, 1979), they are treated with increased scepticism nowadays, largely because their implied origin is in fact very speculative (Thorn, 1992; Thorn and Hall, 2002; André, 2003). But even if we agree that mid-slope terraces are reliable indicators of periglacial environments, their absence on Dartmoor is striking. Granite slopes are instead long and planar, locally interrupted by tors, but not by transverse cliff lines of any sort. However, in north-west Dartmoor the marginal part of the upland is built by metamorphic country rock and there, on the Cox Tor elevation, benched slope profiles are evident (Gerrard, 1988*a*).

A general view emerging from the observational evidence is that Dartmoor is a landscape of long history, significantly controlled by and adjusted to structure and lithology (Gerrard, 1974; Ehlen, 1992), in which certain cold-climate elements are superimposed on older landforms. Block fields and stone stripes are most characteristic, but their presence has not modified the slope form to any appreciable extent. Likewise, slope cover deposits of periglacial origin are widespread but their thickness only locally exceeds 2 m (Fig. 7.2).

The picture provided by Dartmoor is essentially repeated in other European granite uplands (Fig. 7.3). In the Karkonosze Mountains, south-west Poland, the major components of granite relief are elevated surfaces of low relief, tors, and dome-like hills. The origin of the summit plain predates the Pleistocene, as indicated by glacial cirques undercutting the palaeosurface and fluvial valleys formed in response to Neogene uplift of the massif, deeply incised into the plateau (Sekyra, 1964; Migoń, 1999). At lower elevations, closely spaced hills and intervening basins rather than level surfaces exist. Onto these landform elements, many of them evidently controlled by structure (Dumanowski, 1968), an assemblage of cold-climate features is superimposed. Its distinctiveness, though, varies according to the textural variant of granite involved. In the fine- and medium-grained granite of the summit palaeosurface autochthonous block fields and block slopes occur, whereas large stone polygons provide evidence of frost sorting. A few of the hills rising above the summit plain have developed stepped slope profiles, resembling textbook cryoplanation features. Rock glaciers have also been reported from the northern slope of the massif. The long planar slope of the main ridge bears a heterogeneous cover of solifluction and slope wash deposits, derived from weathering of the host granite. Although a few distinct horizons may be distinguished within the head deposit, the total thickness of this sediment seldom exceeds 2 m. Below it, a mantle of *in situ* grus is invariably present. The periglacial legacy is much more difficult to identify in the northern part of the massif, which is built of coarse granite.

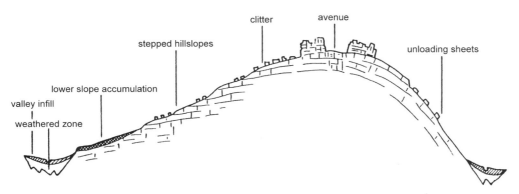

Fig. 7.2 Composite profile of a Dartmoor slope (after Gerrard, 1988b, fig. 13)

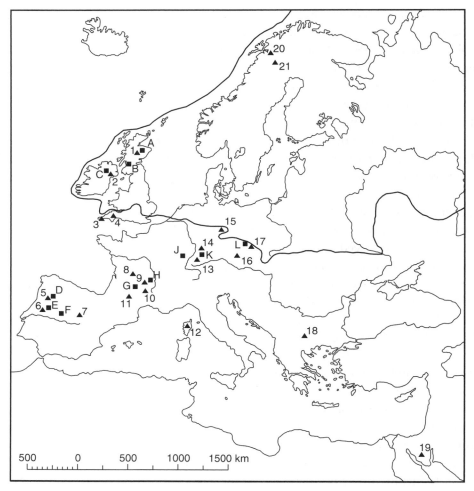

Fig. 7.3 Periglacial and glacial phenomena in granite uplands in western and central Europe (selected examples), including localities from Balkan and Sinai: (A) Cairngorms, (B) Arran, (C) Mourne Mountains, (D) Serra do Gêres, (E) Serra da Estrela, (F) Serra dos Gredos, (G) Margeride and Aubrac, Massif Central, (H) Forez, Massif Central, (J) Vosges, (K) Schwarzwald, (L) Karkonosze; (1) Cairngorms, (2) Mourne Mountains, (3) Dartmoor and Bodmin Moor, (4) Land's End and Scilly Islands, (5) Serra do Gêres, (6) Serra da Estrela, (7) Sierra da Guadarrama; Massif Central: (8) Limousin, (9) Forez, (10) Margeride, (11) Lozère; (12) Corsica, (13) Schwarzwald, (14) Odenwald, (15) Harz, (16) Bavarian Forest, (17) Karkonosze, (18) Vitosha, (19) Sinai

Solifluction deposits do occur on slopes, but distinctive landforms are missing. Slope morphology is dominated by irregularly shaped outcrops of massive granite and scattered rounded boulders, evidently former corestones exposed after the fine material was washed away. In summary, large- and medium-scale geomorphology does not show any clear periglacial imprint. This imprint may be discerned in the suite of smaller landforms and cover deposits, but even their clarity depends upon the lithology.

Other granite terrains in the Bohemian Massif show similar scale relationships. Periglacial inheritance is recognized in frost-riven cliffs in medium and fine granites, localized mid-slope benches, cracked boulders, block fields and block streams on steep slopes (Demek, 1964; Votýpka, 1971, 1974, 1979), but all these are evidently subordinate to the pattern of hills, basins, and level surfaces which has developed through repeated deep weathering and stripping over protracted periods of the Cainozoic. In addition, it is not certain that all geomorphic features described as periglacial do indeed form the cold-climate legacy. Too often little is said about the granite itself, and about the spatial distribution patterns of landforms claimed to be periglacial. Hence, one may wonder how much structure matters in their development. For example, split boulders are not necessarily products of frost cracking, but are more likely to be products of tensile stress build-up in laterally unsupported rock masses (Migoń and Roštinský, 2003).

The impact of periglaciation is even less pronounced in the areas overridden by ice sheets in the Pleistocene. The Cairngorm Mountains in Scotland have a rich assemblage of small-scale periglacial landforms, which includes block fields, boulder lobes, solifluction sheets, patterned ground, and occasional rock glaciers (Ballantyne and Harris, 1994), but their gross morphology is clearly dominated by preglacial palaeosurfaces with pockets of grus saprolite and tors, into which impressive glacial troughs and cirques are incised (Sugden, 1968; Hall, 1996).

The hypothesis of limited periglaciation and the subordinate role of cold-climate landforms arising from the above review has one significant weak point. All the areas mentioned, from the Cairngorms and Dartmoor to the Bohemian Massif, were located in a relatively mild, maritime periglacial environment, where temperature fluctuations were modest. The impact of cold conditions might have been much more severe in the continental interior of Siberia and Mongolia, where granites occur in abundance. What needs to be remembered, though, is that Siberia is also a relatively dry environment, whereas it is moisture availability that is crucial for frost-related geomorphic processes to occur. In the context of cryoplanation, Demek (1969a, 1969b) and Czudek (1989) reviewed the existing Russian literature, and themselves documented many benched slopes, in the Stanove Mountains, the Aldan Upland, the Oymiakon Upland, and other areas. The examples provided show that no uniform cryoplanation relief exists and a variety of forms can be encountered. Terrace treads are of variable width, from a few tens to more than 500 m, whereas risers are either rock cliffs of up 10–20 m high or rather gently sloping (20–25°), debris-covered steps. In addition, slope breaks appear more angular in metamorphic and massively bedded sedimentary rocks than they are in granite. These differences in form are probably related to fracture density and breakdown patterns, but quantitative observations are yet to be presented. The Russian literature contains reports about backing cliffs more than 200 m high (see Demek, 1969b), but these seem to be excessive figures for cryogenic scarp retreat and another origin is more likely. Other cold-climate landforms in Siberia include block fields, for which the term 'kurum' (sing.) is applied and block streams. The Khangai Mountains in northern Mongolia host a similar assemblage of block fields, benched slopes, angular tors, and patterned ground, for which an origin in the cold-climate environment is claimed (Pękala and Ziętara, 1980).

One further aspect of periglaciation of granite uplands requires consideration. Periglacial environments are known as having a very efficient slope transport system of weathering-derived detritus (see Washburn, 1979; French, 1996; Matsuoka, 2001b). Likewise, older deposits and *in situ* weathering mantles can easily be mobilized and eroded away from the slopes. The likelihood of mobilization applies to fine-grained material in particular. The consequence of removal of the cover is the exposure of the weathering front, bedrock projections, and any large boulders that were big enough to resist movement. In fact, the Linton's classic theory of tor origin (Linton, 1955) implied periglacial denudation of grus as the ultimate process of tor excavation, and tors in other European uplands have been considered to be exposed in the same way (Wilhelmy, 1958; Jahn, 1962). Periglacial denudation has also been implicated in the origin of boulder fields and some of the block streams in the central European mid-mountains. Shallow solifluction and slope wash would have removed all fine material, leaving a residual boulder lag, possibly subjected to minor transport afterwards (Flohr, 1934; Büdel, 1937; Wilhelmy, 1958). Big rounded boulders projecting through periglacial cover deposits up to the current slope surface typify many granite uplands and are difficult to link with frost shattering of outcrops. Not uncommonly, rock outcrops do not exist in the vicinity of boulders. One may thus wonder if the most important geomorphic role of periglacial conditions was denudation of an older regolith and exposure of tors and corestones, rather than creation of new landforms.

This brief review of cold climate landform assemblages within granite landscapes permits the following conclusions:

1. Granite landform development in cold environments is guided, as elsewhere, primarily by lithological and structural properties of the rock. Granites vary in their ability to support frost-related landforms. Fine- and medium-grained granites easily disintegrate into angular detritus, although how much of this breakdown can be safely attributed to frost wedging is not clear. This detritus then becomes incorporated into the regolith cover and reworked by frost sorting and creep. In this way, rock cliffs, block fields, and block streams can form. However, comparable landforms on coarse granite

are rare, or even non-existent in many granite massifs. In particular, the contrast between fine and coarse granite is revealed in slope form. As observed by Demek (1969b) in his worldwide survey, benched profiles and cryoplanation terraces are generally rare on the latter.

2. In most maritime environments the periglacial legacy is most evident in near-surface cover deposits: block accumulations and solifluction covers. Only a few new landforms are created and these are evidently superimposed onto older or non-periglacial (i.e. not driven by frost action) landscape elements such as pre-Pleistocene slope and summit surfaces, valley sides, glacial cirques and troughs (André, 2003).

3. Although it is occasionally suggested that the absence of typical periglacial landforms signifies pre-Pleistocene inheritance and increased rock resistance against cryogenic processes, such views are probably incorrect. Rather, the influence of structure in coarse granite is so pervasive and granular disintegration so dominant that distinctive suites of cold-climate landforms are unable to develop. Thick solifluction and slope wash deposits present in almost all granite uplands provide the best evidence that cold-climate geomorphic systems were by no means dormant.

4. No 'standard' periglacial landscape exists on granite. Different landform assemblages occur in areas which are, or were, under the influence of atmospheric cold. This variability reflects rock variability, different geomorphic setting, and probably, climatic conditions. In addition, inherited and non-periglacial elements occur and may even dominate, despite a long history of periglaciation.

Glaciated Granite Terrains

Glaciation has been an important component of geomorphic history for many granite areas. Glaciated landscapes occur in a multitude of form, depending on geographical location, elevation, local relief, geomorphic history, and climatic conditions. Considering spatial scale, we may distinguish mountain glaciation and inland (ice sheet) glaciation, but two comments are necessary. First, there are granite massifs in high latitudes which may have been both overridden by ice sheets and modified, usually subsequently, by local glaciers. Consequently, superimposition of relief features created by both types of glaciations may occur. Second, despite scale differences certain landforms occur regardless of the size of glaciations. Ice-moulded hills, otherwise known as *roches moutonnées*, are an example. They can be found within valley floors occupied by local glacier tongues as well as typifying large areas of high-latitude shields covered by Pleistocene ice sheets.

It is neither feasible nor appropriate to provide a comprehensive review of glacial landforms encountered in granite terrains. Most, if not all, are repetitive across a range of lithologies and cannot be claimed as distinctive for granite. The aim of this section is to highlight selected aspects of glaciation which are particularly important in the context of granite geomorphology.

Mountain Glaciation

Many granite mountain ranges were at sufficient elevation during the Pleistocene to allow local glaciers to develop. The scale of mountain glaciation varies hugely, from extensive ice caps covering tens of thousands of square kilometres as in the Alps or the Sierra Nevada in the western USA, to almost insignificant cirque glaciation in some of the Central European uplands. Consequently, the degree of landscape transformation by glaciers has been widely different. At one extreme, there is an alpine-like topography dominated by serrated ridges (arêtes) and pyramidal peaks (horns) separating deep U-shaped valleys and huge amphitheatres composed of many intersecting cirques (Plate 7.2). On the other hand, the impact of glaciation on the moderately elevated plateaux of Central Europe is restricted to enhanced undercutting of plateau edges by local cirque glaciers, but valley and slope morphology have been little modified (Fig. 7.4). There is a whole range of intermediate situations, such as the co-existence of watershed surfaces of low relief and deep long troughs a few hundred metres deep lined by impressive rock walls and truncated spurs, as epitomized by the landscape of Yosemite Valley in the Sierra Nevada (Plate 5.1).

Plate 7.2 The northern mountains, Isle of Arran, have been completely transformed by glaciers and show an array of horns, serrated ridges, deeply incised cirques, and glacial troughs

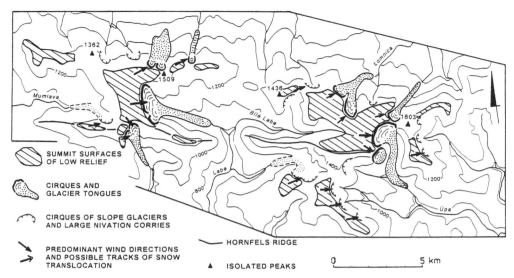

Fig. 7.4 Distribution of glacial landforms in the Karkonosze Mountains, Poland/Czech Republic (after Migoń, 1999, fig. 4)

Geomorphologists investigating the patterns and causes of mountain glaciations tend to agree that these have invariably been much influenced by landform configuration at the onset of the Pleistocene. This pre-glacial inheritance has been highlighted particularly in respect to mountain massifs typified by selective glacial erosion and cirque glaciation (e.g. Partsch, 1894; Sugden, 1968; Daveau, 1971; Veyret, 1978; Migoń, 1999). Many uplands owe their present-day altitude to Neogene surface uplift, which transformed low-lying palaeosurfaces of low relief into undulating summit plateaux. The associated change in relative relief induced river downcutting and the development of deep valley heads along the margins of an uplifted plateau. These in turn, especially if located on the lee-side in respect of prevailing wind directions and exposed to the north and east (in the northern hemisphere), began to act as suitable storage places for snow received both directly from atmospheric snowfall and indirectly, from summit plains. With contributing areas sufficiently large and elevation of the basin floor above the snowline, local glaciers may have formed in the preglacial valley heads, whereas deep fluvial valleys below channelized ice flow downslope. If, however, the pre-glacial plateau relief was of restricted extent, conditions for glaciers to develop along its margins were much less favourable. In the granite massif of the Karkonosze, Poland, it was possible to demonstrate that the magnitude of local glaciation was positively correlated with the surface area and shape of the summit plain above the respective valley heads (Migoń, 1999; Fig. 7.4). Asymmetry of glaciation in respect to regional circulation patterns and aspect is a repetitive feature. In western and central Europe, for example in the Serra da Estrela, the Vosges, and in the

248 Cold-Climate Granite Landscapes

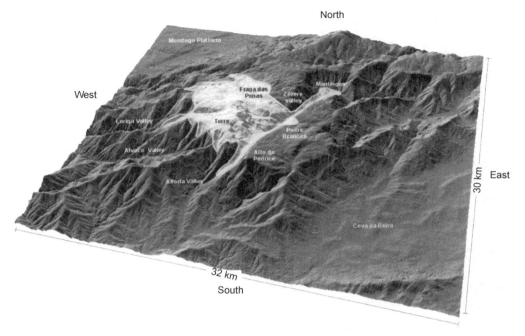

Fig. 7.5 Distribution of glacial landforms in the Serra da Estrela, Portugal. The steep eastern margin of the plateau is a fault-generated escarpment (digital model courtesy of Gonçalo Vieira)

Schwarzwald, cirques facing east and north are much better developed than those facing west. Likewise, U-shaped valleys tend to dissect eastern and northern slopes rather than western and southern slopes (Fig. 7.5). Inheritance is occasionally sought in heavily glaciated mountain ranges too. Klimaszewski (1964) linked the staircase relief pattern within the glaciated valleys of the Tatra Mountains with pre-Pleistocene valley morphology and argued that steps within the longitudinal profiles of glacial troughs have developed at knick points separating 'ancient' and rejuvenated valley sections.

Although geomorphic features such as troughs, cirques, and steps are undoubtedly glacial features, there is abundant evidence that many are guided by structures. An excellent example is the 7 km long trough of Vale do Zêzere in the Serra da Estrela (Plate 7.3). It retains its straight south–north course all along its length and has an impressive depth of 300–400 m. Despite having been fed by tributary glaciers coming exclusively from the west, no eastward deflection of the major valley is observed. The trough is located along the regional fault zone, the Vilariça-Manteigas Fault, and there is little doubt that it was the presence of a weak zone, perhaps preferentially affected by deep weathering in pre-Pleistocene times, that has favoured significant valley deepening. On the much larger spatial scale of the Sierra Nevada, California, structural guidance of many glacial troughs, some tens of kilometres long, can also be demonstrated (Ericson et al., 2005). However, a paradoxical feature of the Sierra Nevada is

Plate 7.3 Glacial trough of the Vale do Zêzere, Serra da Estrela, Portugal, owes its straight course to adjustment to a regionally important fault line

the apparent V-shaped cross-profile of many valleys, for which there is independent sedimentary evidence for glaciation (Huber, 1987; Schaffer, 1997; Ericson *et al.*, 2005). In a few of these cases, glaciers may have been as much as a few hundred metres thick, yet they proved unable to transform valley cross-sections to any appreciable degree. Again, recourse to rock structure may be helpful here. Many valleys, especially major tributaries of trunk valleys, follow structural discontinuities which are probably zones of more dense fracturing along ancient strike-slip zones. Numerical modelling by Harbor (1995) has revealed that U-shaped cross-sections are increasingly difficult to achieve if linear glacial erosion is concentrated along an outcrop belt of weaker rock. The narrower this belt and the more pronounced contrasts in rock resistance, the less distinctive a U-profile of the trough becomes. In the Sierra Nevada we deal with very narrow zones of rock which, because of the high degree of shattering, are much less resistant than massive granite and granodiorite nearby, hence the very limited transformation of valley cross-sections.

The recognition of the geomorphic role of glaciers in shaping mountainous granite topography ought to be extended to the assessment of the impact of glaciation on other landforming processes. Deep glacial erosion leads to the steepening and not uncommonly undercutting of slopes, and therefore changes the state of stress in rock masses. Removal of lateral support, consequent upon both ice wastage and bedrock erosion, increases tensile stress and causes opening of slope-parallel discontinuities which can be either pre-existing or newly formed (Plate 7.4). This, in turn, leads to a variety of mass movement along trough and cirque walls which may take the form of rock falls, slides, and avalanches. Therefore, glacier-modified mountain relief is almost invariably

Plate 7.4 Royal Arches, Yosemite Valley, California—an example of large-scale sheeting, likely to have formed in response to erosional unloading

associated with landforms resultant from high magnitude rock slope failures: slope recesses, rock fall scars, talus aprons, scree cones. Moreover, glacial erosion increases local topographic gradients, thus creating a suitable setting for other gravity processes such as debris flows, gullying, and avalanching.

Ice Caps

Extensive tracts of granite terrains in the high latitudes of northern Europe and North America were covered by ice sheets in the Pleistocene. However, it must not automatically be inferred that they have all suffered from vigorous glacial erosion nor that their present-day topography has been produced solely by the erosive action of ice. In fact, widespread occurrences of pre-Pleistocene saprolitic cover on both the Laurentide and Fennoscandian Shield provide convincing evidence that the depth of erosion caused by ice sheets may have been locally, or even regionally, rather insignificant (Lidmar-Bergström *et al.*, 1999). In other cases, glacial erosion has appeared limited to the removal of sedimentary cover rock and contributed to exhumation of ancient granite landscapes, bringing them to light after millions of years of burial (e.g. Lidmar-Bergström, 1989, 1997).

Sugden's (1974) classification of landscapes of glacial erosion may readily be applied to granite terrains. It includes three main categories:

(1) landscapes of areal erosion, dominated by scour topography, abraded and streamlined bedrock, multitude of basins;
(2) landscapes of selective linear erosion, in which deep troughs produced by fast-moving ice streams are separated by plateaux and flattened interfluves with few signs of glacial remodelling;
(3) landscapes of little or no glacial erosion, which lack glacial erosional features despite ice cover.

These differences are linked to the thermal characteristics at the base of an ice sheet. Cold-based ice, frozen to bedrock, hardly moves and as such does not cause significant erosion, hence there are extensive areas belonging to category (3) over northern hemisphere glaciated shield areas. Location in ice divide zones does not favour considerable erosion either. An inherited preglacial topography with tors, boulder fields, structure-controlled basins, and regolith pockets, preserved under non-erosive cold-based ice sheets has now been recognized in many formerly glaciated granite areas, including northern Scandinavia (Kleman, 1994; Lidmar-Bergström et al., 1997; Stroeven et al., 2002; André, 2004) and Canada (Sugden and Watts, 1977; Gangloff, 1983; Watts, 1986). Survival of preglacial topography typifies interfluves within landscapes of selective linear erosion too, a well-known example being the Cairngorm plateau in Scotland (Sugden, 1968).

Scour topography is occasionally described using the local Gaelic term of 'knock-and-lochan' topography. However, in recent years doubts have been expressed over whether the interpretation of this type of relief exclusively in terms of glacial erosion is everywhere correct (e.g. Thomas, 1994b, 1995; Johansson et al., 2001a). First, some examples are located in the former centres of ice mass dispersal, such as in the big intermontane basin of Rannoch Moor in the Scottish Highlands, where vigorous glacial scour is less likely (Plate 7.5). Second, the irregular topography of an exposed weathering front bears many similarities to the knock-and-lochan relief, hence landforms alone may be a poor guide to their origin. It is probable that landscapes of areal erosion may in fact owe more to relief inheritance than commonly assumed.

In heavily glaciated areas structures may still influence topographic patterns. Mount Desert Island is a predominantly granite island within the Acadia National Park in eastern Maine, USA. It has a mountainous topography, with elevation approaching 500 m a.s.l., and a conspicuous feature is the series of parallel fracture-guided valleys running roughly from north to south. In the Pleistocene the island was overridden by an ice sheet moving from north-northwest. Glacial erosion succeeded in truncating hill slopes exposed to the south-east, but the general pattern of north-south trending valleys has survived until today (Chapman and Rioux, 1958).

Domes are another example of characteristic granite landforms which have complex relationships with glaciation. They are usually considered as components of preglacial landscapes which survived the passage of ice due to their massiveness, but have been

252 Cold-Climate Granite Landscapes

Plate 7.5 Intermontane basin of Rannoch Moor, Scotland

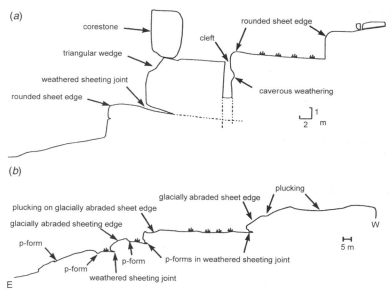

Fig. 7.6 Glacial modification of a pre-Pleistocene etched granite relief, after Johansson *et al.* (2001b, fig. 18): (*a*) unglaciated granite topography at Hampi, India, (*b*) superimposed glacial landforms, Bohuslän, Sweden

modified to various extents. Olvmo *et al.* (1999) and Johansson *et al.* (2001*a*) described truncated domes from Bohuslän in south-west Sweden, which have lost their uppermost parts (Fig. 7.6). Today they have unusually flat hilltop surfaces, onto which many minor features indicative of glacial action are superimposed. Some domes in the Sierra Nevada, in turn, have been re-shaped into giant *roches-moutonnées* and are now distinctly asymmetrical due to efficient plucking on their lee sides. Lembert Dome and Pothole Dome in the Tuolumne Meadows provide excellent examples and, in addition, bear a range of small erosional features such as crescentic gouges and striations (Plate 7.6). Transformation of granite domes into big *roches-moutonnées* has also been described from Scotland (Sugden *et al.*, 1992). In other areas changes to the inherited dome form have probably been negligible, as in the Mourne Mountains, Northern Ireland. The Hen Mountain in the northern part of the massif is a steep-sided cupola which does not show any evident signs of glacial modification.

The most clear evidence of glaciation is the multitude of small-scale landforms produced by ice and meltwater. They have been surveyed in detail in Bohuslän (Olvmo *et al.*, 1999; Johansson *et al.*, 2001*a*; Olvmo and Johansson, 2002). Minor glacial features include striated and polished surfaces, plucked lee sides of granite hills, detached sheeting slabs and corestones, whereas potholes, p-forms, and subglacial canyons constitute the geomorphic evidence of meltwater action. On a larger scale, glacial and meltwater erosion are seen as powerful agents in stripping cover rocks and saprolites to re-expose bedrock, but their role in shaping basement topography has been rather limited.

Plate 7.6 Plucked lee side of a granite dome, Lembert Dome, Sierra Nevada, California

8

Geological Controls in the Evolution of Granite Areas

The preceding chapters have already indicated that granite properties and structures play a key role in the progress of rock weathering, the development of many medium- and small-scale landforms, and in the patterns of mass movement phenomena on slopes. But the influence of geological factors *sensu lato* is by no means limited to these, rather restricted spatial scales. Geotectonic settings, modes of emplacement, and long-term geological histories are all relevant to the understanding of the diversity of granite landforms and landscapes. The aim of this section is show the variety of geological controls and how they are reflected in granite landscapes, moving progressively from large to small spatial scales.

Tectonic Setting

If plate tectonics is used as a framework, then granite intrusions form in two major settings: orogenic, including transitional, and anorogenic (see Chapter 2). Geographically, the former take place at convergent plate margins, whereas the latter take place at divergent plate margins (rift zones) and within continental interiors, at hot spots. However, for the purpose of a geomorphological approach to granite landscapes of the world, the time-independent plate tectonics framework is less useful. This is because many granite intrusions occur in settings different than those in which they formed millions of years ago, and it is their post-emplacement long-term geological history and current location that are crucial to understanding the landscapes that have developed upon them. For example, late Paleozoic granite intrusions in central and western Europe took place within the Hercynian orogenic belt, hence in a convergent plate margin setting, but their present-day morphology is mainly the legacy of long-term evolution in an anorogenic regime and late Cainozoic rejuvenation, including plateau uplift and faulting. In a somewhat similar manner, ancient orogenic granite intrusions have been incorporated into shield interiors and passive margins. Figure 8.1 is an attempt to relate the tectonic settings of granite intrusions to the distribution of granite areas, as we see them today, against the background of global tectonics.

From this point of view, granite landscapes occur within the following five main geodynamic settings: (1) orogenic zones along convergent plate margins, (2) eroded and rejuvenated ancient orogenic belts, subject to geologically recent plateau uplift,

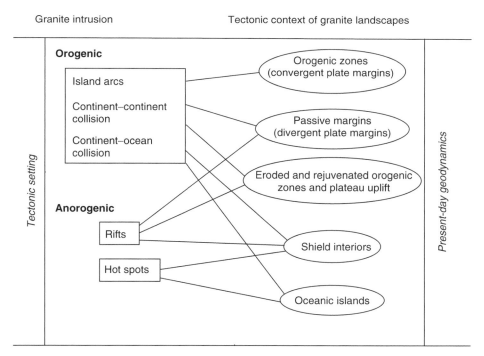

Fig. 8.1 Relationships between tectonic settings of origin of granite masses and the setting of granite landscapes in relation to present-day global tectonics

(3) passive margins at divergent plate boundaries, (4) stable shield interiors, and (5) oceanic islands.

Orogenic Settings

Granite landscapes in recent orogenic settings are perhaps the least distinctive of all granite terrains of the world, particularly in areas subject to the highest surface uplift. There is a notable absence of characteristic medium-scale landforms such as inselbergs, tors, topographic basins, and tracts of planar relief in a watershed position. The dominant morphology is one of a highly dissected terrain and relief energy is very high, which creates a favourable environment for rock weathering and mass movements. Depending on climatic conditions, which in turn depend on latitude, elevation, and aspect, weathering may be predominantly physical or chemical. The former is the case in such high-mountain ranges as the Himalaya, Karakoram, Canadian Cordillera, or the southern Andes. Therefore rock slopes are ubiquitous and develop through particle fall, rock slides, and rock falls. The high efficacy of debris liberation and mass movement is also related to a high degree of rock fracturing, consequent on both strong tectonic stresses and the release of lithostatic stresses. Earthquakes play an additional role in forcing slope instability. Regolith cover is thin and coarse, subject to easy mobilization and downhill movement as debris flows. Valleys may follow regional fault lines which facilitates rapid incision, but their detailed morphology is a combined outcome of weathering, slope evolution, glacial and fluvioglacial processes. Rates of incision are typically very high, well in excess of 1 mm yr^{-1}, and not uncommonly above 10 mm yr^{-1} (Caine, 2004), not uncommonly of the same order of magnitude as the rates of rock uplift, for example in the Nanga Parbat area of the western Himalaya (Burbank *et al.*, 1996). In these circumstances the evolution of topography appears to proceed too fast to allow for any more distinctive manifestation of lithological and structural variations of granite, except for details of rock slope geomorphology such as joint-controlled chutes and ravines or smooth vertical surfaces being joint planes (Plate 8.1). On the other hand, the high degree of fracturing may also mean that the granite mass is poorly differentiated into massive and densely jointed compartments, hence it offers little scope for selective denudation anyway.

Less elevated mountain ranges may show a predominance of rock decomposition over disintegration, while the residence times of weathering products may be significantly longer. The Japanese Alps typify this type of granite relief. They are highly dissected and the slopes are steep, not uncommonly in excess of 30°, but there is also widespread grus mantle (Japanese *masa*) present on these slopes (Ikeda, 1998). Its thickness may be as much as 30 m. The combination of high rates of uplift, up to 5 mm yr^{-1} (Ohmori, 2000), poorly consolidated grus mantle, high annual rainfall (> 2,000 mm), and high frequency of typhoons provides an optimal environment for landslides

Plate 8.1 High-mountain topography developed in granodioritic rock, Karakoram (Photo courtesy of Andrew S. Goudie)

and debris flows. Therefore, the Japanese Alps are a very dynamic area, with denudation rates of the order of 1–5 mm yr^{-1} (Ohmori, 2000). Rates of rock uplift and surface denudation roughly balance themselves and steady-state geomorphic evolution is envisaged for the range. As in the more elevated glaciated mountain ranges, rates of denudation appear too high for a 'typical' granite structural relief with inselbergs and basins developing. But the reason for their absence may be partly structural. As Ikeda (1998: 197) pointed out: 'Granite landforms located in the island arc mobile belt, i.e. as is Japan, show a very high density of fractrures and joints. The density and size of these greatly influence surface conditions'. He maintains that fracture spacing does not exceed 3 m, hence only medium-scale landforms such as tors and boulder fields may occur, especially in less elevated places (Ikeda, 1998).

Eroded and Rejuvenated Ancient Orogenic Belts

This is a very specific category of granite massifs which solidified in an orogenic setting, but since have been significantly eroded and, because of their proximity to recent orogenic belts or rifts, subject to recent plateau-like uplift. Western and central Europe provides numerous examples of such massifs, which date back to both the Hercynian movement in the late Palaeozoic (e.g. Bohemian Massif, Schwarzwald, Massif Central, Armorican Massif, Central Iberian Range, Dartmoor) and the Caledonian orogeny in the Silurian/Devonian (e.g. granite massifs in mainland Scotland). Granites forming mountainous terrains in south-east China and the Malay Peninsula are Far East examples, having been emplaced during the so-called Yanshanian orogeny in the Mesozoic. Sierra Nevada Batholith in North America would also fit here, as it was emplaced in the Cretaceous and elevated to its present-day altitude largely during the late Cainozoic (Wakabayashi and Sawyer, 2001).

Two main types of granite intrusions are present in rejuvenated mountain belts: syn-kinematic and post-kinematic (Marmo, 1971). The former were emplaced during the main deformation phases, typically along regional thrust and strike-slip faults in a compressive stress field, whereas the latter postdate these major deformation events. These contrasting temporal relationships to the timing of principal deformation have important implications for the structure of intrusions. Syn-kinematic granites are usually much deformed and fractured, show evidence of foliation and local mylonitization, and form concordant laccolithic bodies. Therefore, they are generally weak, even if their composition is dominated by quartz and potassium feldspar. By contrast, post-kinematic granites typically form extensive oval bodies discordant to structures in the country rock. Fracturing is regular and, on average, less dense than in syn-kinematic granites. Because of the high degree of fracturing and the presence of gneissic textures, syn-kinematic granites rarely support distinctive landscapes, although deep weathering, corestones, tors, and medium-scale features of selective denudation are usually present. South-western Brittany, France, hosts a series of syntectonic leucogranites, which were emplaced at shallow crustal levels along the South Armorican Shear Zone in the Carboniferous (Román-Berdiel et al., 1997). The geomorphology of these massifs is not very distinctive and basins rather than uplands dominate (Garreau, 1985).

Much more characteristic is the geomorphology of post-kinematic, less deformed granites. In Europe, a repetitive motif is an undulating upland surface with broad elevations and wide, saucer-like basins, interconnected by narrower defiles (Plate 8.2). Depending on fracture density, tors may be frequent, as they indeed are in the Cairngorms (Ballantyne, 1994), Dartmoor (Linton, 1955; Gerrard, 1974; Ehlen, 1992, 1994), Bodmin Moor (Gerrard, 1978), Serra da Estrela (Migoń and Vieira, in preparation), Harz and Fichtelgebirge (Wilhelmy, 1958), Šumava (Votýpka, 1979), and Karkonosze (Jahn, 1962). Locally, more distinctive residual hills, likened to inselbergs, occur, giving the landscape a multi-convex appearance (Czudek et al., 1964; Migoń, 1996). Some of these granite uplands such as the Karkonosze, Schwarzwald, different parts of the Massif Central, and the Serra da Estrela are bound by impressive fault-generated escarpments (Fig. 7.5). Where the uplift of former surfaces of low relief has been more than a few hundred metres, it has typically been accompanied and followed by significant fluvial incision. Hence the typical contrast between gently rolling watersheds with tors and deeply incised valleys (Plate 8.3). Some of these rejuvenated massifs have been sufficiently elevated to cross the Pleistocene snow line and their relief has been modified by glaciation. Cirques and glacial troughs are the most evident signs of

Geological Controls 259

Plate 8.2 Upland topography of Bodmin Moor upland, south-west England, dominated by saucer-like basins and broad ridges with occasional tors

Plate 8.3 Dissected plateau of the Cairngorms, Scotland. Valleys have been significantly reshaped by glaciers during the Pleistocene

glacial history, whereas intervening plateau relief usually suffered much less reshaping, because either little relative relief did not favour glacial erosion or glaciers were confined to valleys (Sugden, 1968; Daveau, 1971; Veyret, 1978).

How much of the upland topography is really inherited from the distant past is a matter of persistent debate. Deep weathering is common and may reach a few tens of metres, but its dating and temporal relationship to uplift is controversial. Whereas argillaceous mantles are most probably relict and predate the late Cainozoic, arenaceous residuals are likely to form concurrently with uplift and dissection (Migoń and Lidmar-Bergström, 2001; Migoń and Thomas, 2002). Thermochronological studies may shed some light on this problem, but surprisingly there are not many available. Nevertheless, very distant (Palaeogene and earlier) ages seem rather unlikely, unless a scenario involving denudation into a plain, subsequent burial, and relatively recent exhumation is applicable. Such scenarios have been hypothesized by Lidmar-Bergström (1986) in relation to Dartmoor and by Hall (1991) in relation to Buchan, north-east Scotland. In each case, burial by Cretaceous marine deposits is envisaged. In subsided areas, though, end-Palaeogene ages of hilly relief are not unreasonable and are confirmed by the occurrence of Miocene deposits (Migoń, 1997c).

The Rocky Mountains in North America show a similar granite plateau relief, despite their uplift having been much stronger. Extensive high-altitude surfaces of low relief with tors, domes, and widespread deep weathering typify the Laramie Range (Cunningham, 1969; Eggler et al., 1969). In the Rocky Mountains National Park the elevation was sufficient to allow for mountain glaciation to develop, hence the relief along the main watershed is dominated by steep slopes, but at lower elevations gently rolling surfaces reappear.

Not all rejuvenated mountain ranges host the type of granite topography described above. Granite landscapes in Corsica, Serra da Guadarrama, and Galicia in Spain show little evidence of the former existence of planation surfaces (Klaer, 1956; Pedraza et al., 1989; Vidal Romani, 1989). Their relative relief is much higher, well in excess of 1 kilometre, and can hardly be classified as a dissected plateau. Huge domes with impressive sheet structures and vertical jointing protrude from steep valley sides and watershed ridges. The mountainous topography in these areas may have evolved from uplifted plateaux, but their remnants are difficult to identify unambiguously (Plate 8.4). A growing interest in the thermochronological history of rejuvenated Palaeozoic mountain belts of Europe and elsewhere will undoubtedly help to elucidate the relationships between their landscape evolution, uplift, and subsidence. Deep weathering assists in the transformation of a former plateau into a mountainous topography. Therefore, low-latitude examples of this class of granite massifs are typically strongly weathered and deeply dissected, evolving into multi-convex topography with massive rock domes adjacent to deeply weathered hilly terrain. Examples can be found, among others, in south-east China (Xu, 1996) and Malaysia (Swan, 1970; Tjia, 1973).

Plate 8.4 All-slopes topography of Yanshan Mountains, north of Beijing, China. Numerous massive compartments protrude from the slopes

The Sierra Nevada, California, is a special, complicated, and controversial case. The most widely accepted scenario envisages the existence of a gently rolling mid-Cainozoic landscape—a product of advanced degradation of the Cretaceous batholith, within which the predecessors of the famous Sierra domes occurred (Bateman and Wahrhaftig, 1966; Huber, 1987; Wakabayashi and Sawyer, 2001). In the late Cainozoic the area was differentially uplifted along the major north-south running fault system and tilted westwards. The maximum uplift occurred in the east, where altitudes exceeded 4 kilometres. Uplift induced fluvial incision, but erosion of the granite mass was interrupted by a period of burial by volcanic rocks in the Neogene. In the southern and central part of Sierra Nevada the removal of volcanic rocks was largely complete by the Pliocene, thus deep weathering and dissection of the underlying granite may have started anew. It proceeded selectively, exposing massive domes at different topographic levels between 4,000 and 2,000 m a.s.l. (Plate 5.1, Plate VIII) and creating stepped topography in the less elevated parts (Wahrhaftig, 1965). During the Pleistocene significant deepening was accomplished by extensive mountain glaciation, which may have induced further climatically driven erosion (Brocklehurst and Whipple, 2002). In effect, the present-day topography of the granite part of Sierra Nevada includes landscape facets of different types: remnants of moderate relief at various altitudes, high-altitude topographic basins (Plate 8.5), fault-generated mountain fronts, all-slopes relief, domes of variable geometry, U-shaped valleys and cirques (Matthes, 1950, 1956; Huber, 1987).

262 Geological Controls

Plate 8.5 Intermontane basin of the Tuolumne Meadows, Sierra Nevada, California

Passive Continental Margins

Passive margins, and their Great Escarpments in particular, provide a very specific setting for granite massifs. In general, they are very controversial features because they do not fit easily into a conventional view of relief being built in effect from plate collisions (Ollier, 2004). Therefore numerous theories as to the causes and course of uplift have been suggested and not uncommonly there are at least several competing models for each passive margin. Passive margins include four principal landscape facets: hinterland plateau, marginal swell, Great Escarpment, and coastal plain, which may or may not have a sedimentary cover (Ollier, 2004). Granite massifs may crop out within each of these facets, but only those located within Great Escarpments and coastal plains are considered here. Those present further inland will be described later, within the context of continental interiors.

The Serra do Mar and Serra Mantiqueira in subtropical south-east Brazil, both containing abundant granite and massive gneiss, are examples of relief developed within a Great Escarpment during the late Cainozoic (Thomas, 1995). Differential uplift and subsidence along regional WSW-ENE faults created elevation contrasts in excess of 2 kilometres, contributed to setting up high tensile stresses in rock masses, and induced additional fracturing. However, numerous extremely massive bedrock compartments have withstood these stresses and have survived as poorly jointed cupolas within an otherwise highly fractured bedrock. High topographic gradients resulted in high hydraulic gradients, facilitating deep circulation of groundwater and enhancing deep weathering. The resultant thickness of the weathering mantle may exceed 100 m (Branner, 1896), but is very uneven. Deeply weathered hilly terrain is juxtaposed with massive domes, which typify the surroundings of Rio de Janeiro and Petropolis (Plate 8.6). Weathered terrain is highly dissected and subject to frequent landsliding and debris flow activity (Coelho Netto, 1999; Fernandes *et al.*, 2004), but there is a constant renewal of saprolite at the weathering front and many thick weathering profiles are arenaceous (Power and Smith, 1994; Thomas, 1995). The resultant topography is one of highly dissected terrain, in which deep weathering, mass movements, and fluvial incision conspire to maintain, and probably increase, relief over time.

The Bega batholith in south-east Australia is also located within the Great Escarpment, but it lacks most of the features present in south-east Brazil. High domes and deeply dissected terrain are missing and there is an extensive topographic basin instead, with widespread arenaceous deep weathering, occasional tors, and boulders (Dixon and Young, 1981). Dixon and Young suspect that granite suffered from early hydrothermal changes, which made the rock highly susceptible to further subaerial weathering and contributed to the absence of more massive compartments. Subsequent weathering and erosion, acting upon an already weakened rock, have been particularly effective, hence a basin has been hollowed within the escarpment. However, Ollier (1983) disagrees with the idea of hydrothermal alteration and the absence of bold relief in the Bega basin may have other causes.

Plate 8.6 Co-existence of deeply weathered hills (foreground) and monolithic domes (background) near Petropolis, Serra do Mar, south-east Brazil

Plate 8.7 Domes around the Guanabara Bay, Rio de Janeiro, Brazil

By contrast, a classic granite area located within the coastal plain in the Valley of the Thousand Hills in Natal, South Africa, shows many similarities to the Serra do Mar (Thomas, 1994b), although relative relief does not exceed a few hundred metres. Again, there are massive domes scattered across the area, tors and boulders, whereas other hilly compartments are weathered throughout. The granite here is Precambrian, but has remained under the cover of the Palaeozoic Table Mountain Sandstone until the late Cainozoic. Current stripping of the sandstone strata has re-exposed the granite, but incipient weathering has probably taken place prior to the exhumation (Thomas, 1978).

In conclusion, great escarpments appear to provide a very favourable setting for granite landscape development through deep selective weathering and saprolite removal, and for relatively rapid exposure of massive compartments as residual landforms. If in addition high levels of rainfall and high temperatures occur, as in south-east Brazil, geomorphic evolution would lead to and maintain a very spectacular topography (Plate 8.7).

Cratonic Continental Interiors

Granite relief typical of cratonic continental interiors is an extensive plain, above which some scattered inselbergs rise. Thus, large tracts of the Laurentide Shield in northern Canada and the Fennoscandian Shield in central Sweden and Finland have very low relief. It is sometimes erroneously maintained that the flatness of shield areas is the result of vigorous glacial scouring during repetitive inland glaciations in the Pleistocene. In reality, granite relief in formerly glaciated shield areas can be very varied, as will be shown below, whereas flatness, where it occurs, significantly predates the Pleistocene. In addition, many low-latitude shield surfaces, in Australia, the Middle East, southern Africa, and southern America, are remarkably flat over tens and hundreds of kilometres, with a total absence of any signs of glaciation.

Planar relief typified many cratonic areas in the Precambrian. One of the regional surfaces in Scandinavia is called the Sub-Cambrian Peneplain (Lidmar-Bergström, 1996), but given its astonishing flatness the term 'ultiplain' coined by Twidale (1983) would be more appropriate. A similar surface developed at that time across the Laurentide Shield (Ambrose, 1964). In the Middle East the end-Precambrian planation surface was covered by a thick series of Cambrian sandstone and has been re-exposed only in the late Cainozoic (Plate 8.8). Exhumation of the plain has been caused by the development of the Dead Sea Transform and graben shoulder uplift. Hence the Precambrian surface is now being rapidly reshaped into a rugged all-slope topography seen in the Sinai Peninsula, Saudi Arabia, and south-west Jordan (Plate 8. 9). Extensive planar relief is present in shield areas within the former Gondwana Land and this too had formed in the Precambrian and Palaeozoic, and was subsequently buried under sedimentary strata (Fairbridge and Finkl, 1980; Twidale, 1994).

Plate 8.8 Precambrian planation surface preserved as an unconformity beneath the Cambrian sedimentary succession, south-west Jordan

Plate 8.9 Mountainous topography of the Dead Sea Transform shoulder, developed through dissection of an ancient exhumed plain, north-east of Aqaba, south-west Jordan

However, if exhumation of ancient planation surfaces was of an earlier date and happened to coincide with the period of humid climate, favouring deep weathering and fluvial incision, geomorphic evolution followed a very different pathway. This is best documented in southern Sweden, where differential structure-controlled etching in the Mesozoic transformed the primary peneplain into an array of dissected uplands, undulating hilly landscapes, fracture-aligned basins, and massive domes (Lidmar-Bergström, 1995; Johansson *et al*., 2001*b*). These south Swedish etchsurfaces and their specific topographies are discussed more thoroughly in Chapter 9.

Not all flat surfaces in cratonic settings are exhumed. A long-term aridity, the resultant absence of deep weathering, and poorly developed drainage do not favour differentiation of relief, but planation instead. This is the case of the Central Namib Desert, underlain mainly by the Precambrian Salem Granite, which is remarkably flat. Evidence of ongoing superficial weathering, especially of salt weathering, is abundant (Ollier, 1978*b*; Goudie and Migoń, 1997) and these various weathering mechanisms seem capable of reducing outcrops of coarse Salem Granite relatively quickly. The impressive inselbergs of the Namib are all built of massive, far more resistant rocks, including younger Jurassic granites and marbles (Selby, 1977, 1982*b*).

The geomorphology of cratonic terrains in more humid environments, such as in the Guyana Shield, southern Deccan, equatorial Africa, and Madagascar, is more diversified. They show relief of at least a few tens of metres, carry a blanket of thick weathering

Plate 8.10 Inselberg landscape, Kora area, Kenya (Photo courtesy of Andrew S. Goudie)

mantle, and may contain tors and inselbergs of various morphological types. Residual hills are typically built of very sparsely fractured granite and may cluster into larger inselberg groups (Thomas, 1965, 1966; Meyer, 1967; Eden, 1971; Büdel, 1977; Godard *et al.*, 2001) (Plate 8.10). Some of them may attain impressive heights of a few hundred metres.

The crystalline basement of the cratons is Precambrian in age, but in Africa there also occur the so-called Younger Granites, emplaced during the Mesozoic and Cainozoic. Their emplacement is connected with hot spot activity. Hot spots are regions of significant thermal anomaly within the earth crust, caused by the occurrence of unusually hot mantle at the base of the litosphere. This leads to melting of the lower litosphere, its replacement by low density mantle material, and an isostatic uplift of the hot material which may force its way up to the surface. The Younger Granites are typically potassium-rich and occur as either individual batholiths and stocks, or more complex volcano-magmatic structures. The latter are common in Africa, for example Erongo and Brandberg in Namibia (Goudie and Eckardt, 1999).

Hot-spot-related anorogenic granites suffered from very little subsequent deformation. Their common massive appearance, paucity of fractures, combined with an elevated content of more resistant potassium feldspar, account for a distinctive geomorphology. The Younger Granites tend to support higher ground, whether uplands or mountainous relief, and domed inselbergs are typical. The height difference between watershed geomorphic surfaces developed in the Younger Granites and the adjacent basement complexes may attain hundreds of metres (Plate V). Internal differentiation of these complex intrusions is clearly reflected in landform patterns, as documented by Thorp (1967b, 1969, 1975) in the Kudaru and Liruei Hills, Nigeria and in the Air Mountains.

Lithological Variation

The family of granitoid rocks is very broad and its members differ considerably in terms of texture, mineralogy, and geochemistry. The nature of and the reasons for these differences have already been reviewed in Chapter 2. Here, the geomorphological implications of these variations are explored more thoroughly.

Lithological characteristics have profound significance for the efficacy of geomorphic processes, particularly for weathering. However, in different environments and for the different rock breakdown processes involved, different aspects of lithology assume major importance. Thus, mineralogical composition is understandably one of the key controls of chemical weathering, but it matters less if physical disintegration is concerned. On the other hand, the size of crystals and related porosity have a direct bearing on the strength of rock fabric, which may be weakened by both chemical and physical processes.

However, to establish definite relationships between rock properties, geomorphic processes, and landforms is not easy. The underlying problem is that the role of individual lithological factors is very difficult to isolate, hence assessment of their relative weights becomes an arduous task. Not uncommonly, certain lithological features make the rock more susceptible to weathering and erosion (e.g. high porosity), but at the same time high content of quartz and potassium feldspar contributes to increasing resistance. Moreover, the local weathering environment and water availability are important. Biotite-rich granites may be expected to be weak in moist environments, because biotite may hydrate and expand, but this is not necessarily so in arid lands. The effects of variable fracture density are superimposed on those related to lithology, complicating the issue even further. Therefore, inevitably, rock–landform relationships tend to be site-specific and few general comments may be offered.

Texture

At the most general level, granite textures may be classified according to the size of constituent minerals, or to their arrangement within the rock. Hence, granites may be fine, medium, and coarse grained, and may show a directional or apparently chaotic arrangement of individual minerals. Some granites, especially those emplaced at greater depths in orogenic settings during the main phase of crustal deformation, may show certain textural similarities to gneiss, such as foliation (Buddington, 1959).

Other characteristics being equal, fine-grained granites should show an increasing resistance against most weathering processes because they are less porous, have a tighter fabric, and their crystals are often intertwined. Hence, meteoric water cannot access the rock mass as easily as it can penetrate into coarse, porous masses. The elevated strength of fine granite is indicated by the results of laboratory experiments on freeze-thaw weathering (Martini, 1967; Swantesson, 1989) and inferred indirectly from field relationships. An excellent example is provided by the summit plateau of the Serra da Estrela, Portugal, which is cut across different textural variants of Carboniferous granite (Ferreira and Vieira, 1999). The most elevated (> 1,600 m a.s.l.) parts of the plateau, the Torre Massif in the south-west, Alto da Pedrice and Curral do Vento in the south-east, are developed in fine and medium granite, whereas the lower parts of the plateau are underlain by much coarser, in places strongly porphyritic granite of Seia and Covilha variants (Fig. 7.5). Distinct topographic steps 100–200 m high have developed between higher and lower ground. Similar relationships to those in Serra da Estrela have been found within the summit plateau of the Karkonosze Massif, south-west Poland (Dumanowski, 1968). In the same area there occur stocks of fine-grained, aplitic granite which intruded into an older, coarser variant. These stocks typically form hills, which occasionally rise very abruptly above the surface of low relief eroded in coarse granite (Migoń, 1996, 1997c; Plate 8.11). However, a higher resistance of fine granite must not be universally assumed, and there are field examples

Plate 8.11 Residual hills associated with stock-like intrusions of fine-grained granite, Jelenia Góra Basin, south-west Poland

Plate 8.12 Rock shelter at Haytor, Dartmoor, signifying differential weathering of fine (below) and coarse (above) variants of granite

to the contrary. In Dartmoor, England, the main coarse granite appears more resistant than the fine Blue Granite (Plate 8.12). The explanation of this situation probably lies in the fact that the Blue Granite is both more closely jointed and contains more biotite than its coarse counterpart, and these two features may offset any positive effect of tighter fabric. On a much larger spatial scale, there is a reverse relationship between crystal size and topography in the Isle of Arran, Scotland (Godard, 1969). Fine-grained granite, which occurs in the central part of the granite massif, forms a topographic low of subdued relief, which contrasts with rugged relief in the outer coarse granite.

Another aspect of crystal size and packing is that fine granites tend to support less varied topography than coarse variants often do, whatever their elevation. Again, the Serra da Estrela may be cited in this context. The summit plateau in fine granite is rather monotonous and consists of long planar slopes which only locally host tors. These tors, in turn, are not very impressive and fields of scattered rounded boulders are uncommon. By contrast, lower surfaces in coarse granites are dotted with domes, tors, boulder piles, and shallow basins are frequent (Plate 8.13). The effect of rock texture on surface roughness at the spatial scale of a few tens of kilometres is also evident in the Karkonosze Massif (Fig. 8.2). The reasons relate to the variability of the host rock. Domes, tors, boulders, and basins are in most cases, and certainly in the Serra da Estrela and the Karkonosze, geomorphological manifestations of selective deep weathering followed by stripping. Fine-grained granites, however, offer limited potential for

Plate 8.13 Planar granite landscape developed upon medium-grained Curral do Vento granite, Serra da Estrela, Portugal. Tors on the skyline are angular and may have formed in periglacial conditions. Compare also with Plate VI

selective weathering as they have crystals of similar size throughout the rock mass. Moreover, fractures are typically rather densely and uniformly spaced. Therefore, the rock mass is relatively homogeneous in respect to weathering and conditions for highly differential descent of weathering front are not met.

The internal arrangement of crystals is also important. The presence of linear gneissic textures decreases the strength of the rock and makes it more prone to weathering and erosion. Foliation planes offer continuous lines of weakness which can be readily penetrated by water when weathering starts. Subsequently, foliation planes may develop into parallel lamination, called micro-sheeting (Folk and Patton, 1982; Chigira, 2001), and may facilitate large-scale gravitational movements on steep slopes (Varnes et al., 1989). Because of the weakened fabric, many granite massifs which show directional textures have not evolved into a type of topography most readily associated with this rock. For example, sheared Precambrian granite massifs in the Bohemian Massif support gently rolling uplands, which do not differ very much from the relief developed on laminated metamorphic rocks. However, there are spectacular examples of massive domed inselbergs in granite gneiss and even gneiss, of which sugar loaf cupolas in Rio de Janeiro are the prime example (Freise, 1938; Thomas, 1978). This again highlights the complexity of structural and lithological factors in the development of granite relief.

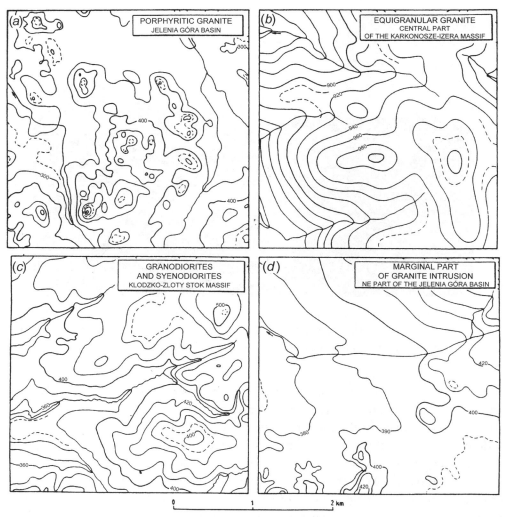

Fig. 8.2 Contrasting surface roughness in different parts of the Karkonosze Mountains, south-west Poland, depending on structural and textural characteristics of granite (after Migoń, 1996, fig. 3)

Mineralogy and Geochemistry

Granite is made of minerals which differ in hardness, chemical composition, internal structure, and the ease with which they can be broken down. Quartz is normally very durable, whereas biotite and plagioclases are 'weak links in the chain'. Therefore, depending on the proportions between different minerals, unequal resistance against weathering and erosion may be inferred for different granites. Again, field relationships help to formulate general propositions, but exceptions are common and must not be ignored.

Two mineralogical features that matter most are the proportion of potassium feldspar to plagioclase and the percentage of biotite. The higher the content of plagioclase and biotite in granite, the less resistant the rock may be expected to be. These relationships are implied by the geochemical and textural characteristics of the relevant minerals, which contain abundant weatherable elements such as iron, calcium, sodium, and magnesium, and, in the case of biotite, show distinct cleavage. Studies of chemical weathering, whether experimental or field-based, confirm that their rates of weathering are indeed much higher than for potassium feldspar. If these observations are scaled up to the landscape level, a consistent pattern emerges. Granites rich in quartz and potassium feldspar tend to underlie higher ground and to support impressive dome clusters and inselberg groups. By contrast, a typical granodiorite countryside is gently undulating, with topographic basins, perhaps with tors and boulders, but distinct inselbergs are rather few. Weathering mantles are thicker on granodiorites than on granites, but their characteristics are also very much controlled by fracture patterns. If granites *sensu stricto* and granodiorites occur side by side, the former are usually in elevated position. But there are clear examples to the contrary. In the Sierra Nevada, California, the spectacular topography of deeply incised valleys, rock slopes, and domes has developed in apparent disrespect of lithological differentiation of the batholith. There is no observable change in morphology between granites and granodiorites in the Yosemite National Park, both variants being massive and evidently very resistant. It is probable that lithological differences become geomorphologically significant only if a general trend of landscape development has been maintained over very long timescales, thus allowing for protracted and uninterrupted selective weathering and denudation.

Numerous examples can be provided to illustrate the general statements above. Lagasquie (1984) analysed the granite geomorphology of the French Pyrenees and used landforms and altitudinal relationships to rank rocks according to their resistance. Granodiorite invariably appeared as less resistant than monzogranite and in many places has been hollowed out to form topographic basins of various dimensions. In another French example, Flageollet (1977) compared granites from Limousin in the north-west Massif Central and found that higher altitudes correlate with decreasing plagioclase, and especially biotite content. More resistant leucogranites have a biotite content of less than 6 per cent and a plagioclase content typically within the range of 30–40 per cent. Biotite granites, in turn, may have as much as 20 per cent of biotite and 35–45 per cent of plagioclase. In the Jelenia Góra Basin, south-west Poland, granites with an elevated content of potassium feldspar build higher ground and support more diversified, inselberg-like relief, whereas higher percentages of sodium and calcium, relevant to the content of plagioclase, occur in less elevated areas (Migoń, 1996; Fig. 8.3). In the nearby Lusatian Granite Massif, a distinctive hilly morphology with relief up to 200 m is associated with two-mica granite, whereas granodiorite supports surrounding surfaces of rather low relief (Möbus, 1956).

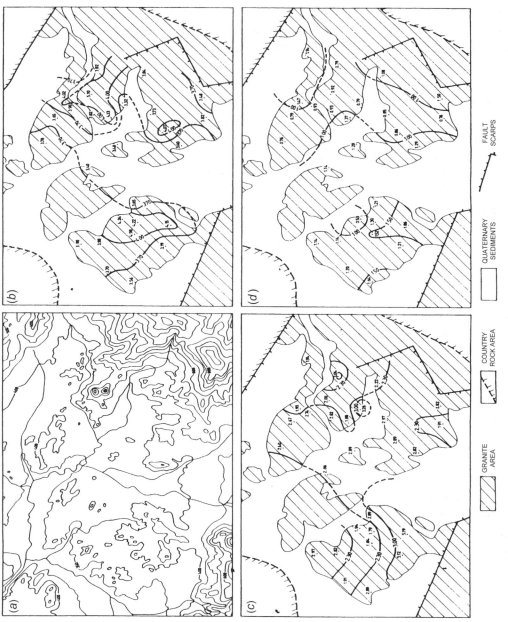

Fig. 8.3 Relationships between relief and mineralogical composition of granite, Jelenia Góra Basin, south-west Poland (after Migoń, 1996, fig. 4): (a) relief (contour lines every 50 m), (b) potassium content in per cent, (c) sodium content in per cent, (d) calcium content in per cent

Among North American examples, that provided by Eggler *et al.* (1969) from the southern Laramie Range is very informative, but it is also in a sense atypical. This area, located on the border between Colorado and Wyoming, is underlain by the Precambrian Sherman Granite, which occurs in two variants. Trail Creek granite is rich in potassium feldspar (42% on average) and depleted in biotite (2%), whereas the Cap Rock quartz monzonite has 32 per cent of potassium feldspar and 8 per cent of biotite. These relationships might suggest that the Tail Creek granite would be more resistant and form higher ground, but the situation is actually the reverse. It is deeply weathered into grus, up to 60 m, and forms an extensive level surface called the Sherman Surface. The Cap Rock monzonite, in turn, gives rise to a very distinctive tor and parkland topography, with widespread boulder fields, also described by Cunningham (1969). Eggler *et al.* (1969) attribute this unexpected situation to the early hydrothermal weakening of the Tail Creek granite fabric, now exploited by subaerial weathering. The Cap Rock monzonite disintegrates into grus too, but despite its higher biotite and plagioclase content the progress of disintegration does not keep pace with that experienced by the Tail Creek granite. The granite geomorphology of the Joshua Tree National Park, southern California, shows lithological control too. The main variant, Tank Rock granodiorite, is coarse and rich in biotite, but also differentially jointed, hence it is very prone to selective weathering. The extensive topographic basin in the middle of the park is underlain by the Tank Rock granodiorite, but numerous domes, tors and boulder piles rise from the flat basin floor and marginal pediments (Plate VII). The Northern Appalachians is another area where the patterns of landforms show sympathy to lithological differentiation of the basement (Birot *et al.*, 1983). Granodiorites typically underlie topographic lows, which is attributed to both their composition and high degree of microfracturing. The widespread Conway granite is very differentiated itself in respect to plagioclase and biotite content, which is reflected in the presence of uplands and basins alike. However, wherever it is intruded by hastingsite and riebeckite granite, such as in the White Mountains, New Hampshire, these younger stocks occupy the most elevated position in the landscape.

In Africa a repetitive motif is the presence of inselberg groups and massive domes in intrusions of the so-called Younger Granite. These Younger Granites are typically potassium-rich and poorly jointed, hence reasons for their higher resistance against weathering are complex. Examples abound in all parts of the continent, regardless of regional climate, and have been reported from South Africa (Brook, 1978), Zimbabwe (Whitlow, 1979), Namibia (Selby, 1982*b*; Migoń and Goudie, 2003), Kenya (Pye *et al.*, 1986), Nigeria (Thomas, 1965), and Sierra Leone (Thomas, 1980) among others (Plate 8.14). On the other hand, Pye *et al.* (1984) have not found any lithological control on the shape of granite residuals in the Matopos area, Zimbabwe, which appears to be governed entirely by fracture patterns.

Plate 8.14 Massive granite domes, Nigeria (Photo courtesy of Michael F. Thomas)

In the Eastern Desert of Egypt Dumanowski (1960) observed how mineralogical zonation within a small pluton, probably a stock intruding older metamorphic rocks, is reflected in landform patterns. Its central part is a prominent hill, Gebel Harhagit, that rises about 350 m above the surrounding desert floor, which is an erosional surface of variable width, from 0.2 to more than 1 km, also cut in granite. Scattered low granite residuals rise from the floor of the depression. On the outer side of the depression the terrain rises again, towards the contact between granite and country rock. The situation is repeated in the yet smaller stock-type pluton of Little Harhagit, located nearby. The reasons for differential erosion are to be found in mineralogical differences between the outer and inner parts of the pluton. The inner zone is depleted in biotite and lacks lamination.

On a larger spatial scale, Lageat (1978) was able to show certain relationships between rocks and landscapes and to rank different suites of crystalline rocks according to their resistance in the Barberton region, East Transvaal, South Africa. Leucocratic granites and the group of Younger Granite are situated at the top of the scale, supporting highlands, whereas topographic depressions such as the Kaap Valley in front of the Drakensberg escarpment are excavated in porous granodiorite (Fig. 8.4).

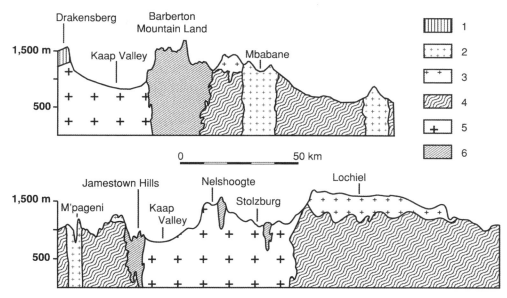

Fig. 8.4 Bedrock–topography relationships in the eastern Drakensberg, South Africa (after Lageat, 1978, fig. 9): (1) quartzites, (2) Younger Granites, (3) Hood granites, (4) gneiss and migmatites, (5) granodiorite, (6) older metamorphic formations

Veins, Enclaves, Borders, and Roof Pendants

Small-scale lithological contrasts between the host granite and enclosed veins and enclaves may be responsible for a range of minor geomorphic features on rock surfaces. However, it is not possible to offer any general statements because veins may have widely different compositions and, likewise, the host granite may be lithologically and structurally varied. Fine-grained aplite veins in most cases protrude above the rock surfaces, giving rise to a few to 10 or so centimetres of relief. In recently deglaciated areas, such as the High Sierra, California, and northern Scandinavia, projecting aplite veins may still retain glacial polish and striations (André, 2002). But in other places, where the adjacent granite is very massive, aplite can be found in recesses and forms negative relief. Quartz veins are also typically protruding and the resultant features are called *nerviaciones* in Spanish (Pedraza *et al.*, 1989). Pegmatites are coarse vein rocks, usually internally fractured, therefore they tend to weather at a faster rate than the surrounding granite.

Enclaves, in the majority of instances, are made of more mafic rock than the granite itself and contain abundant minerals which weather rather easily, such as biotite and hornblende. Therefore they decompose at a faster rate and are normally associated with shallow hollows on rock surfaces. Enclaves are occasionally suspected of offering initial points of weakness, upon which selective weathering begins to operate to produce an end-form of a weathering pit or tafoni. This may indeed be the case in specific situations, but it is difficult to accept as a universal explanation.

The phenomenon of contact metamorphism, typically associated with epizonal intrusions into a cool country rock, also has its geomorphic expression. This is because the metamorphosed rock, called hornfels, acquires a very tight fabric while heated, but also because marginal parts of granite intrusions are not uncommonly weakened by late-magmatic hydrothermal and pneumatolitic alteration. Hence, two rock types of contrasting resistance occur next to each other, offering potential for selective erosion. Dumanowski (1963) analysed the border zone of the Karkonosze granite batholith in Poland and the Czech Republic and found that along 90 per cent of its circumference there exists a country rock escarpment facing towards the granite mass (Fig. 2.3). Where massive hornfels occurs, the height difference between the two rock masses may approach 400 m. Contact metamorphism may also affect the roof of an intrusion, again strengthening the country rock against granite. If differential denudation exposes both granite and thermally altered country rock, the former may then erode faster and the latter, known as the roof pendants, may survive in water divide position for a long time.

Ring Dykes

Ring dykes tend to have very clear geomorphic expression and whenever the granite that forms dykes is more resistant, for whatever reason, than the rock it cuts through, an elongated ridge following the strike of a dyke will occur. Thorp (1967*b*) described a perfect example of ring dyke geomorphology from the Kudaru Hills in Nigeria. There is an almost continuous circle of elongated hills around the central granite upland, separated from it by a 1–2 km wide stretch of lower ground underlain by rocks belonging to a Precambrian basement complex (Fig. 8.5). Ring dyke ridges are 100–200 m high and steep sided. Numerous streams cut through the ring in narrow gorges, but there are also wider openings towards the surrounding plain. Ring dykes are also a part of the Erongo volcano-magmatic structure on the fringe of the Namib Desert. In north-west Erongo the high plateau built of lava and ignimbrite slopes down to an inner plain, some 4–5 km wide, enclosed from the outer side by a semi-continuous crescentic ridge of massive granite, which rises up to 500 m above the plain.

Discontinuities

Master Fractures

Regionally extensive structural discontinuities, traditionally described as master joints but in reality ancient fault zones, may be traced across tens of kilometres. They occur in different structural settings, from shields to relatively young mountain systems such as the Sierra Nevada, California, and exert an evident influence on large-scale landform

Geological Controls 279

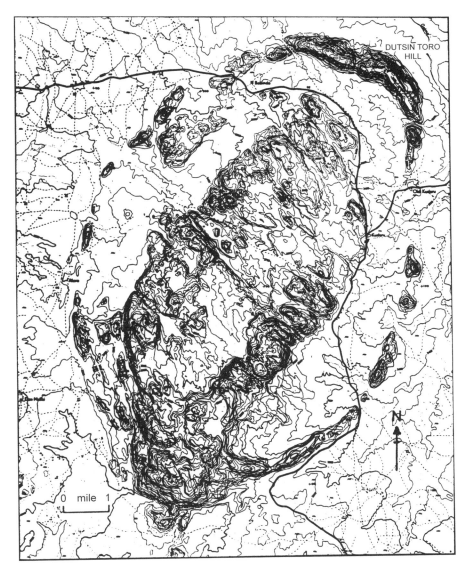

Fig. 8.5 Geomorphological expression of a granite ring dyke, Kudaru Hills, Nigeria (after Thorp, 1967b, figs. 1 and 2)

patterns. As has been shown in Chapter 2, 'master fractures' in close-up are zones of closely spaced partings, locally affected by cataclasis, with significant alteration of primary minerals. Preferential erosion of such zones of weakness gives rise to a range of distinctive geomorphic features.

In certain shield settings one may observe that domes are arranged into quasi-regular geometric patterns, most often orthogonal and rhomboidal. Massive compartments occur next to each other, standing in rows and forming uplands separated by linear troughs. Where two lines of depressions cross, troughs are locally enlarged to intermontane basins. Some troughs are used by streams, but this is not necessarily always so and the pattern of negative relief elements is essentially independent of drainage lines. The presence of such linear depressions results in the compartmentalization of relief into individual units of higher ground, particularly well visible on aerial photographs and satellite images. Examples of basement relief controlled by master fractures have been forwarded, among others, from South Africa (Mabbutt, 1952), Zimbabwe (Whitlow, 1979; Pye et al., 1984), and Nigeria (Jeje, 1973, 1974; Thomas, 1974) (Fig. 8.6).

Master factures can be very prominent in high-latitude shields as well. Bohuslän in south-west Sweden is an example of an area where preferential etching of these zones has given rise to an intricate pattern of basins of variable size and outline, interconnected by narrow defiles (Lidmar-Bergström, 1995; Johansson et al., 2001a, 2001b; Migoń and Johansson, 2004; Fig. 8.7). Basins are associated with N-S to NW-SE trending zones, probably related to the evolution of two major regional tectonic structures: the Tornqvist-Teisseyre line in the south-west and the Oslo Rift in the west. It has been hypothesized that the etching of the master fractures proceeded geologically over a rather short timescale and was brought to a halt by marine transgression in the late Cretaceous, otherwise a wider extension of basin floors would have been expected.

The geomorphic role of master fractures in high mountains is somewhat different, as demonstrated in the Sierra Nevada (Ericson et al., 2005). Two principal systems of such fracture zones exist, one trending SSW-NNE and another WSW-ENE. Along most of their lengths they are used by mountain rivers draining the western slope of the mountain range, but the location of some valley side ravines and passes within water divides is also related to these structural discontinuities. Late Cainozoic uplift of the High Sierra triggered vigorous fluvial incision and currently it is stream erosion which exploits the master fractures. Hence, there is little scope for excavating larger basins at fracture intersections, which is a characteristic feature of more stable cratonic settings where weathering prevails over incision.

Jointing

Joint control is perhaps referred to most often in the context of relationships between landforms and structures. Several aspects have already been mentioned in preceding chapters; hence they will be summarized rather briefly here. Joint control appears dominant at smaller spatial scales than those influenced by lithology and magmatic history. First, it is medium-sized landforms such as residual hills and tors whose outlines show the clear influence of jointing patterns. In fact, the variability of form of residual hills (Table 3.1) results precisely from the variability of jointing patterns.

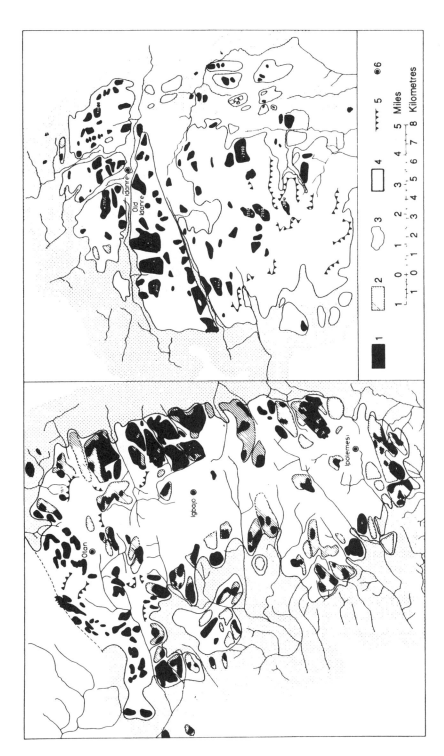

Fig. 8.6 Master fractures and granite relief, southern Nigeria (after Thomas, 1994b, fig. 10.18): Streams and avenues trending east–west, north-east–south-west and north–south indicate the direction of master fractures. The left-hand map shows the Igbajo Plateau, the right-hand one shows the Idanre Hills: (1) bornhardts and convex rock terrain, (2) convex summits with soil cover, (3) dissected terrains, (4) deeply weathered Basement Complex rocks, (5) major breaks of slope, (6) towns

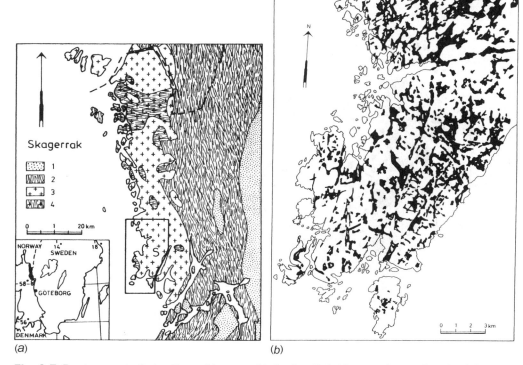

Fig. 8.7 Fracture-controlled pattern of topographic basins, Bohuslän, south-west Sweden (after Johansson et al., 2001a, figs. 2 and 4): (a) Location and simplified geological map: (1) orthogneiss, (2) supracrustal gneiss, (3) Bohus granite, (4) Bohus granite with abundant gneissic xenoliths, S–Sotenäset Peninsula. Inset shows the location of detailed map (b). Extent of basin floors is marked in black

Second, a range of minor bur nevertheless distinctive facets within residual hills are joint-controlled, for example hill-top avenues, clefts, caves, and basal re-entrants. Third, joint characteristics assume crucial importance for slope form and evolution, including the pattern of mass movement on both rock slopes and in weathered terrains because joints control groundwater flow and provide ready slip planes. Fourth, joint spacing controls the progress of deep weathering and there is usually a positive correlation between joint spacing and the depth of weathering profiles. Fifth, the pattern of physical breakdown at the surface is joint-controlled, hence the relationship between average joint spacing and the nature of superficial deposits.

Generally speaking, the presence of joints weakens the rock mass and makes it more prone to weathering, mass movement, and erosion. Various lines of evidence may be used to validate this statement. Residual hills of all morphological types are typically developed in granite, which is less jointed than in the surrounding terrain

(Twidale, 1964, 1982; Thomas, 1965; Gerrard, 1974; King, 1975; Brook, 1978). Domes are typically virtually unjointed rock masses, the only discontinuities observed being tight sheeting planes (Plate V; Plates 3.10, 3.11, 8.7). An inspection of deep weathering profiles reveals that corestones form where joints are widely spaced and if this spacing is large, they may survive even in the upper horizons of very thick mantles. On the other hand, grus mantles may offer clues as to why granite has disintegrated so thoroughly. They may contain closely spaced relict joints, visible owing to the iron staining along them. Observations of landslide scars in weathered granite demonstrate that inclined joint surfaces acted as slip planes (Chigira, 2001), and geotechnical tests provided conclusive evidence that strength along joint surfaces is significantly reduced (Au, 1996). Inclination and long-term behaviour of rock slopes depends on joint density and it is evident that more closely jointed rocks support gentler slopes. In addition, they tend to fail through particle fall rather than voluminous but rare rock slides and rock falls. Joint orientation also plays its part in dictating the ways of subaerial degradation of residuals and patterns of slope collapse. The domination of inclined sheeting joints tends to favour sliding, whereas cleft opening, block collapse, and toppling are associated with steeply dipping and vertical joints.

However, Twidale and Bourne (1993) have sounded a cautionary note, arguing that relationships between joint density and rock susceptibility to degradation are more complicated than one might think. Fractures, in their opinion are a 'double-edged sword', in the sense that more joints does not automatically mean more efficient weathering and erosion. It is the flushing rate and the residence period of water which are important. Thus, if joints cutting a rock mass are open and permit free drainage towards valley sides, they will actually help to keep the rock relatively dry and therefore more resistant. Furthermore, precipitation of iron compounds supplied by capillary moisture flow on joint planes may locally strengthen the rock mass. Such highly indurated joint surfaces may prove more resistant than the rest of the rock, to give a curious effect of a lattice of raised rims (Plate 8.15).

Microfractures

These minor discontinuities, typically around 100 μm long and too miniscule to be seen by the naked eye, are nevertheless very important in the context of granite weathering. Their presence facilitates the breakdown of individual minerals, especially of otherwise rather resistant quartz, and the loosening of the fabric of the entire rock mass (Whalley *et al.*, 1982). Therefore, their role in the origin of grus must not be ignored (Folk and Patton, 1982; Pye, 1986, Power *et al.*, 1990). Opening of microcracks allows for easier access of water into the rock, which then drives other weathering processes associated with crystallization of salts and freezing and thawing. Likewise, enhanced biological weathering may be inferred for granites with higher

Plate 8.15 Raised, iron-enriched rims left after weathering of a granite boulder, Gobabeb, Namib Desert

density of near-surface cracks as these may be penetrated by lichen hyphae and provide living space for endolithic algae.

Hierarchy of Geological Controls

If the various relationships between landforms and geological features are put into a spatial and temporal context, then a certain hierarchy appears to emerge. Different structural and lithological properties control landforms of different sizes, and the same rock characteristics may be of the utmost importance for some landforms and largely irrelevant for others. For example, to explain the present-day altitude of granite masses in relation to the country rock, the tectonic setting, uplift rates, and related long-term denudation rates need to be considered first, whereas such information matters little if patterns of mass movements on rock slopes over decadal time scales are to be deciphered. For the latter, the characteristics of jointing patterns will be crucial.

On the largest scale, tectonic setting and its changes over time as well as the mode of emplacement are the crucial variables to consider if the long-term evolution of entire batholiths and plutons is in focus. Altitude differences within granite massifs of the order of a few hundred metres and more are typically produced by differential tectonics and straight escarpments tens of kilometres long are in most instances fault-generated. Inactive fault lines of regional extent ('master fractures') are no less significant, as they

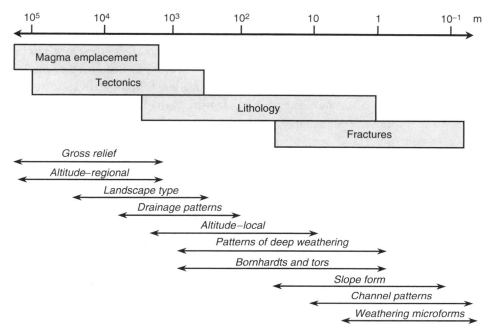

Fig. 8.8 Hierarchy of geological controls in relation to spatial scales and relevant components of granite landscapes

may focus erosion and control regional drainage patterns. At medium spatial scales (~10–10^3 km^2) lithological variation appears to play a key role in differentiating granite landscapes. Going into more detail, fracture control becomes crucial for the shape of residual and erosional landforms (inselbergs, tors, valleys, basins, bedrock channels), the progress of deep weathering, and the patterns of mass movement on rock and weathered slopes alike. However, minor lithological contrasts may guide selective weathering of exposed rock surfaces and account for the variety of microrelief of granite surfaces (Fig. 8.8).

The two examples below, taken from areas differing in respect of tectonic setting and the temporal context of landform evolution, illustrate the concept of hierarchical geological control. Krumlovský les is built of Proterozoic granite too, but this was subsequently incorporated into an orogenic zone in the late Palaeozoic and deformed again in the Neogene. The Serra da Estrela is a representative of a Palaeozoic mobile belt, rejuvenated by plateau uplift in the Cainozoic.

Krumlovský Les, Czech Republic

Krumlovský les is a relatively small upland area in the south-east part of the Czech Republic, in the marginal part of the rigid basement block of the Bohemian Massif. It rises to an altitude of 415 m a.s.l. and overlooks adjacent lower ground by 100–150 m. The gently undulating upland surface is elongated north to south and is 12 km long and up to 5 km wide. It is poorly dissected; there is only one active river gorge and a parallel wind gap in the northernmost part of the massif. Granite is Proterozoic, medium-grained, with predominant plagioclase, and occurs in a few lithological variants, which are however not very dissimilar. The unifying feature is the high degree of fracturing, which is a superimposed result of many episodes of crustal deformation. The hierarchy of structure-controlled landforms of the Krumlovský les has been presented by Migoń and Roštinský (2003).

The reasons for the existence of an upland are tectonic. Krumlovský les is a horst, separated from the west by a prominent, rectilinear, and poorly dissected steep scarp which exactly follows the course of a fault. The uplift of the block is associated with Neogene vertical and strike-slip movements along the boundary between the rigid Bohemian Massif and the Carpathian orogen (Roštinský, 2004). The upland surface is little differentiated and lacks residual hills, impressive tors, and extensive boulder fields. Few existing excavations reveal a mantle of grus, with little evidence of selective weathering. Interestingly, lithological change within the plateau does not correspond to any appreciable change in landform patterns. This situation is likely to reflect certain common lithological and structural characteristics of granite under the plateau. It is medium grained and densely fractured, hence offering little scope for selective weathering. It is also rather poor in potassium content, which is 4.38 per cent on average, and plagioclase very much prevails over potassium feldspar. The monotony of the plateau is only locally interrupted by the presence of parallel low convexities elongated east-west, built of aplite veins. Joint control is discernible in minor landforms. There are occasional tors, but these are low (< 5 m high) and have a ramp-like appearance due to dipping rather than vertical joints. There are scattered boulders on the plateau and these seldom exceed 1 m long, reflecting dense joint spacing. Topographic lows between aplite ridges and tor-crowned elevations lack tors and boulders entirely, suggesting increased joint spacing. The hierarchy of geological controls in the Krumlovský les upland is presented in Fig. 8.9.

Serra da Estrela, Portugal

The mountains of Serra da Estrela in central Portugal owe their present-day elevation, reaching 1993 m a.s.l. at the summit of Torre, to differential uplift in the Cainozoic along SW-NE trending faults (Brum Ferreira, 1991). An impressive fault-generated escarpment runs along the eastern shoulder of the massif, creating an elevation drop of almost 1,000 m, and a similar, although more dissected, escarpment occurs in the west. Above the marginal escarpments there extends a plateau-like relief, sloping gently to the north. Its general geomorphology, with a focus on glacial history, has been described by Daveau (1969, 1971), and much new information has recently been added

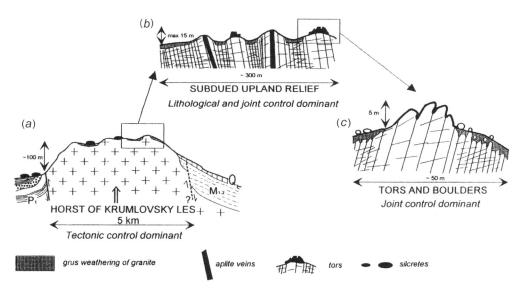

Fig. 8.9 Hierarchy of geological controls in the granite upland of Krumlovský les, Czech Republic (after Migoń and Roštinský, 2003, fig. 7). P_1—Permian infill of the Boskovice Graben, M_{1-2}—Miocene sediments, Q—Quaternary deposits, mainly loess

by Vieira (1998, 1999). An impressive geomorphic feature related to fault tectonics is the system of two valleys in the middle of the plateau, sloping respectively to the south (Vale da Alforfa) and to the north (Vale do Zêzere). Both are deeply incised into the upland surface, up to 500 m (Fig. 7.5, Plate 7.3), and hosted thick valley glaciers in the Pleistocene.

Within the plateau no fewer than six distinctive variants of granite crop out (Ferreira and Vieira, 1999), each supporting its own distinctive assemblage of landforms (Table 8.1). From the table it appears clear that fine- and medium-grained, more lithologically uniform granites are associated with less diversified relief, although their altitudinal position is higher. The scarcity of tors is particularly striking on the summit plateau, because it has been largely unglaciated and theoretically cold-climate weathering processes acting over protracted time spans would have had an opportunity to produce abundant frost-riven cliffs. However, one small area of the plateau where the Covilhã granite crops out is dotted with domes and boulder tors. The explanation seems to lie in the apparent uniformity of the summit granites, unfavourable to selective weathering. On the other hand, coarse and differentially fractured coarse variants give rise to an extremely varied topography regardless of altitude, from 1,600 down to 1,200 m a.s.l., and the proximity to deeply incised valleys (Plate 8.13; plate VI).

The shape of individual residual landforms is controlled by fracture patterns. Domes and tors which have developed in coarse granite on the intermediate plateau provide

288 Geological Controls

Table 8.1 Rock–landform relationships in Serra da Estrela

Granite variant (name)	Lithological characteristics	Topographical setting and altitude	Characteristic medium-scale landforms and weathering mantles
Estrela	medium-grained	summit plateau	planar surface, rock cliffs, scarce angular tors, vertical walls in cirques, shallow grus
Pedrice	fine-grained	summit plateau	planar surface, straight slopes, block fields, and block slopes
Curral do Vento	medium-grained	summit plateau	planar surface of very low relief, sporadic trapezoidal tors
Covilhã	coarse	intermediate plateau	rough surface, medium-sized basins, domes and shield-like elevations, tors, boulders, weathering pits, deep grus mantles, locally clayey grus
Seia	coarse, with large megacrysts	intermediate plateau	rough surface, medium-sized basins, domes and shield-like elevations, tors, boulders, deep grus mantles, locally clayey grus
Manteigas	medium-grained, biotite-rich	valley slopes and Manteigas Basin	deeply weathered slopes

Source: After Migoń and Vieira, in preparation.

Plate 8.16 Massive domes in the coarse Seia granite, Serra da Estrela, Portugal

the best examples. Dome shapes are related to sparse jointing, paucity of vertical partings, and predominance of regularly arranged upward-convex sheeting joints. Tors are mainly of the boulder tor type and show an irregular pattern, but joint spacing is typically more than 1 m, and not uncommonly more than 2–3 m. Superimposition of widely spaced sheeting and vertical joints allows for efficient free drainage of rainwater and has made the occurrence of impressive castellated domes possible (Plate 8.16). Unfortunately, it is not possible to determine the reasons for the excavation of basins and pockets of grus over the plateau. This may relate to higher joint density or mineralogical differences.

Typology of Natural Granite Landscapes

Approaches to Granite Landscapes

There are two major recurrent themes in geomorphological research into granite landscapes. On the one hand, there is a recognition of the extraordinary diversity of landforms supported by granitic rocks, on a variety of scales, from microrelief on exposed rock surfaces to regional landscape types. On the other hand, there are striking geomorphic similarities between basement regions across the world, noted again at the scale of individual, almost omnipresent landforms, as well as in larger landform assemblages. To account for both diversity and similarity, various attempts have been made to produce a typology of granite landscapes.

One of the early systematic approaches was that presented by Wilhelmy (1958) in his *Klimamorphologie der Massengesteine*. In line with the dominant paradigm in German geomorphology, he was an advocate of strong climatic control on the development of landforms and, accordingly, used climatic zonation of the globe as a basis for his classification system. Seven major morphoclimatic zones with allegedly distinctive phenomena of granite weathering and denudation have been distinguished (Table 9.1).

Table 9.1 Climate–landforms relationships according to Wilhelmy (1958)

Climatic zone	Characteristic landforms and weathering patterns
Humid tropical, equatorial	deep chemical weathering, spheroidal weathering and corestone development, karren on exposed rock surfaces
Seasonally humid, with long rainy season (including monsoon countries)	deep chemical weathering, boulder fields as residual deposits, bornhardts with massive sheeting
Seasonally humid, with long dry period	inselbergs, piedmont angle only if distinct contrasts between dry and wet period, deep arenaceous weathering, two-stage boulder fields, tafoni
Arid	granular disintegration and boulder development
Mediterranean	deep arenaceous weathering, two-stage boulder fields, bornhardts and castellated tors, tafoni
Humid temperate	arenaceous weathering, castellated tors, block fields and block streams
Polar	angular block fields

In addition, Wilhelmy emphasized climate-controlled change in landform inventories with altitude, citing examples from Corsica, the Sinai Peninsula, and Korea. Following Lautensach (1950), he specifically mentions mountain ranges in the Korean Peninsula, where a dissected landscape typified by deep ferruginous weathering gradually gives way to assemblages typical of harsh periglacial environments as altitude increases. The problem with the climatic approach is the likely co-existence of landforms and weathering patterns of different ages, hence formed in different environmental conditions, especially in middle and high latitudes. In fairness, it has to be said that inherited components have been recognized by Wilhelmy, but the evidence for inheritance is not always conclusive. Selected controversial examples have been presented in the previous chapters. These difficulties also raise a wider methodological issue, if climatic zonation of the globe is to be an appropriate framework to analyse granite landforms. Stoddart (1969) and more recently Twidale and Lageat (1994) offered insightful reviews in this respect, concluding that the uncertainties and limitations of climatic geomorphology are far too many to make it a preferred paradigm.

Thomas (1974) adopted a different approach, a morphological one, attempting to identify characteristic granite landform systems. These include:

(a) multi-concave (basin-form) landscapes, with individual basins being enclosed, partially enclosed, or dissected;

(b) multi-convex (dome-form) landscapes, which can be further subdivided into ones with dominant weathered convex compartments and ones with rock-cored compartments;
(c) stepped or multi-storey landscapes;
(d) plains.

There is a certain degree of structural control behind this division in that plains are preferred end-landforms if jointing is closely spaced, whereas multi-convex relief with rock-cored compartments is favoured in massive rock, with individual fractures spaced 10 metres or so apart. Thomas envisaged specific pathways for the development of granite landscape, depending on climate change and tectonic instability, but also makes a strong point that: 'Many granite landscapes have undergone prolonged sub-aerial development that has involved the formation, destruction and reformation of characteristic forms such as basins, tors and domical compartments. There is thus no single starting point and no inevitable end product in the sequence of landform development in granite' (Thomas, 1974: 33). One implication of this statement is that classifications of granite landscapes should probably emphasize the contemporary patterns of landforms above any other factors and controls. Such a rationale will be adopted by the present author later on.

The most comprehensive typology so far, however, has been offered by Godard (1977). In the first instance, he attempted to distinguish landforms characteristic of different spatial scales, thus relating geomorphic objects of widely different sizes (Table 9.2). This is a very helpful proposal since different sizes usually imply different lifetimes of relevant landforms, a factor which is not always addressed in studies of granite terrains. He then goes on to identify characteristic landscapes (*paysages*) and in doing so, he simultaneously considers morphotectonic setting, timescales of relief evolution, and environmental conditions (Table 9.3).

Table 9.2 Characteristic granite landforms according to Godard (1977)

I. order	plains
	great escarpments
	all-slope topography
II. order	degraded (dissected) plateaux
	multi-convex relief
	rock domes
	sharp-crested ridges
III. order	boulders
	tors and boulder fields
IV. order	tafoni
	weathering pits
	karren

Table 9.3 Characteristic types of granite landscapes according to Godard (1977)

Tropical shields	
(a) in the humid tropics and rainforest environments	deep weathering abundant
	gullying and debris slides on weathered slopes
	multi-convex topography
	topographic basins and alluviated valley floors
(b) in savanna and desert environments	superficial rock disintegration
	planation surfaces with inselbergs and tors
	pediments
Ancient massifs and younger intrusions in monsoon environments of South-East Asia	deep weathering abundant
	gullying and landslides
	multi-convex topography
	sharp-crested relief at higher altitudes
Ancient massifs in recent orogenic belts	
(a) Mediterranean environments	arenaceous weathering
	tors and boulder fields
	basins
	altitudinal morphological belts
	tafoni as a characteristic microform
(b) Alpine-type high mountain relief	sharp-crested ridges
	talus accumulation
	glacial landforms abundant
Ancient massifs in middle latitudes	elevated relict planation surfaces
	residual hills and massifs
	basins
	relict deep weathering
	tors and block fields
Shields and great escarpments in high latitudes	
(a) shields	shallow grus weathering
	landforms of glacial erosion
	knock-and-lochan landscape
	inherited, including exhumed planation surfaces
(b) great escarpments	elevated relict planation surfaces
	sharp-crested ridges of former nunataks
	block fields
	strandflat

In contrast to the elaborate classification by Godard (1977), a very simple morphological framework has been chosen by Twidale (1982) in his *Granite Landforms*. Four categories of major landforms are distinguished: boulders, inselbergs, all-slopes topography, and plains. Then a range of minor landforms, subordinate to the major ones, are discussed. However, this approach has two significant limitations as a general framework. First, there is a problem of scale, which has not been consistently addressed. Although plains and all-slopes topography may be considered as major landforms of the same order, inselbergs are much smaller, and boulders may occur on inselberg' slopes, on mountain slopes, and on plains. Second, several distinctive landscape types developed on granite are omitted, such as the joint-valley topography and multi-convex relief. Even the classic granite landscape of Dartmoor can hardly be described in terms of these four fundamental landforms, as it is neither a true plain nor a highly dissected upland, except for some marginal parts, and inselberg-like hills are missing.

The comprehensive relief classification developed by Lidmar-Bergström (1995) for Sweden is also worth mentioning alongside other approaches (Table 9.4). Although it has not been specifically designed for granite terrains, the abundance of granite in the study area makes it a potentially valuable tool in analysing granite geomorphology.

Table 9.4 Relief classes in Sweden according to Lidmar-Bergström (1995)

Relief classes	Relative relief (m)
Plain (peneplain)	0–20
Dissected plain	20–50
Plain with residual hills	0–200
Undulating hilly land	20–200
Joint-valley landscape	20–100
Large-scale joint-valley landscape	50–400
Deeply incised highlands	400–1500
Tectonic horst-and-graben relief	

Note: Some class names have been modified in order to eliminate specific regional context.

Etching Concept and its Application

The recognition of the profound role of deep weathering in shaping granite landforms, from minor to medium-sized, raises the question of whether an etching concept can be of use in describing, classifying, and explaining granite landscapes as a whole. However, it is not feasible to offer a comprehensive review of the concept and its history here (for this, see Thomas, 1989a, 1989b; 1994b; Bremer, 1993; Twidale, 2002).

Etching and Etchplains

The word 'etching' means corroding a surface by aggressive reagents and is used in geomorphology to describe progressive rock decomposition which occurs within deep weathering profiles, at the weathering front. In particular, it is applied to situations where rocks differ in their resistance to chemical decay and consequently where the thickness of a weathering mantle becomes highly variable over short distances. Removal of products of deep weathering will expose bedrock surface, the topography of which is the direct result of differential etching, thus it is an 'etched surface'. At an early stage of development of geomorphology, when the focus on planation surfaces and peneplains was pre-eminent, etched surfaces were visualized as surfaces of low relief and thought of as a special category of peneplain, produced by subsurface rock decay followed by stripping of the weathering mantle. Although geomorphic surfaces originating in this way had been recognized as early as the end of the nineteenth century (Twidale, 2002), the very term 'etchplain' was proposed by Wayland (1933) and Willis (1936), both working in East Africa in the 1930s. Accordingly, the process of producing an etchplain through weathering and stripping later became known as 'etchplanation'.

The impact of the concept of etching and etchplanation on general geomorphology was initially limited, but the situation began to change with the arrival of a paper by Büdel (1957), which was noteworthy for a number of reasons. Although the term 'etchplain' was not used in this publication, written in German, yet the argument leaves no doubt that 'die doppelten Einebnungsflächen' (translated as 'double planation surfaces') have all the characteristics of an etchplain. First, Büdel made it clear that he was applying the weathering/stripping concept to entire landscapes rather than to limited areas within them or to individual landforms. Second, he pointed out that transition from the phase of dominant weathering to the phase of dominant stripping might be associated with major environmental changes, whose profound impact for landform development was realized only later. Third, he suggested that many upland surfaces in middle and high latitudes are inherited Tertiary etchplains, hence extending the applicability of the concept outside the Tropics.

Realization of the crucial role of deep weathering and saprolite development in shaping most tropical landscapes, which was achieved in the 1960s, led to the expansion of the original ideas of Wayland and Willis. Different types and categories of etchplains were proposed by individual authors (Ollier, 1960; Thomas, 1965; Finkl, 1979), and ultimately Thomas (1989a, 1989b) offered the following classification:

(a) mantled etchplain—weathering mantle is ubiquitous and virtually no bedrock is exposed. Weathering progressively attacks solid rock at the base of the mantle, shaping the etched surface which is to be exposed later, but the mantle can also be relict;

(b) partly stripped etchplain—develops from mantled etchplain through selective removal of the weathering mantle and exposure of the bedrock surface, but part of the original saprolite remains. The proportion of areas still underlain by saprolite may vary from 10 to almost 100 per cent;

(c) stripped etchplain—most of the bedrock is exposed from beneath a weathering mantle and only isolated patches of saprolite are left (< 10 per cent of the area). These characteristics conform with the original definition by Wayland;

(d) complex etchplain—includes a few variants, in which deeply incised valleys may be present (incised etchplain), or removal of saprolite is accomplished by pedimentation (pedimented etchplain), or a new generation of weathering mantles begins to form (re-weathered etchplain);

(e) buried etchplain—one which has been covered by younger sediments or lava flows;

(f) exhumed etchplain—one which has been re-exposed after burial.

One important terminological problem has been noted, that a stripped surface is rarely a plain but tends to show some relief, which reflects differential rock control on the progress of etching. In granite areas in particular, stripped surfaces are typified by inselbergs, domes, tors, basins, and boulder piles, and to call them 'etch*plains*' would be both inappropriate and misleading. Therefore the term 'etch*surface*' has been recommended for use wherever evacuation of weathering mantles reveals a varied topography.

Etchplanation, and specifically the transition from the weathering phase to the stripping phase, is commonly linked with major external changes experienced by a landscape, related either to a change in tectonic regime or to environmental changes. It is envisaged that mantled etchplains form and exist during long periods (up to 10^9 yr) of stability, whereas stripping is initiated by uplift, or climatic change towards drier conditions, and is accomplished over much shorter timescales (10^5–10^7 yr) (Thomas, 1989a, 1989b). In this scenario, major external disturbances are essential for the formation of etchplains and the static nature of planate landsurfaces during the 'phase' of deep weathering might be implied. However, in many low-latitude areas subject to efficient deep weathering there is also extensive field evidence of geomorphic activity at the surface, hence an idea of 'dynamic etchplanation' has been introduced to emphasize ongoing landscape development through etching and stripping (Thomas and Thorp, 1985). Key points made are simultaneous weathering and removal of its products, lowering of both interfluves and valley floors, continuous sediment transfer, redistribution and temporal storage of weathering products, and the importance of minor environmental disturbances. An important part played by episodic events is emphasized in the parallel term of 'episodic etchplanation' (Thomas and Thorp, 1985).

From the 1980s onwards, following progress in weathering studies, the etching concept has been extended from tropical plains, or their inherited equivalents, to a much wider range of settings. Emphasis on the process of deep weathering rather than on the ultimate form of a plain has made it possible to see geomorphic development of many low-latitude mountain ranges of moderate relief as being accomplished by differential etching. In European middle and high latitudes the origin of many palaeosurfaces has recently been re-assessed and their etched nature made evident (e.g. Söderman, 1985; Hall, 1991; Lidmar-Bergström, 1995; Migoń, 1997b; Migoń and Lidmar-Bergström, 2001). Today, the idea of etching and etchplanation has the status of an autonomic concept in geomorphology, capable of accounting for both planate and topographically complex landsurfaces, integrating tectonic and climatic controls, linking historical and process geomorphology.

Regional Examples

The occurrence of assemblages of minor landforms of etched origin, the partial survival of deep weathering mantles, and evidence of the ongoing advance of weathering front shows that many lowland and upland granite landscapes are of etched origin too (Fig. 9.1). However, there is no one universally present topography of granite etchsurfaces, although some medium- and small-scale landforms such as boulders and tors are certainly repetitive. On the contrary, depending on the duration and intensity of deep weathering, bedrock control, and environmental conditions, the resultant landscapes will show a wide range of relief types.

In the southern part of the Fennoscandian Shield a few distinctive types of exhumed etchsurfaces of Mesozoic age exist on granite bedrock. Major variants include planate surfaces with isolated residual hills (inselbergs), multi-convex hilly topography with relief up to 200 m, and joint-valley landscapes, in which deep valleys were etched into the upland surface along regional fracture lines (Lidmar-Bergström, 1995). Subsequent studies along the Bohuslän coast north of Göteborg (Johansson et al., 2001a, 2001b; Migoń and Johansson, 2004) have revealed that the etched joint-valley topography includes a more specific type of relief, namely that of interconnected basins cut into the upland surface. Basins are of variable area and depth, from a mere 0.01 to as much as almost 2 km^2, and form an intricate spatial pattern (Fig. 8.7). Fracture lines exert a dominant control over the shape and size of basins. Lidmar-Bergström (1995) suspects that topographical differences between etchsurfaces reflect primarily the duration of deep weathering acting upon a sub-Cambrian peneplain in the Mesozoic. There was a specific time window for basement etching in the southern part of the Fennoscandian Shield. It commenced after the Palaeozoic sedimentary cover had been stripped from the sub-Cambrian peneplain and was interrupted by the marine transgression in the late Cretaceous. Longer duration of etching resulted in a more evolved hilly topography,

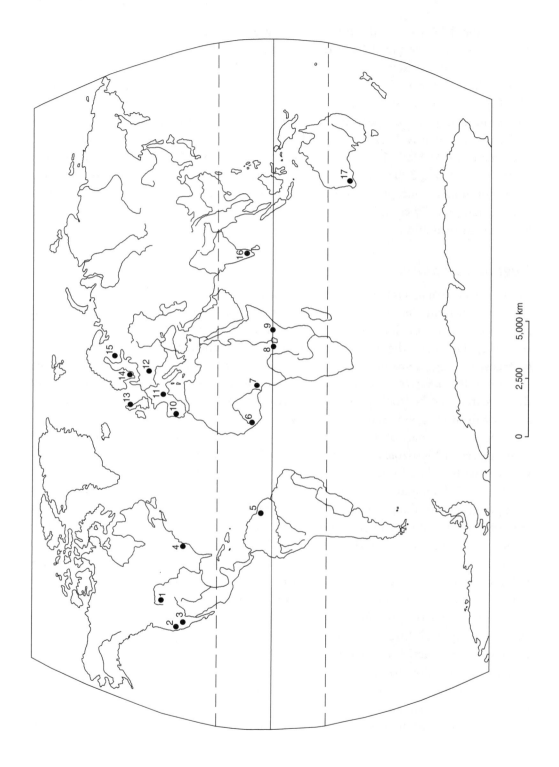

whereas a relatively late exposure allowed only for etching the fractures and the origin of joint-valley topography. Late Cretaceous transgression caused burial of Mesozoic etched topography, which has only been re-exposed by the end of the Cainozoic. Exhumation is in fact not yet everywhere complete and the hilly etchsurface in southern Skåne is best described as a partly stripped surface (Fig. 9.2).

Since the publication of important, but internationally little known, papers by Bakker and Levelt (1970) and Czudek and Demek (1970), many European palaeosurfaces have been reinterpreted as Cainozoic etchsurfaces, and granite landscapes are no exception. It appears that the most typical topography for a European granite etchplain is one of gently rolling relief, with scattered boulders and occasional tors (Plate 9.1). Prominent inselbergs built of less fractured or less coarse rock may occur, for example in the Žulova Highland in the Czech Republic (Czudek et al., 1964; Ivan, 1983) and the Jelenia Góra Basin in south-west Poland (Migoń, 1997b), but these are not very common regionally. In more elevated areas, such as Harz and Fichtelgebirge in Germany, Šumava in the Czech Republic, and Karkonosze at the Czech/Polish border, rolling summit surfaces of etched origin abound in impressive castellated tors. It has to be noted that the term 'etchplain' to describe surfaces produced by deep weathering and stripping is not used everywhere, but the geological and geomorphological evidence presented in many papers speaks clearly for an etched origin of other undulating granite landscapes as well. This is the case with granite uplands in the French Massif Central (Godard, 1977; Coque-Delhuille, 1979; Simon-Coinçon et al., 1997). In other localities, including Dartmoor, the geomorphic legacy of Pleistocene periglacial environments has been superimposed on an older etchsurface (Waters, 1964; Gerrard, 1988a).

In North America the term 'etchplain' has not been used much, but this must not obscure the fact that the key role of stripping a saprolite in granite landscape development has been recognized by many authors in many settings. Pavich (1986, 1989) presented examples of ongoing lowering of weathering front, saprolite production, and removal from the Appalachian Piedmont. The relief on the interfluves is not very much

Fig. 9.1 (opposite) Etchplains and etchsurfaces—a global view: (1) Laramie Range (Cunningham, 1969; Eggler et al., 1969), (2) Sierra Nevada (Wahrhaftig, 1965), (3) Mojave Desert (Oberlander, 1972, 1974), (4) Appalachian Piedmont (Pavich, 1986, 1989), (5) Guyana and Surinam (Eden, 1971; Kroonenberg and Melitz, 1983), (6) Sierra Leone (Thomas and Thorp, 1985; Teeuw, 1991), (7) southern Nigeria (Thomas, 1965), (8) Uganda (Ollier, 1960), (9) Kenya (Veldkamp and Oosterom, 1994), (10) eastern Portugal (author's observations), (11) Massif Central (Godard et al., 2001), (12) Bohemian Massif (Czudek and Demek, 1970; Migoń, 1997b), (13) Scotland (Hall, 1986, 1991), (14) southern Sweden (Lidmar-Bergström, 1995), (15) Finland (Söderman, 1985), (16) southern Deccan (Büdel, 1977; Johansson et al., 2001b), (17) south-west Australia (Mabbutt, 1961a; Finkl and Churchward, 1973; Finkl, 1979)

300 Typology of Granite Landscapes

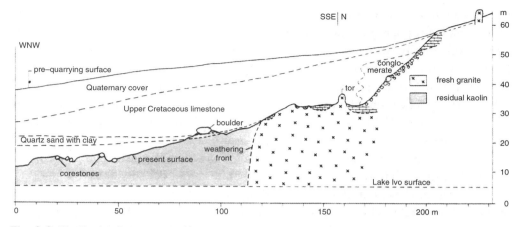

Fig. 9.2 Weathering front exposed in a quarry at Ivön, south Sweden (after Lidmar-Bergström, 1989, fig. 3)

Plate 9.1 Etchsurface topography, near Sortelha, eastern Portugal

varied and gently rolling surfaces dominate. High-elevation stripped surfaces with tors and boulder clusters typify granite uplands and mountain ranges in the northern Rocky Mountains (Cunningham, 1969; Eggler *et al.*, 1969) and the evolutionary model proposed is clearly one of etching/stripping (Fig. 9.3). A rolling plain with numerous inselbergs, tors, boulders, and basins stretches in southern California, from the San

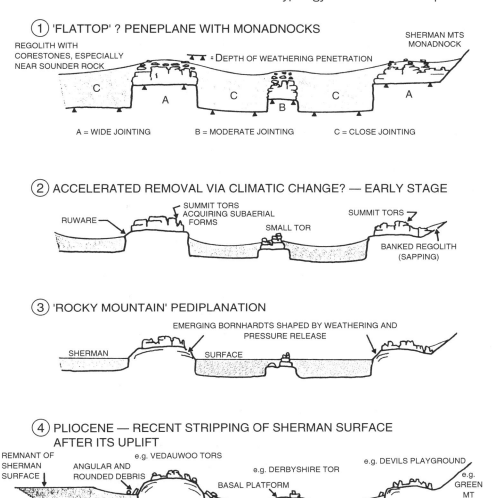

Fig. 9.3 Etchplain development in the Laramie Mountains (after Cunningham, 1969, fig. 5)

Bernardino Mountains east of Los Angeles across the Mojave Desert to the north and east (Oberlander, 1972, 1974). Remnants of the saprolite, from which residual landforms have been excavated, are locally preserved under Miocene lava flows. Finally, an example of stepped topography formed by alternating deep weathering and stripping is offered by the southern part of Sierra Nevada, California (Wahrhaftig, 1965).

Low-latitude etchsurfaces are as varied topographically as their mid-latitude counterparts, although the most common variant appears to be a low relief surface carrying a discontinuous blanket of saprolite, protruded by bedrock hills ranging in morphology from boulder piles to massive bornhardts. A few of these low-latitude landscapes of etched origin have been described in detail. The Koidu etchplain in eastern Sierra

Leone, developed upon Archaean granites, gneiss, and migmatites, has become the type locality for dynamic etchplanation, hence it also offers an insight into how etchplains form today (Thomas and Thorp, 1985). This is a low relief (< 100 m) topographic basin located at an altitude of 350–450 m a.s.l. and surrounded by higher hills and mountain ranges built of schist and granite. Mean annual rainfall is in excess of 2,300 mm, most of it (80%) falling in 6 months. The geomorphology of the basin floor is dominated by interconnected channelless swamps ('bolis') filled with a lag deposit of quartz gravel. Most of the runoff takes place in the subsurface and export of solutes and fine particles occur via this subsurface flow, resulting in ground lowering of both valley floors and interfluves. Over time, most resistant bedrock compartments emerge along drainage lines and on slopes as boulders and tors. Temporal storage of material occurs in headwaters as colluvial aprons and along major valleys in the form of river terraces, but evacuation of these deposits caused by episodic high-energy events allows the process of bedrock lowering to continue. Environmental changes throughout the Quaternary must have contributed significantly to episodic changes in the balance between weathering and stripping, as evidenced by sedimentary characteristics of alluvial deposits (Thomas and Thorp, 1980, 1985). Using the example of Koidu, Thomas and Thorp (1985) list a few characteristic geomorphic features of an active etchplain. These are:

(1) low but otherwise varied relief;
(2) an intricate pattern of valleys which includes flat-floored, saucer-shaped 'bolis', deeply incised (15–30 m) valley heads, and major valleys with permanent streams;
(3) the presence of alluvial and colluvial sedimentary infill of the valleys;
(4) the presence of bedrock channels in large rivers;
(5) glacis-like surfaces linking valley floors and duricrusted hills.

A subsequent detailed study of this area (Teeuw, 1991) highlighted significant differences within the regional Koidu etchplain, dependent on bedrock variability and different drainage histories. Two adjacent drainage basins conform to the characteristics of mantled etchplain and partly stripped etchplain, respectively, and this reflects a relatively recent river capture in the second basin considered which triggered fluvial incision and saprolite stripping. On the other hand, Teeuw (1991) observed that etched topography on granite is dominated by numerous rock exposures, including inselbergs up to 200 m high, whereas on granitic gneiss it is much more planate, with few residual hills and deep and almost continuous weathering mantle.

In South America, an etchplain developed upon Precambrian basement rocks in southern Guyana has been identified by Eden (1971). It is typified by undulating lowland relief (170–200 m a.s.l.), within which bedrock exposures are common.

Many are of whaleback and boulder pile type, but castellated and domed inselbergs are present too. Summits of these inselbergs are located at different altitudes, so that the existence of a former plain is difficult to hypothesize. Shallow saprolitic mantle is widely distributed between the hills. Using this evidence, Eden (1971) considers the topography of the Rupununi Surface as an example of a stripped etchplain, although the term 'partly stripped' is perhaps more appropriate. He also observed that the higher Kaieteur Suface (> 220 m a.s.l.) is still blanketed by weathering mantle and infers that the Rupununi surface has evolved through advanced stripping of saprolite.

Further evidence of the etched nature of geomorphic surfaces in the Guyana Shield has been offered from adjacent Surinam. Bakker (1960) described inselbergs and deep weathering profiles, whereas Kroonenberg and Melitz (1983) emphasized that topographic steps separating different summit levels follow lithological boundaries and their occurrence is consistent with differential bedrock-controlled etching. Likewise, the mountains are best explained by rock resistance to etching. They are of two types: domed inselbergs built of particularly massive granite and lateritized plateaux. Multiple phases of etching and stripping, reflecting major climate alternations from humid to semi-arid, are advocated for this part of South America and regional relief is proposed to increase through time. A lowland erosional surface in the north-eastern part of the Roraima, Venezuela, is described as 'associated with savanna, waterlogged flats, closed lakes and lateritic outcrops; all interspersed with domed granitic inselbergs with boulders' (Schaefer and Dalrymple, 1995: 18). The absence of rock-cut benches around inselbergs is taken as evidence that etching/stripping has been a dominant mode of landscape development. In fairness, diverging opinions about the origin of the Guyana Shield basement landscapes, emphasizing cyclic pediplanation, have to be mentioned (McConnell, 1968; Zonnenveld, 1993), but the weight of evidence clearly favours the etching concept, at least for the timescale of the late Cainozoic.

Surfaces of etched origin are widely developed in the southern part of the Deccan (Büdel, 1977). A fine example of a stripped etchsurface of very varied relief exists near the famous archaeological site of Hampi, in the state of Karnataka (Johansson *et al.*, 2001*b*). It consists of residual hills occurring in two main variants: domed and boulder-strewn, tens of metres high, whose geomorphology in detail is controlled by fracture pattern characteristics. The lower slopes are mantled with boulders and remnants of detached sheeting slabs, while the floors of topographic basins between the hills are commonly protruded by rounded boulder piles.

Granite landscapes of etched origin have also been reported from many parts of Australia (Jutson, 1914; Mabbutt, 1961*a*, 1965; Finkl and Churchward, 1973; Finkl, 1979), although most of these accounts have adopted a broad, regional approach rather than focusing on topographic details or contemporary dynamics.

The Concept—Relevance and Limits

From the above review it appears clear that etchsurfaces supported by granitic rocks are indeed common. They are also highly variable and can be characterized using two parallel approaches. One places an emphasis on relief and may include the landscape type categories introduced earlier. Thus, the Koidu etchplain in Sierra Leone is best described as multi-concave (Thomas, 1980), whereas deep weathering within a tilted block of Sierra Nevada, California, proved crucial in the origin of stepped topography (Wahrhaftig, 1965). Büdel (1977) in turn emphasized the planate geomorphology of the Tamilnad etchplain in south-east India. In another approach, the proportion of saprolite-covered surface versus bedrock surface is important. Thus, a certain area can be subdivided into a finite number of units, whose characteristics would conform to one of the etchsurface types outlined earlier. This has been attempted by Söderman (1985) for the whole of Finland (Fig. 9.4). He distinguished poorly stripped, partly stripped, and dominantly stripped, landscape facets, as well as valley areas with preserved kaolinized saprolites.

A point to consider is whether etching is continuing today, or whether a relict etched morphology is observed. The Koidu etchplain is active (Thomas, 1980; Thomas and Thorp, 1985; Teeuw, 1991), and so is the Appalachian Piedmont (Pavich, 1985, 1989). Gunnell (2000) showed how weathering and stripping intensities vary in relation to each other along a climatic gradient in Karnataka, southern India. Other etchsurfaces, especially in medium and high latitudes, are largely inherited from pre-Quaternary times and over the last million years have been sculpted by processes other than deep weathering. Therefore, geomorphic effects of glacial erosion or mechanical weathering may dominate in mesoscale, but the gross morphology is produced through protracted etching in the more distant past. A study of palaeosurfaces in southern Sweden (Johansson et al., 2001b) provides excellent examples of superimposition of glacial landforms on the pre-existing, etched morphology.

Although differential weathering is ubiquitous and occurs in all possible relief types and climatic zones, to see *all* landscapes as variants of etched topography would certainly be a step too far if the concept is to remain meaningful. First, deep weathering is a necessary contributor to relief evolution and if it does not operate, and has not operated in the relevant geological past, then the concept cannot be applied. Many, though by no means all, contemporary arid landscapes lack evidence of past deep weathering and their genetic interpretation must be different. Likewise, the geomorphology of heavily glaciated mountains is incompatible with etched landscapes. Second, the residence times of saprolites appear to be a crucial discriminating factor. If local relief is high and processes of mass wasting on slopes are active, then the residence time is reduced to a minimum and weathering mantle in slope settings is very thin and

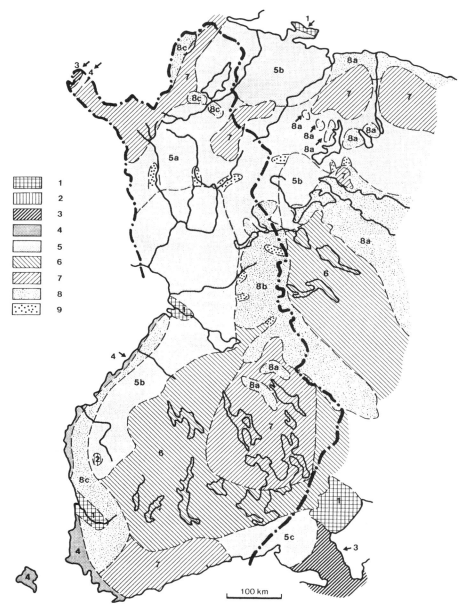

Fig. 9.4 Spatial distribution of different types of etchplains in Finland (after Söderman, 1985, fig. 2): (1) sub-Jotnian stripped surface, (2) sub-Eocambrian surface, (3) stripped surfaces covered by Palaeozoic sediments, (4) exhumed Precambrian surface, (5) poorly stripped Pliocene etchplain (a—with remnants of earlier meso-Cainozoic etched surfaces, b—with remnants of earlier Cainozoic etched surfaces, c—with remnants of Pleistocene etched surfaces), (6) partly stripped Pliocene etchplain, (7) dominantly stripped Pliocene etchplain, (8) stripped Pliocene etchplain (a—stripping level below weathering front, b—with remnants of older etched surfaces, c—stripping level approximately at weathering front), (9) valleys with preserved kaolinitic saprolites

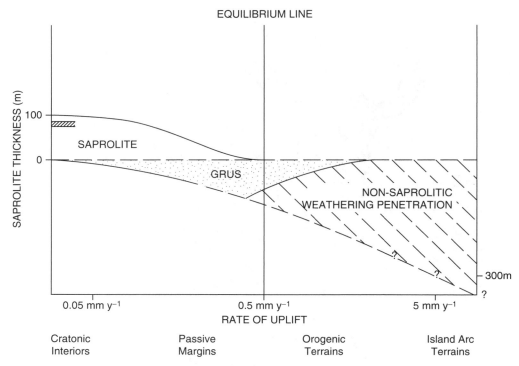

Fig. 9.5 Model relating weathering types and products to the rate of uplift in different tectonic settings (after Thomas, 1997, fig. 1)

unstable. For recent mountain belts in collision settings a geomorphic steady-state is often inferred (Willett and Brandon, 2002), with the rate of descent of the weathering front roughly balanced by the rate of slope lowering and the rate of fluvial incision. Their all-slopes topography does not fit any of the categories of etched landscapes. Consideration of rates of uplift and denudation places constraints on the applicability of the etching model. If they exceed $0.5 \text{ mm} \cdot \text{yr}^{-1}$, weathering tends to be non-saprolitic (Thomas, 1997; Fig. 9.5) and steady-state dissected landscapes develop. On the other hand, the etching concept is highly relevant to shields, most passive margin areas, and rejuvenated mountain massifs. Among the latter, relict etchplains may sit on water divides, whereas active etching would typify their less elevated parts.

Characteristic Granite Landscapes

Given the limitations in the application of the etching concept and the previously highlighted deficiencies of a climatic approach, a morphological classification of granite landscapes seems most appropriate. The co-existence of landscape facets of different origins and different lifetimes, both often uncertain and poorly constrained, further

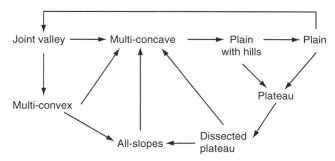

Fig. 9.6 Tentative pathways of granite landscapes evolution. Stepped topography is omitted in the diagram

explains why preference is given to simple relief description as a unified framework for any further analyses. Nine principal types of regional granite landscapes are distinguished and Figure 9.6 shows tentative pathways of their evolution. The relevant spatial scale is from several tens of square kilometres up or, referring to Godard's (1977) typology, his first and second order landforms. However, no attempt is made to propose any numerical values pertinent to relative relief or slope, which theoretically might help to separate one category from another, as these would be inevitably arbitrary.

Plains

As Twidale (1982: 186) remarked, 'Despite the understandable interest evinced in the positive relief features developed on granite ..., plains are, measured by their areal extent, by far the most characteristic landform developed on granitic bedrock'.

Plains are, simply speaking, extensive monotonous surfaces of very low to non-existent relief, located close to the regional base level, with drainage lines in the level of the plain. The gradient of plains may be significantly less than 1°, but even this visually imperceptible inclination of the surface would result in a few hundred metres of altitude difference over tens of kilometres. Plains of large extent tend to occur in shield settings, where they are cut across Precambrian granites and other basement rocks. Examples include the Namib Desert, Namaqualand in South Africa and the north-west Saharan Desert in Africa, the Yilgarn craton in south-west Australia, the sub-Cambrian peneplain in Fennoscandia, and the palaeoplain of Arctic Canada. The absence of residual hills (inselbergs) over large tracts of these plains is particularly striking.

Plains can be subdivided into two main variants: rock-cut and weathered. In the latter case, the topography of the weathering front may be quite variable and stripping of the saprolite would reveal a landscape of some relief, possibly hilly, of the *Grundhöckerrelief* type envisaged by Büdel (1957; Fig. 9.7). The examples named above are mostly rock-cut, although a mantle of weathered rock is present on the Yilgarn craton. Weathered plains are abundant in more humid environments, such as the Guyana Shield and the Deccan.

308　Typology of Granite Landscapes

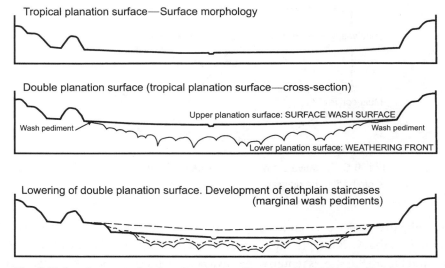

Fig. 9.7 Development of etchsurfaces according to Büdel (1957, fig. 5)

Rock-cut plains may be of different origin. An etched origin is possible, although explanation is required as to why deep weathering did not act selectively, as it usually does. Therefore, an epigene origin is more likely and arid conditions may be particularly favourable for plain development. In the hyperarid Namib Desert there is abundant evidence of a multitude of superficial weathering processes working towards the reduction of any more conspicuous rock outcrops (Ollier, 1978b; Goudie and Migoń, 1997). The plains of the Namib are considered to be actively forming today (Plate 9.2). Many rock-cut plains exist in the geological record as regional unconformities. Over large areas they have been exhumed from beneath sedimentary covers in the Cainozoic and are relict features. Most of the plains on the shields of Europe, Greenland, and North America are exhumed, as are the plains in the Middle East and northern Africa. Northern plains may carry evidence of glacial reshaping, such as polishing and striations, but their gross morphology is inherited from the very distant past. Timescales of plain formation are unknown, but a few tens of millions of years are probably required to arrive at the stage of a featureless landscape cut across structures. Plains need to be viewed as products of protracted denudation under generally stable base level conditions.

Plains with Residual Hills

The monotony of many plains is interrupted by the occurrence of residual hills. These may be dramatically high and visible from a long distance, fulfilling the criteria for an

Plate 9.2 Plains cut across the Salem Granite, Central Namib Desert, Namibia

inselberg, or low and subdued, akin to the African *ruware*. Again, shield settings abound in plains with residual hills, although such landscape facets of limited extent (< 100 km^2) may be present within rejuvenated ancient mountain belts.

Perhaps the best and most widely known examples of inselberg landscapes come from Africa. They have been reported from the Namib Desert, the coastal strip of Angola, parts of South Africa and Zimbabwe, the savanna plains of Tanzania and Kenya, as well as from various localities in west Africa (Plate 8.10). Inselbergs rising above extensive plains typify the Eyre Peninsula in South Australia and basement terrain in south-west and north-west Australia. Further examples are known from southern Deccan, where inselbergs are common within the Tamilnad etchplain. The occurrence of this type of granite relief is not limited to warm environments. Mid-latitude analogues are known from the Asian steppes of Mongolia and Kazakhstan, the uplands of the Bohemian Massif, and the Fennoscandian Shield. The South Småland Peneplain in southern Sweden owes its distinction to the presence of scattered residual hills, non-existent within the sub-Cambrian peneplain.

The origin of plains with inselbergs has been controversial. King (1953, 1962) and his followers argued that these plains are pediplains, formed through long-term scarp retreat and pedimentation, whereas the survival of inselbergs is mainly related to their large distance from lines of dissection. They may be associated with more resistant rocks (King, 1975), but this is not a necessary prerequisite. Clustering of inselbergs in front of escarpment has been cited as an argument in favour of the hillslope retreat hypothesis, but such a distribution pattern is also compatible with differential etching in the zones of crustal upwarp (Bremer, 1971, 1993; Büdel, 1977). According to an alternative model, inselbergs dotting the plains are built of rock which is more resistant by virtue of either its mineralogical composition or, more commonly, its wider fracture spacing. It is significant that the majority of inselbergs are massive bornhardts or poorly jointed castellated forms (Plates V, VII). The plains between the inselbergs are etched surfaces and typically carry a layer of weathered bedrock at least a few metres thick.

Like their inselberg-free equivalents, plains with hills are products of long-term landform evolution and have considerable lifetimes. Their rates of lowering, determined in Australia by means of cosmogenic isotopes, are very low, locally less than 1 m per one million years (Bierman and Caffee, 2002). They may occur in exhumed variants too. East African plains with inselbergs are essentially facets of Cretaceous landscape, revealed again after their Cretaceous cover has been stripped away.

Multi-Convex Topography

The essential characteristic of the multi-convex topography is the occurrence of closely spaced and irregularly distributed hills, convex in outline and circular or oval in plan, with very little intervening flat land. Surface drainage is accomplished by streams winding between the hilly compartments, although many concave landscape facets are channel-less and drained in the subsurface. Thomas (1974) distinguishes two morphological variants. One is the multi-convex relief with weathered hilly compartments, while the other has rock-cored convex compartments.

Weathered multi-convex relief is a widespread type of landscape in the humid and seasonally humid areas, and is not restricted to granite rocks. It is known in Portuguese as *meias laranjas* (meaning half-orange) and *demi-orange* in French. Thomas (1994b) cites numerous examples, from south-east Brazil, Sierra Leone, Cameroon, and Hong Kong. Godard (1977) mentions further occurrences in the Ivory Coast, Brazil, and Guyana. The hilly relief of the granitic parts of south-east China may also be mentioned in this context. The relative relief within weathered multi-convex relief is generally small, hardly exceeding 100 m, and typical slope inclination is 20–25°. Many hills are weathered throughout (Plate 9.3), but a solid rock core may exist at depth. Isolated boulders and boulder piles may be visible on hillslopes, especially in their lower parts, but very rarely on the summits. Landslides and gullying are the main geomorphic processes shaping this kind of landscape, hence many hills have shallow hollows and gully networks incised into their slopes.

Rock-built compartments typify the other variant of multi-convex relief, although these usually occur alongside weathered hills. Hurault (1963) described a landscape of this sort from Guyana, whereas the marginal parts of Serra do Mar in Brazil are another example. Relative relief is higher, due to the higher elevation of rock domes.

Plate 9.3 Multi-convex topography in weathered granite, Cameroon (Photo courtesy of Michael F. Thomas)

Thomas (1974) identifies 'dome and cleft' terrain as a subtype, in which massive bornhardts are divided along fracture lines by deep clefts used by drainage. Because of clear fracture control, such a landscape is described here as a variant of joint-valley topography (see below), but surely transitional cases exist.

Multi-convex relief appears to originate through dissection and differential lowering of an upland or an escarpment, therefore it is characteristic of passive margins and continental interior areas subject to large-radius epeirogenic uplift. It is maintained through the constant renewal of the saprolitic cover. Interestingly, no landscapes of this kind have been reported from high latitudes, which is probably related to the subordinate role of deep weathering in their contemporary geomorphic systems.

Multi-Concave Topography

In contrast to the multi-convex terrain, dominated by hilly compartments, topographic basins assume the status of a characteristic landform in the multi-concave topography (Plate 9.4). The variability within this type of relief is considerable. They may be of different sizes, from much less than 1 km² to 10–100 km², and nested basins may occur. Some basins may stand in isolation and these are drained via narrow breaches in their rocky rims, whereas others are interconnected and form spatially complex patterns. Flat basin floors are often marshy and poorly drained. Elongated channelless alluviated variants are widespread in the tropics and are known under a variety of local names,

Plate 9.4 Multi-concave topography, Serra da Estrela, Portugal

such as *dambo* in central Africa or *bolis* in Sierra Leone (Mäckel, 1985; Meadows, 1985; Thomas and Thorp, 1985; Boast, 1990). Fully enclosed basins, resembling karst dolines, have been reported from the humid tropics, including Colombia (Feininger, 1969) and Brazil (Thomas, 1994b). By contrast, the 'knock-and-lochan' landscape is a distinctive high-latitude variant, present in formerly glaciated terrains of Scotland, Greenland, and Canada. It is thought to be the product of areal scouring by ice sheets, although a possibility exists that it may be derived from an exposed weathering front and only slightly reshaped (Thomas, 1995). Multi-concave topography has worldwide distribution and has been reported from the tropics, middle and high latitudes alike. French geomorphologists were particularly concerned with basins, which have long been a rather neglected topic, overshadowed by tors and inselbergs. They described numerous examples from African and European localities (Grelou-Orsini and Petit, 1985). Upland terrains in the Massif Central abound in topographic basins and there is a transition between multi-concave topography and plateau relief.

How and why the multi-concave topography evolves is not always clear and the reasons for its development are likely to be complex. Basins are surely products of selective bedrock weathering which exploits fractures and mineralogical differences (Thorp, 1967a; Bremer, 1975; Johansson *et al.*, 2001a). However, over time, they may extend into adjacent higher ground due to enhanced scarp-foot weathering and the primary structural control becomes obscure. Further development of concave landforms in granitic terrains may be facilitated by the contrasting behaviour of granite in the presence and absence of moisture. Once relief differentiation into upland and basin compartments takes place, the former become areas of runoff and moisture deficit, and hence are more durable. Several pathways of long-term basin development leading to their coalescence have been hypothesized by Johansson *et al.* (2001a) in respect of Bohuslän, south-west Sweden.

The occurrence of flat-floored, undissected basins indicates suppression of fluvial incision and, probably, a longer period of relative stability of the local base level. By contrast, uplift and associated increase in relative relief would tend to enhance fluvial erosion and rejuvenate weathering systems, leading to the development of multi-convex topography. Indeed, African examples tend to come from continental interiors, whereas nearer the coasts multi-convex relief becomes more common. Multi-concave relief may, and does, exist at higher altitudes, for instance in many places in the French Massif Central (Veyret, 1978; Etlicher, 1983), but only as long as headward erosion has not reached the area.

Plateaux

Plateaux differ from plains in that they occupy a different position in respect of the regional base level. In brief, they are elevated surfaces of low relief separated from their surroundings by steep marginal slopes. These may be of tectonic origin, as is in the Serra da Estrela in Portugal, Margeride Plateau in the French Massif Central, or in the Rocky Mountains, and attain the height of many hundreds of metres. In other cases uplands slope down via less steep denudational escarpments, related to the apparently higher resistance of granite in relation to the country rock. The granite plateaux of south-west England belong to this category. Plateaux are repetitive landforms in western and central Europe as well as in North America, but they appear not to be widely represented at low latitudes. This is probably related to two factors. First, European plateaux are clearly associated with rejuvenated ancient mountain belts, which are notably absent in Africa and most of Australia. Second, apparently more efficient deep weathering in warm and humid low latitudes tends to transform areas subjected to uplift into a multi-convex type of relief. It is thus possible that low-latitude multi-convex landscapes and high-latitude plateaux are equivalent landscape types, the development of each being controlled by the characteristics of the weathering system.

Plateaux usually have some relief and are hardly ever flat, as are base-levelled plains. Some are gently rolling and the dominant landscape facet is a long planar slope connecting flattened summits and floors of shallow troughs. Much of Dartmoor and Bodmin Moor presents this type of geomorphology (Plate 8.2). Others, like the Aubrac Plateau in the French Massif Central, are typified by shallow basins occupied by marshy ground (Veyret, 1978). In turn, the relief of the Serra da Estrela plateau is considerably rough in detail, especially in coarse granite variants (Plate VI). Tors are common on many plateaux and may occur both on summits and in less elevated parts, separating minor topographic basins. The parkland topography of the Laramie Range in the Rocky Mountains serves as an example (Cunningham, 1969). Altitude differences within the plateaux may be inherited, if there has been a geologically young uplift involved. Alternatively, they may testify to ongoing differential surface lowering, favoured by the elevated position of a plateau and the resultant combination of an increased topographic and hydraulic gradient, free drainage, and joint dilatation.

Many plateaux in middle and high latitudes bear the clear imprint of Pleistocene cold-climate conditions and possess a suite of periglacial landforms superimposed on older, mainly structure-controlled landforms. These include rock cliffs, block fields and block streams, patterned ground, solifluction sheets, and dry valleys. Representative suites of periglacial landforms on plateaux have been described, among others, by Gerrard (1988a) and Campbell *et al.* (1998) in respect to Dartmoor, Coque-Delhuille (1979) for the Margeride Plateau in France and Hövermann (1953) for the Harz Mountains in Germany. Ice caps may have existed in sufficiently elevated plateau sectors (Veyret, 1978).

Dissected Plateaux

Dissected plateaux are a variant of the above category. Their prominent geomorphic feature is a close juxtaposition of watershed surfaces of low relief and deeply incised valleys. There may occur a zonation within an upland, with a dissected outer zone and a more compact inner zone, not yet reached by headward erosion. The dissected plateau belt may, in turn, give way to all-slopes topography in the marginal parts of the elevated area. Serra do Gêres in northern Portugal shows elements of this tripartite topographical pattern.

The Cairngorms, Scotland, are a high elevated plateau (1,310 m a.s.l.) built of Silurian granite. Protracted deep weathering and stripping in pre-Quaternary times resulted in a subdued topography, with gently rolling watershed surfaces, smooth slopes, wide and shallow trough valleys, abundant tors and pockets of regolith (Sugden, 1968; A. M. Hall, 1996). Late Cainozoic uplift brought the plateau near its present-day altitudinal position, triggering stream incision, which has gradually led to the dissection of the summit surfaces of the Cairngorms. However, dissection became truly spectacular during the Pleistocene, when valley glaciers deepened former fluvial tracts and a few glacial breaches cut through the entire plateau (Plate 8.3). Lairig Ghru is the most striking example of these breaches, incised into the summit surface by more than 400 m. Another example of a dissected granite plateau in the Scottish Highlands is the Lochnagar Massif.

In the American Rockies dissected granite plateaux are also common. Dissection is both fluvial, as along the eastern boundary of the mountains near Boulder, Colorado, and glacial in more elevated parts. Plateau remnants have distinct signs of continuous shaping in cold climatic conditions, including extensive block fields and patterned ground.

Joint-Valley Topography

Joint-valley topography, named after a category introduced by Lidmar-Bergström (1995) into the analysis of regional relief of Sweden, is a specific morphological type, distinguished by the close relationships between fracture patterns and negative relief. There is a recognizable lattice-like pattern of linear concave landforms overprinted on major (master) fractures. These landforms may be fluvial valleys, narrow clefts and gorges, defiles, and highly elongated basins (Fig. 8.7). Larger basins, although less regular in plan, occur at fracture intersections. Intervening higher ground may be either plateau-like, with little relief, or consist of closely juxtaposed domes. There are several salient differences between joint-valley topography and dissected plateaux. First, dissection is fluvial or glacial, whereas fracture exploitation occurs predominantly through selective weathering, with a subordinate role for fluvial erosion. Therefore, many defiles are not occupied by permanent streams and are dry. Second, dissection proceeds from the outer parts of a plateau inward, resulting in a hierarchical pattern of drainage lines, whereas the pattern of fracture-guided valleys does not show a hierarchy. Third, the adjustment of dissection lines to bedrock structures is imperfect, reflecting the key role of regional slope. Fourth, relative relief in joint-valley topography is typically 50–200 m, hence much less than in dissected plateaux. However, joint-valley landscapes may evolve into dissected plateaux, as will be shown later.

The type area for the joint-valley topography, Bohuslän in south-west Sweden, has been presented elsewhere in this book. Another region of similar topography is Blekinge in southern Sweden, where major concave features follow the general north-south trend and are lined by steep rocky cliffs up to 50 m high (Plate 9.5). These are connected by a series of narrow defiles developed along minor fractures trending west-east (Lidmar-Bergström, 1995). There are also low-latitude examples, for instance in South Africa (Mabbutt, 1952) and Sri Lanka (Bremer, 1981b). Pye *et al*. (1984) mention joint and fault guidance of valleys in the Matopos Batholith in southern Zimbabwe and enclose an aerial photograph showing an excellent example of joint-valley landscape. There is a distinct rhomboidal pattern of concave relief features, adjusted to NW-SE, WNW-ESE, and W-E fractures. Massive domes and enclosed shallow basins typify higher ground between the valleys and defiles.

Despite the profound role of fractures in the evolution of granite landscapes, the joint-valley topography as defined above is not a very widespread variant of granite morphology. The probable reason is that it requires a specific combination of efficient, but selective deep weathering acting over a geologically short timescale. This has been proposed by Lidmar-Bergström (1995), after analysing the spatial relationships between different landscape types and cover rock occurrences. Noting that both Blekinge and

Plate 9.5 Steep slopes of Dalhejaberg, Blekinge, southern Sweden, one of the features of etched joint-valley topography

Bohuslän are located close to the limits of the sub-Cambrian peneplain and that a more open landscape occurs further away, she argues that:

The joint valleys are interpreted as the first stage in the transformation of the sub-Cambrian peneplain to an undulating hilly land. There is a sequence from the Blekinge type, over the Bohuslän type to the undulating hilly land of Halland and NE Skåne with increasing relief. The differences in relief amplitude are explained by the weathering intensity and the length of time of exposure between the erosion of the Palaeozoic cover and deposition of the Cretaceous cover. (Lidmar-Bergström, 1995: 52)

Consequently, a longer period of etching would not only deepen fracture-aligned concavities, but also result in their widening, gradual reduction, and fragmentation of intervening upland surfaces. With time, the distinctive character of joint-valley topography would fade away, up to the stage where it is no longer recognizable. Indeed, joint-valley topography may grade into other relief types, notably into multi-convex and multi-concave relief. The Idanre Hills in southern Nigeria exemplify an advanced stage of etching of the fractures and their present-day landscape is dominated by domes and dissected slopes. Nevertheless, tracing structural discontinuities which outline the dome clusters and were likely to be exploited first is still relatively easy (Jeje, 1974; Thomas, 1994b).

There are also granite terrains combining certain characteristics of joint-valley relief with those of other types. The northern part of the Yosemite National Park, Sierra Nevada, California, is an example of a landscape which is transitional between joint-valley, all-slope, and dissected plateau (Ericson et al., 2005). The main tributaries of the trunk stream of Tuolumne River closely follow the regional fracture lines striking SSW-NNE and WSW-ENE, and show straight courses, sharp bends, and confluences at acute angles. However, their depth of incision is up to 500–600 m and there are few other relief features which show clear fracture control. Minor plateau streams debouch from hanging valleys and have long stepped profiles. The intervening areas are either plateau-like, with hills and basins filled by lakes, or steeply sloping ridges. Fluvial incision into the western slope of Sierra Nevada was probably concurrent with the Cainozoic asymmetric uplift of the range and is likely to have accelerated in the last few million years (Wakabayashi and Sawyer, 2001). To keep pace with rapid uplift, eroding rivers focused on existing lines of structural weakness, deepening them while leaving inter-stream areas relatively untouched. The preparatory role of deep fracture-guided etching within the gently rolling surface of the ancient Sierra is unknown; however, the absence of clear fracture-controlled features on the plateau remnants speaks against the existence of a well-defined joint-valley topography.

All-Slopes Topography

The all-slopes topography is characterized by Twidale (1982: 177) in the following way: 'Some granite uplands ... consist of sharp-crested ridges bounded by essentially rectilinear slopes. The latter vary in inclination from area to area but they are characteristically steep. There are no significant areas of flat land either on ridge crests or in valley floors, and on this account such topography is called *all-slopes*'. For the all-slopes landscape type the presence of valleys dissecting the relief appears essential, to further underline the difference with the multi-convex morphology of closely spaced domes. This restriction has implications for the origin of all-slopes relief. Fluvial and glacial erosion are crucial components in its long-term geomorphic evolution.

All-slopes topography is thus intimately associated with mountain regions, although heavily dissected marginal parts of modestly elevated plateaux may also fulfil the criteria. All-slopes relief occurs in many variants, depending on the relative intensity of fluvial and glacial processes, the depth of incision, the presence of regolith on slopes, and mechanical properties of granitic rocks. In the most elevated mountain ranges on Earth, such as the Himalaya, Karakoram, the Andes, the Alps, or the highest parts of the Sierra Nevada, California, steep, almost vertical rock slopes dominate. Their dynamics is very high, with rock falls and rock slides both frequent and voluminous. Crests are serrated, with individual joint-bound compartments sculpted into pinnacles, turrets, and angular towers, separated by deep clefts. Powerful glacial erosion may result in significant slope undercutting and vertical rock walls many hundred metres high then form.

The less elevated but heavily dissected mountain massifs of Scotland and Ireland are best described in terms of all-slopes topography too. A good example is provided by the Isle of Arran, where the Northern Granite area has been shaped into an array of sharp-crested ridges, partly undercut by cirques, steep but generally straight slopes with a thin regolith cover, and glacially moulded U-shaped valleys (Plate 7.2). Further examples from the region include the Red Hills in the Isle of Skye, the north-eastern part of the Mourne Mountains in Northern Ireland, and granite massifs in Donegal. Dissection of these massifs was probably well under way in pre-Pleistocene times, but was additionally augmented by glacial action during the Pleistocene.

Low latitudes abound in all-slopes topography, which has evolved without a contribution from glacial processes. Deeply dissected mountain terrains extend across the Korean Peninsula (Lautensach, 1950; Tanaka and Matsukura, 2001) and the Japanese Alps (Ikeda, 1998; Onda, 2004). Examples have been described from small Pacific islands off the coast of Papua New Guinea (Pain and Ollier, 1981), Puerto Rico (Monroe, 1979), and Sierra Madre del Sur in Mexico. In the latter region, the Veladero Mountain towering above the Acapulco Bay is a highly dissected massif, typified by bouldery crests, deeply incised amphitheatre valley heads, steep slopes in excess of 30°, and narrow valley floors (Lugo *et al.*, 2002; Fig. 9.8). Slopes are underlain by a few metres thick layer of saprolite, usually of geochemically immature grus type. For most of these occurrences a long-term balance between fluvial incision, slope degradation accomplished mainly by shallow mass movement in weathered material, and uplift may be inferred. In other places, such as the Serra do Mar in south-east Brazil, all-slope relief cut in weathered material co-exists with massive domes protruding from watershed ridges and spurs (Plate 8.6). Given the continuous uplift of the area and ongoing deep weathering, a gradual relief increase and eventually a two-storey topography may be hypothesized, with the upper storey of residual domes rising to different heights, and the lower storey of dissected weathered terrain maintaining the steady-state.

All-slopes topography also has its desert representatives, for example the granite mountains of the Sinai Peninsula in Egypt and in south-western Jordan, near Aqaba (Plate 8.9). It is true that fluvial processes today are of marginal importance, but one needs to realize that most of the geomorphology of these mountains is inherited from more humid, pre-Quaternary times.

Stepped Topography

Numerous granite mountain ranges show stepped topography. Its topographic expression is a step-like rise in elevation from the foothills towards the crest, accomplished by the alternating occurrence of steep scarps and gently sloping surfaces in between. The stepped, or benched, appearance of mountain massifs, including those built of granite,

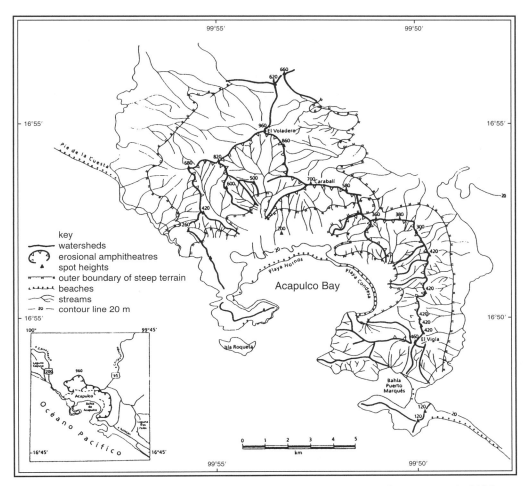

Fig. 9.8 Granite amphitheatre of the Veladero Massif, Acapulco, Mexico (after Lugo et al., 2002, fig. 13.2)

was noted at the beginning of the twentieth century by German geomorphologists, notably Walter Penck (1924), and explained in terms of his somewhat elusive concept of *Piedmonttreppen*. However, this theory was not meant to be specific to granite, nor were other explanations of that time which implied relatively simple stories of uplift and planation, proceeding with variable intensities.

The credit for recognizing stepped topography as an important variant of granite morphology goes to Wahrhaftig (1965), who presented his hypothesis by referring to the southern part of the Sierra Nevada, California. The part of this mountain range between the Yosemite Valley in the north and the Kern River in the south can be resolved into a series of topographic steps separating surfaces of low relief, described as treads. Steps are of variable height, from as little as 30 m to as much as 1,300 m, but heights within the range of 90–600 m are typical. Likewise, the steps vary in steepness, from 5 to 75°, but again, there is a class of predominating values, between 15 and 35°. Steps generally follow a NNW-SSE direction, hence they are parallel to the elongation of the entire range, but show much sinuosity in detail, including projecting points and re-entrants, and may even enclose basin-like features from almost all sides. As many as 11 steps have been mapped in the area studied by Wahrhaftig (Fig. 9.9). The treads are up to 8 km wide and their length to width ratio varies between 1 and 10. Their surfaces are not exactly planar, but may show local relief of the order of tens of metres. Their overall gradient is commonly less than 5°, but interestingly, the inclination is in many places towards the next highest step rather than outward, giving the appearance of back tilting.

The key to the origin of stepped topography is the distribution of bedrock outcrops and weathering products. Step faces abound in granite outcrops, which are either extensive clusters of massive boulders or bare surfaces of sheeting joints, occasionally in the form of half-exposed domes. By contrast, bedrock outcrops are rare on treads and tend to concentrate near their fronts. Further inward, a grus weathering mantle of up to 30 m thick becomes ubiquitous (Plate 2.12). Stream channel morphology shows much sympathy to the outcrop pattern, in that valleys assume a boulder-choked, gorge-like form within steps, whereas they are wide, shallow, and have sandy beds on the treads.

Wahrhaftig suggested that it was the differential weathering of granite and the influence it exerted on streams' ability to erode that was responsible for the origin of stepped topography. He envisaged the evolutionary sequence as follows (pp. 1166–7):

The stepped topography is believed to be caused by differences in the rate of weathering in the two environments to which granitic rocks in the Sierra Nevada are subject. Where buried by overburden or gruss, the solid granitic rocks are moist most of the year, and disintegrate comparatively rapidly to gruss; where exposed, the solid granitic rocks dry after each rain and therefore weather slowly. Small streams, even overland flow, can transport the gruss, but the unweathered rock is jointed into blocks so huge than not even the largest streams can move them. Consequently, any exposure of solid rock is a temporary base level, to which the country upstream or upslope can be leveled, but below which it cannot be reduced. Irregular lines of such outcrops, formed during periods of downcutting or accelerated erosion, grow into escarpments irregular in height and ground plan. These are the step fronts. The areas between them become flattened by weathering and erosion to form the treads of the steps. Thus, the stepped topography is caused by the development and propagation of perturbations in the rate of weathering and erosion. (Wahrhaftig, 1965: 1166–7)

Wahrhaftig dismissed alternative hypotheses, such as those implying faulting, the influence of lithological contacts, or scarp retreat, for which there was no evidence. He also rejected the overwhelming influence of jointing density on the location of steps, arguing for their independence from master joints. However, he did admit that some steps may be controlled by non-uniform fracture density and that direct testing is virtually impossible due to the scarcity of outcrops on treads. The claim that the

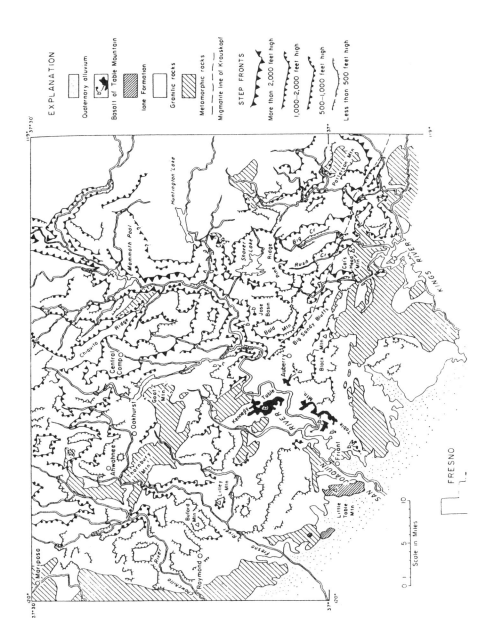

Fig. 9.9 Stepped topography of the Sierra Nevada, California (after Wahrhaftig, 1965, fig. 4)

influence of fractures is subordinate at best is particularly interesting in the context of very clear and prominent fracture control in the topography of the part of granitic Sierra Nevada north of the Yosemite Valley (Ericson et al., 2005).

The above explanation has important implications for the search for palaeosurfaces in higher parts of the Sierra Nevada. Clearly, Wahrhaftig considered the stepped topography to be time- and altitude-independent. It would have begun to evolve after tilting of the range commenced at the Oligocene/Miocene boundary, at the expense of a former hilly topography. Given abundant grus exposures, high relative relief, and current annual precipitation of 500–800 mm, differential weathering and step formation may well be under way today.

Further possible examples of stepped mountainous topography in the western hemisphere include the Peninsular Ranges in southern California, parts of the Front Range in Colorado, and coastal ranges in south-east Brazil, but no detailed studies adopting Wahrhaftig's model have been produced. On the other hand, Thomas (1974) noted similarities between stepped topography described in the Sierra Nevada and relief developed upon the Separation Point granite on South Island, New Zealand.

Stepped topography appears typical of many European mountain and upland areas, but whether they conform genetically to the model introduced by Wahrhaftig, or are down-faulted blocks, is not clear everywhere. For example, in the Cairngorms there are fewer elevated surfaces present below the summit level (Sugden, 1968; Ringrose and Migoń, 1997), but a geomorphic interpretation is yet to be advanced. A further problem with this category of granite landscape is that individual steps may show their own relief that fulfils the criteria for multi-concave or perhaps multi-convex relief. In the Serra da Estrela, Portugal, the summit plateau appears to descend via a series of steps on which, depending on lithology, different relief types occur (Plate VI).

Rock versus Climatic Control

In the history of geomorphology diverging views have been expressed about the relative weight of two principal factors controlling the evolution of erosional landscapes, that is bedrock properties and climatic conditions. Since the advent of the 'quantitative revolution' in geomorphology in the early 1960s, the focus on process-oriented research and, inevitably, on short timescales has generally turned attention away from material properties. Contributions to 'Process geomorphology' have often been strikingly enigmatic about the characteristics of the rock which supports a particular slope section or valley reach. In German geomorphology, the emphasis on environmental conditions as a controlling variable is even more deeply rooted and has evolved into a distinctive climato-genetic approach to landform analysis (Bremer, 1989, 2004), epitomized through the concept of climate-dependent relief generations expounded by Büdel (1977). How the climatic paradigm has influenced inquiries into granite geomorphology can be observed while consulting two excellent monographs from the late 1950s, one regional, focused on Corsica (Klaer, 1956), another truly global in scope (Wilhelmy, 1958). In each the issues of climate-driven processes, vertical zonation of process domains, and inheritance are much more widely discussed than the variability of the rock. Godard (1977) has graphically shown contrasting type-examples of granite terrains, attaching them to the presentation of granite landscapes in different climatic zones. In effect, an impression may have arisen that the geomorphic variability of granite terrains is primarily controlled by climatic conditions and that specific climate–landform relationships can be identified.

On the other hand, Thomas (1976) warned that criteria for recognizing climatically induced landforms are often uncertain, whereas Twidale (1982) provided examples of different individual granite landforms, from major to minor, coming from almost all climatic zones of the world. More explicitly, strong structural control has recently been made evident by French geomorphologists (Godard et al., 2001).

Are granite landscapes rock- or climate-controlled? First, the very way of formulating the question is flawed, as it implies that one of these variables might be immaterial. This is not true. Each landform, whether large or small, is a product of a process (or processes) acting on a geological substrate. Second, this question ignores other variables which are no less important, although less exposed. These are the timescales of landform development and the geomorphic stability of rock surfaces. Many examples presented so far were chosen to demonstrate their utmost significance. In particular, Lidmar-Bergström (1995) has shown that the prime control on the evolution of etched granite topography in the Mesozoic was the duration of bedrock exposure. In turn, granite landscapes in recent mobile belts, where rates of uplift and denudation are both very high, look very different from their shield counterparts, where the respective rates are very low and have probably been so over millions of years. But can we attribute the geomophological variability within granite areas to one of these general factors more than to the others?

It is clear that there is a whole range of granite landforms, at all possible spatial scales, which originate due to certain structural and lithological predispositions. Furthermore, many of them have their outlines in all three dimensions controlled by, and adjusted to, structures existing in the rock, chiefly to fracture planes. The following representative examples may be cited. Plains with residual hills often signify the juxtaposition of two granite variants, differing in resistance to weathering and erosion. In the Central Namib Desert the gravel plains are cut across the fractured Precambrian Salem Granite, whereas impressive inselbergs rising above them are built of very poorly jointed, locally extremely massive, potassium-rich Mesozoic granites. Environmental factors can be discounted as an explanation of the form variability among residual hills. Domed inselbergs require massive rock compartments with well-developed sheeting, while castellated tors form in orthogonally fractured granite, and both types are known to co-exist in different climatic zones, from the humid tropics to hyperarid deserts. Slope form in detail can usually be explained by applying Selby's (1980) rock mass strength approach, which places greatest emphasis on fracture pattern characteristics, thus a structural feature. Specific types of granite intrusions, or characteristic zonal patterns therein, support specific landform assemblages, of which arcuate ridges related to ring dykes are the most striking.

Likewise, direct rock control on the course of geomorphic processes can be demonstrated. It is most evident in the progress of deep weathering, corestone development, and descent of weathering front and there is usually good correlation between rock characteristics and properties of weathering mantles, unless the duration of profile deepening has been long enough to erase bedrock variability in its upper part. Patterns of mass movements show the influence of bedrock structures too. Fracture planes act as slip planes in both fresh rock and in saprolite, whereas fracture density and rock porosity control groundwater circulation and are decisive for the distribution of landslides and debris flows. Further examples may be cited from the domains of glacial and coastal processes.

It is also interesting to note that even if patterns of deep weathering and characteristics of saprolites are different and related to climatic variability, stripping of saprolite may reveal landforms which are strikingly similar. Thus, corestones may form and exist in both grus mantles in high latitudes and clayey saprolites in the humid tropics, and therefore boulder fields and streams resultant from saprolite removal are known from both humid tropical Puerto Rico and Hong Kong and cool temperate northern Europe. A similar argument can be advanced for other two-stage forms, such as flared slopes.

On the other hand, the influence of climatic conditions on the formation of certain granite landforms cannot be denied. Leaving aside the obvious example of glacial landforms, which are in fact not distinctively granitic, there is still a group of medium-sized features apparently developed under specific environmental conditions. This includes angular block fields produced by mechanical weathering in cold climates, stone stripes and other patterned ground phenomena indicative of frost sorting, and pseudokarstic closed depressions in the humid tropics. Further, there is a category of landforms which are not exclusively climate-dependent, but nevertheless develop preferentially in certain climatic conditions. Thus, pediments defined as rock-cut surfaces with a thin sedimentary veneer are features of arid and semi-arid regions rather than humid tropics. Karren are indeed underrepresented outside humid regions, as are tafoni in these humid regions.

But the most evident climatic control is on the rates and intensity of deep weathering (Thomas, 1994a; Oliva et al., 2003), as well as on the frequency of landslides removing products of deep weathering (Thomas, 1994b, 1997). Among climatic parameters rainfall appears the most important, especially the annual total and distribution throughout the year. Most catastrophic events in humid low latitudes which transform the landscape dramatically in a mere few hours or days are associated with abnormally high rainfall episodes. As far as weathering profiles are concerned, the origin of a tropical saprolitic mantle in excess of 50 m thick appears possible during 1 million years, a figure difficult to match in cool temperate lands. The process of weathering penetration into the rock is probably self-limiting, hence similar thicknesses may be achieved if there is ample time available and little surface lowering. But given the history of late Cainozoic environmental change and tectonics, it is unrealistic to assume surface stability over sufficiently long (> 5 Ma or so) timescales, except for extremely planate shields. Moreover, the thickness of the saprolite appears inversely proportional to the probability of its total removal and exposure of the solid rock beneath. With these constraints in mind we may argue that climate becomes a key variable in the context of saprolite renewal after stripping, which is much more rapid in the humid low latitudes than anywhere else. Constant renewal of the weathering mantle would help to maintain characteristic low-latitude landscapes, whereas complete stripping and exposure of the weathering front might significantly slow down further landscape evolution.

The importance of the latter factor may be considerable for one of the very characteristic granite landscapes, the multi-convex relief with weathered compartments. It occurs in many parts of the tropics and subtropics, but it is difficult to find comparable examples in middle and high latitudes. In uplifted and dissected zones such as south-east China and Serra do Mar, Brazil, constant renewal of weathering profiles to compensate for mass loss due to landsliding and gullying emerges as a key component of the geomorphic system of multi-convex relief, crucial to maintaining regional topography. In this context it is worth noting that exhumed mid-Miocene relief along the south-eastern margin of the Bohemian Massif, Czech Republic, and Austria, shows similarities to the multi-convex relief of the present-day warm and humid, low-latitude environments (Ivan and Kirchner, 1994; Roštinský, 2004).

To conclude this discussion, the following points need to be highlighted:

1. The geomorphic evolution of granite terrains is controlled by a multitude of factors and they all need to be simultaneously considered if these landscapes are to be properly explained in their wider environmental, geographical, and geotectonic context. The unique appearance of many granite landscapes and the specific development of individual landforms at individual sites usually results from a unique combination of these controlling factors rather than from a single unique cause.

2. Notwithstanding the above, geological control *sensu lato* assumes major importance in granite landscapes. Consideration of geotectonic setting and specific rock characteristics not only helps to explain the distinctiveness of granite morphology if compared with other common rock types, but also geomorphic variability within the granite areas. Moreover, certain landforms, for instance domed inselbergs, would not

have developed in the absence of structural predisposition. Large massive compartments are required, whatever the climate. Geological control is hierarchical and scale-dependent.

3. The temporal context of relief development is important. Many granite landforms, even more so their regional assemblages, are relatively slow to evolve because of the strength and resistance of the rock. Before conclusions about the inability of certain climates to support certain landforms are reached, timescales need to be considered. In particular, the zone of temperate climate in the mid-latitudes of the northern hemisphere has only been established in the latest Cainozoic, hence it is perhaps premature to argue about granite landform assemblages in equilibrium with this environment.

4. Granite surfaces, once exposed, are very durable. By inference, rock-cut landforms such as inselbergs and domes, often built of massive rock compartments, are very long-lived. In shield areas of Africa and Australia their mid-Cainozoic form may not have been very much different from that observed today. In specific circumstances even more distant ages are envisaged. Burial and exhumation may bring landforms of widely different ages and origins into close juxtaposition. Thus, granite landscapes are expected to have a significant inherited component and if this is not properly recognized, spurious conclusions can be made about some alleged climate–landform relationships.

5. In terms of formative processes, granite landforms are good examples of equifinality. Likewise, different evolutionary pathways are possible for larger landform assemblages. The form, especially of tors and inselbergs, is a poor and unreliable key to the process that shaped the landform in the past. Likewise, the relationships between climate and weathering products are not as simple as was once thought. Palaeoenvironmental reconstructions, if based solely on erosional landforms and the visual characteristics of saprolites and not supported by independent evidence from palaeontology and isotope geochemistry, should be avoided.

10
Granite Landscapes Transformed

An analysis of granite landscapes would not be complete if the modifying human factor were ignored (Godard, 1977). Over the millennia humans have used the resources provided by granite, whether in a solid or weathered state, taken advantage of the spatial configuration of granite landforms, or mimicked natural granite features for various purposes. The combination of rock outcrops, regolith-mantled surfaces, and soil characteristics has acted as a significant constraint on many human activities, especially in the past. Hence many granite areas have very specific histories of human impact. The monumentality of many granite landforms has inspired fear, awe, and spiritual experience, whereas in the modern era the distinctiveness of many granite terrains has become a magnet for tourism. Each of these activities has left its imprint on granite landscapes, to the extent that some of them easily fall into the category of 'cultural landscapes', while in others, man-made features have evidently overwhelmed the natural configuration of the land.

In this closing chapter of the book a few aspects of human transformation on natural granite landscapes will be briefly addressed. The coverage, and particularly the selection, of examples are by no means exhaustive, and the historical context has not been explored. The intention is rather to review some of the most characteristic relationships between humans and granite landscapes and to show that the characteristics of natural granite landforms dictate very specific adjustments, uses, and strategies of landscape change. Therefore, extensive referencing has also been avoided.

Megalithic Granite Landscapes

The middle and late Neolithic in western Europe (3500–1700 BC) was a period of extraordinary construction activity using local and imported stone. It was not limited to granite lands, but the availability of durable monumental stone was certainly important. Therefore, uplands and rolling plains underlain by granitoid rocks abound in a variety of megalithic structures, including standing stones, stone circles and rows, passage tombs, simple dolmens, burial mounds (cairns), and stone enclosures. Extensive assemblages of Neolithic monuments occur on the Alentejo plain in southern Portugal, in western Spain, in Brittany, France, and on the uplands of south-west England, from Dartmoor through Bodmin Moor, Carnmenellis to Land's End. Ireland also has its collection of ancient granite monuments, including a huge megalithic tomb called Browne's Hill Dolmen, near Carlow. It is made of an enormous granite capstone, the mass of which is estimated to be over 100 tonnes and the length is about 8 m, supported by a few smaller blocks.

Among these areas, Brittany assumes prime importance. It hosts thousands of documented megalithic sites and tens of thousands of individual granite boulders must have been excavated, transported, and erected to form them, but it is undoubtedly the area around the little village of Carnac, on the southern coast of Brittany, where their presence has transformed the entire landscape. According to Michell (1982: 64): 'Around Carnac... is the most amazing collection of megalithic monuments to be found anywhere in the world'. Among them the most impressive and most intriguing are the famous stone alignments which stretch for several kilometres. The biggest and the best preserved Menec alignment consist of hundreds of massive granite boulders arranged in twelve parallel rows trending south-west to north-east (Plate 10.1). The largest ones stand at the western end of the structure and are as much as 5 m high, whereas to the north-east their height diminishes to less than 1 m. The entire structure may seem to survive unscathed from antiquity but in fact many boulders seen standing today were re-erected in the nineteenth century. As with many other megalithic structures, the purpose of building the alignments of Carnac remains a mystery, although the majority of scientific authorities agree that they served as places of cult and astronomical observations. The occurrence of several minor or less well-preserved stone alignments, numerous dolmens, and mounds nearby adds to the distinctive human flavour of the gently rolling granite landscape around Carnac.

Plate 10.1 Neolithic stone alignments in Carnac, Brittany, north-west France (Photo courtesy of Alfred Jahn)

The standing stones of Carnac are impressive due to their congested occurrence, but in respect of their size they are easily surpassed by a number of individual structures dispersed over the Breton countryside. Some of these isolated stones, otherwise known as *menhirs* (from the Celtic), exceed 6 m high and the largest ones are more than 10 m high. Had it not fallen down either at some time in antiquity or at the turn of the seventeenth century, or had it not been deliberately broken as others believe, the *grand menhir brisé* at Locmariaquer would be the largest granite standing stone ever known as it was 20.5 m high. Despite its fracture into four pieces it remains an astounding example of purposeful incorporation of natural granite boulders into the human-transformed countryside. It is even more astounding to realize that it was not made of a native stone but had to be brought from a quarry located tens of kilometres away. Without doubt it was the physical properties of granite and its relatively widely spaced jointing that allowed the ancient Bretonese to erect this and many other monuments across their land.

Smaller granite boulders and slabs were used to build less grandiose structures such as dolmens, covered alleys, and stone circles. It seems almost certain that there must have been more of these in the past. Many were destroyed as the stones were re-used for building purposes, others may have fallen victim to developing agriculture, and yet others could have been demolished because of their pagan origin. Some standing stones, however, have been 'christianized' (Plate 10.2) and provide interesting examples

Plate 10.2 Standing stone (menhir) in St Uzec, Brittany, north-west France. Long flutings have developed on the right side of the stone in the last 5,000 years

of the transformation of the cultural landscape brought about by religious changes. A few thousand years of exposure of megaliths to atmospheric agents have left their legacy too. Rillenkarren and incipient alveolar weathering have affected some of the structures, including the aforementioned cross-crowned menhir at St Uzec (Lageat *et al.*, 1994; Sellier, 1998).

Linkages between natural granite upland landscapes and Neolithic communities were at least twofold. From the practical point of view, abundant granite outcrops provided raw building material, which had many advantages from the point of view of megalithic builders. Preferential weathering along fractures contributed to easier quarrying of rock blocks, whereas former corestones and components of Pleistocene block fields were already ready for transport.

On the other hand, natural granite outcrops of sheer size and unusual shape may have been the source of fear, awe, inspiration, and power for prehistoric people living in close proximity. This issue has recently been examined with respect to the granite upland of Bodmin Moor, famous for both numerous tors (see Chapter 4) and abundant megalithic structures (Tilley, 1996). The early and middle Neolithic (*c*.3500–2300 BC) has left a legacy of long tombs located at a certain distance below prominent tors and pointing towards them; a pattern interpreted as evidence of a symbolic relationship based on veneration and respect. By contrast, the abundant tombs of the late Neolithic and Bronze Age (*c*.2300–500 BC) occur much closer to the tors, occasionally incorporating natural outcrops of granite as part of their structures or simply standing on them. Tilley argues that this documents a change in the perception of the landscape in Neolithic societies. The newly constructed tor cairns were meant to resemble the tors, replace them, and assume their spiritual role in a 'cultural triumph over the sleeping powers of the rock' (1996: 175). Similarly, Bradley (1998) argues that the specific shapes of tors in south-west England have probably provided inspiration for tomb and enclosure builders to choose particular designs. However, he also speculates that tors were not necessarily regarded as natural features possessing power to fear or capture, but may have been regarded as ruined tombs left by past generations. In line with this interpretation, large hilltop enclosures incorporating many tors and boulders, such as on Carn Brea in the Carnmenellis Massif, may have been constructed as special monuments to the ancestors.

Origin of Moorlands

Dartmoor and Bodmin Moor in south-west England are for many the typical granite landscape: open, treeless, grassy, with abundant rock outcrops including the famous tors and extensive block and boulder fields, locally called clitter (Plate 10.3). Blanket peats and poor, difficult-to-use gley and podzolic soils, add to the impression of little human interference. In fact, the granite uplands of south-west England have a long history of human impact and their present-day landscape contains few natural ingredients other than tors and boulders. The shaping of this land began in Neolithic times, more than 5,000 years ago, but accelerated during the Bronze Age, from 4000 BP onwards (Bell and Walker, 1992). Neolithic and Bronze Age stone monuments may appear to be the main witnesses to human presence, but it is the land cover that is the most impressive legacy of the human impact.

Human transformation included, in the first instance, widespread forest clearance and the introduction of grazing and subordinately agriculture. The long, gentle slopes of the uplands allowed for easy extension of agricultural land, although widespread clitter must have been an obstruction, especially at higher elevations. Woodland may only have survived in deeply incised valleys in the marginal parts of the Dartmoor

332 Granite Landscapes Transformed

Plate 10.3 Bare granite upland of Dartmoor bears a long history of human impact and lost its forests as early as the Bronze Age

upland, such as the Webburn and Dart valleys. Bronze Age settlements soon covered much of the granite areas and their expansion involved the construction of stone boundary walls (reaves) between 3300 and 3100 BP, which now provide an unmistakable sign of human transformation of the upland landscape. The walls, for which local granite stone could have been used without difficulty, separate individual fields and delimit enclosures, forming 'perhaps the largest area of preserved prehistoric landscape in Europe' (Bell and Walker, 1992: 178). An interesting feature of the spatial pattern of the walls is that it basically disregards the topography, but follows a master layout plan. Land use changes were followed by vegetation changes and these in turn by changes in soil characteristics, leading to the late Bronze Age landscape dominated by grasslands. The human-induced changes were probably exacerbated by climatic deterioration towards wetter conditions, which occurred in the first millennium BC and prevented forest regeneration even in the absence of intense exploitation of the land. Grasslands were replaced by heather and a short-term return of agriculture in medieval times contributed to maintaining the open character of the landscape (Atherden, 1992). Perhaps ironically, it is the long-term human impact that has helped to reveal to our eyes many of the characteristic minor features of the Dartmoor granite scenery, such as valleyside tors, block fields, and stone stripes.

Rural Landscapes

The shaping of rural landscapes is a subject in its own right and enjoys a voluminous literature, which it is impossible to refer to properly in a book focused on natural landforms. However, there are aspects of granite geomorphology which may help us to understand the rural landscapes in granite terrains and provide a link between natural and cultural landscapes.

First, granite as a readily available stone is widely used for construction of not only farmhouses and barns, but different minor objects as well. Specific local styles have developed in different regions, so that these objects have become inherent components of the cultural landscapes. For example, the northern granite regions of Portugal are known, among others, for their elaborately carved stone granaries. The use of granite in the construction of buildings, not uncommonly set amidst natural outcrops, has resulted in an apparent monotony and merging of natural and built features (Plate 10.4), especially if looked at from a distance. According to Whittow (1986: 37), the reasons for this are as follows:

First, its colour and texture never change with weathering, making it impossible to acquire the patina of a sedimentary freestone; second, unless it is polished it fails to reflect the light as well as sandstone or limestone; third, it does not yield easily to the chisel so that its lack of ornamentation renders it incapable of throwing crisp shadows.

Plate 10.4 Granite village of Monsanto, Portugal

Further on, he comments about rural landscapes in granite areas as a whole:

> Many of these rural granitic buildings look as if they have grown out of the ground, as indeed some of them have, for they had to be raised on the spot wherever the surface blocks were too massive to be moved. Perhaps more than most landscapes of stone, granite lands exhibit a remarkable affinity between man and environment, possibly because the solemn contours of the intractable stone evoke visions of a primeval relationship that has scarcely changed. A visit to the mist-shrouded standing stones of Dartmoor or Penwith will confirm such feelings. (p. 37)

Second, the same wide availability of stone has allowed for construction of stone boundary walls, which are a repetitive motif of rural landscapes all across western Europe, from Portugal to Scotland. The granite boulders used in the wall building are mainly derived from cold-climate mixed slope covers and must have appeared widely on ploughed land as soon as vegetation clearance opened the way for soil erosion. The typical size of granite boulders in the cover deposits, between 25 cm and 1 m, explains the massiveness of boundary walls in granite areas, if compared with terrains built of more densely jointed rock. If even larger boulders had to be removed from the field, they were piled up into irregular heaps of up to a few metres high. In the Aubrac Plateau, French Massif Central, 'man-made tors' have been created in this way and

Plate 10.5 Boulders collected from the field have been piled up to form an artificial tor. A natural granite outcrop in the background, Aubrac Plateau, Massif Central, France

Plate 10.6 Great Zimbabwe (Photo courtesy of Michael F. Thomas)

these are not obviously distinguishable from adjacent natural outcrops (Plate 10.5). A unique example is Great Zimbabwe in south-east Zimbabwe, a group of monumental constructions considered to be the largest ancient structure south of the Sahara (Plate 10.6). It is famous for its massive enclosures built of walls as much as 10 m high and 250 m long, set amidst granite outcrops which provided building stone.

Third, natural granite landscapes provided certain obvious locations for villages. On the one hand, isolated rocky hills were favoured as sites of enhanced defensive potential and the aforementioned blending of natural and man-made features rendered the recognition of an inhabited area more difficult. On the other hand, settlements retreated to rock and boulder-covered slopes, in order not to occupy valuable agricultural land.

Godard (1977) and then Valadas (1987) commented about how granite landscapes support an agricultural economy and what the characteristic associated changes are, referring particularly to the French Massif Central. The combination of usually higher elevation and poor, erodible soils has always made granite areas less competitive 'marginal lands' from the point of view of agriculture. If a choice could be made, grazing was preferred over a more intensive land cultivation, whereas ploughed land, if it occurred, was restricted to small areas with the best soils and hydrological conditions locally. However, overgrazing is known to have led to severe land degradation, deep

336 Granite Landscapes Transformed

gully erosion, and significant vegetation loss (Cosandey *et al.*, 1987). There typically occurs a mosaic of land uses in rural granite terrains which makes them scenically attractive but less valued economically. Today, many former agricultural areas in western Europe have been abandoned and are either seasonally used grazing lands, or are subject to aforestation, or have turned into moorlands.

Defensive Aspect of Granite Geomorphology

Isolated granite inselberg-like hills, ridges, and upland spurs have long been favourite places to build defensive settlements, castles, and watching posts. The occurrence of numerous tors and scatters of boulders on slopes and summit surfaces was not neces-

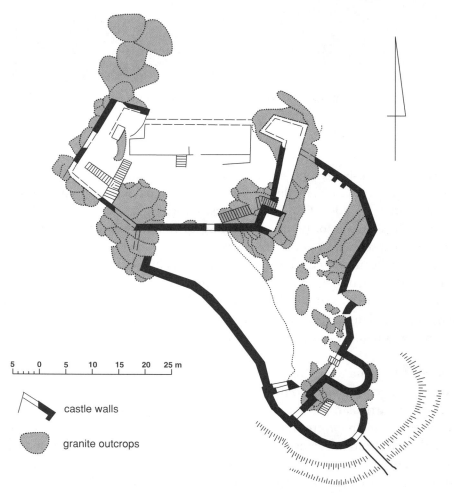

Fig. 10.1 Ground plan of the Bolczów Castle, south-west Poland

sarily considered an obstacle to construction activity. To the contrary, they were often incorporated into building designs in a variety of ways and served as foundations for watchtowers and inner castles (Fig. 10.1), helped to reduce the length of stone walls, and influenced the layouts of winding town lanes, enhancing thereby their defensive character. The world-famous Great Wall of China north of Beijing winds across an all-slope type granite mountain range and follows its narrow crests and spurs. Precipitous rounded outcrops and massive boulders have occasionally been used as foundations for watchtowers, offering particularly wide vistas of the surrounding area (Plate 10.7).

One of the regions which abounds in fortified villages on slopes and tops of granite hills is the eastern part of Portugal, known as Beira Baixa. Perhaps the village most frequently visited is Monsanto (Plate 10.4), strategically located on the northern slope of a prominent inselberg rising 250 m above the surrounding gently rolling plain, which has been used as agricultural land since Roman times. The craggy top part of the hill hosts the ruins of the castle, whereas the block-covered hillslopes below are the site of the village itself. It is not uncommon to observe enormous granite boulders providing partial support for house walls and roofs, more clever examples of the use of natural landscape features to minimize building effort. In addition, numerous natural rock shelters have been converted into stores by the simple addition of a front wall. Although Monsanto is unique in the extent to which architecture and landscape are blended, many more villages in the region occupy poorly accessible granite promontories

Plate 10.7 The Great Wall of China, built on granite outcrops, Badaling, China

and inselbergs, such as the picturesque Sortelha, set amidst an etched granite terrain east of the Serra da Estrela.

The most impressive site of all, however, is the Sigiriya inselberg in north-central Sri Lanka, a huge monolith that rises 200 m above the surrounding countryside. It has a peculiar shape, not very common among granite inselbergs, with steep cliffs and a flattish top. Bremer (1981a) sees this morphology as the result of long-term slope steepening due to long-term basal attack, but there is a structural influence too, namely widely spaced sub-horizontal fractures are very distinctive. The topography of the hill attracted builders of the citadel in the fifth century AD, providing a wide hilltop platform to erect constructions and precipitous slopes to enhance the defensive potential. In 1982 Sigiriya was inscribed on the UNESCO World Heritage Site list and is one of the most popular tourist destinations in the country.

Impact of Mining

Granite areas provide a variety of raw materials which humans can use, and indeed have used, since antiquity. These are the solid granite itself, which serves as a durable building stone, products of granite weathering, from little altered grus to highly decomposed clay-rich kaolin, and various mineral deposits associated with granite intrusions. The latter include tin, wolfram, and, in more recent times, uranium ores. It is not feasible to review all aspects of mining and subsequent use of the materials, but its effect on the landscape needs to be briefly addressed because in specific areas they have been truly enormous. Although for various domestic purposes it sufficed to collect granite stones lying on the ground, any more sophisticated building activity required fresh dimension stone, so quarrying was needed. The origins of granite quarrying are very ancient and in ancient Egypt date back to at least the third millennium BC. However, large-scale extraction of granite stones out of the rock has a 200-year history.

Today, granite quarries can be found worldwide and many of them attain gigantic dimensions. Many claim to be the largest in the world and because of ongoing mining and difficulties in establishing unequivocal criteria it is perhaps difficult to pinpoint which is indeed the largest. However, many granite quarries in the northern Appalachians are big holes, in excess of 100 m deep, and a few of them approach 200 m deep. The now disused Rubislaw quarry west of Aberdeen, Scotland, is often claimed to be the largest in Europe, being 150 m long and 90 m deep. It was in operation from the end of the eighteenth century and eventually closed down in the 1970s. It is now flooded and little of its grandeur remains to be seen. Recognition of a very good quality of particular variants of granite resulted in the concentration of quarrying over relatively small areas, effecting significant anthropogenic transformation of the whole local landscape. In central Europe such areas include the eastern part of Lusatia (Lausitz) in south-east Germany, the environs of the town of Skuteč in the Czech-Moravian

Highland, and the Strzegom Hills in south-west Poland. In Great Britain, the area around Aberdeen has been extensively quarried, in well over 100 quarries, and the city of Aberdeen itself bears the informal name of the 'Granite City', alluding to the widespread use of the local stone. The long-term imprint of quarrying on the landscape varies, depending on the topographical position of the pits. Those cut into hillslopes and hilltops remain visible scars in the countryside for long after quarry operation has ceased, whereas those located on relatively flat ground and subsequently flooded or filled with waste, like the Rubislaw quarry, quickly become almost invisible.

The legacy of granite quarrying does not only involve the quarries and adjacent heaps of waste material themselves. The landscape of Dartmoor, a site of widespread quarrying in the nineteenth century (Plate 10.8), has been enriched by a range of small-scale man-made features, such as causeways and ramps to carry granite blocks, platforms to store the blocks, tramway routes winding along contours across the Dartmoor countryside, and granite-built buildings to serve quarrymen. Among them, the Haytor tramway is unique, its rails having been made of the granite stones themselves rather than of typical iron and wood. Today, the industrial heritage of Dartmoor is as much a part of its scenery as are the natural features, long deforested moorlands, and granite-built villages such as Widecombe in the Moor.

Likewise, kaolin extraction has left its lasting imprint on many areas as it usually involves the opening of huge open-cast pits and the construction of big heaps of waste material. The latter, clearly visible from a distance and not uncommonly shining white,

Plate 10.8 Granite quarries, Dartmoor, south-west England

may not have any equivalents in the natural landscape. In the St Austell granite massif in south-west England kaolin (China clay) mining has been carried out since the turn of the nineteenth century in a landscape that had originally been somewhat similar to that of nearby Dartmoor and Bodmin Moor. A gently rolling countryside prevailed, with a few occasional tors and boulder fields, but apparently lacking inselberg-like relief. Kaolin mining led to the opening of a few dozen deep pits, some as much as 2 km across, scattered over the western part of the massif (Fig. 10.2). Waste material from the pits, which remains in an approximate 9:1 proportion to the clay used in further technological processing, forms high conical and trapezoidal artificial hills. They currently cover around 17 km^2 and provide a very distinctive flavour to the St Austell countryside. Many have been overgrown by planted bush and rhododendron, but the more recent ones remain oddly white. Other kaolin mining areas have been transformed in a similar way, for instance in south Germany near Hirschau, Bavaria, or near Karlový Vary, in the Czech Republic, where kaolin has been mined since the end of the eighteenth century.

The impact of mineral extraction and aggregate production from weathered granite is usually less dramatic. Nevertheless, characteristic landforms may also originate in association with these activities. For example, in the Karkonosze Massif, south-west Poland, there are a dozen or so man-made cave-like features resulting from exploitation of pegmatite for the local glass industry. Some are more than 10 m long. The lasting evidence of tin mining in the Cornubian granite massifs in south-west England includes gully-like looking trenches following mineral lodes, shallow surface excavation areas, as well as deep clefts in the marine cliffs of the Penwith Peninsula, near Land's End.

Large-Scale Granite Carvings: From Sri Lanka to Mt Rushmore

Granite is not an easy rock to work with because it is hard and tough. Therefore, it has never been favoured by artists if the objective was to produce an elaborate sculpture, with as many tiny details as possible. For this purpose, less durable stones such as sandstone or tufa have been chosen. However, if the carved objects were intended to last for a long time, then it was advantageous to select granite. Ancient Egyptian obelisks, remarkably well preserved despite having more than a 3,000-year-long history, were cut out of massive granite cropping out near Aswan, southern Egypt.

However, the most impressive examples are undoubtedly precipitous rock outcrops, the faces of which have been transformed into monumental sculptures, ranking today among the most visited tourist attractions. An ancient example, dating back to the seventh century AD, is the group of open air rock carvings in Mahabalipuram (Plate 10.9), an ancient city in the southern Indian state of Tamil Nadu, some 50 km south of Chennai (Madras). Among them, the Arjuna's Penance is the representation,

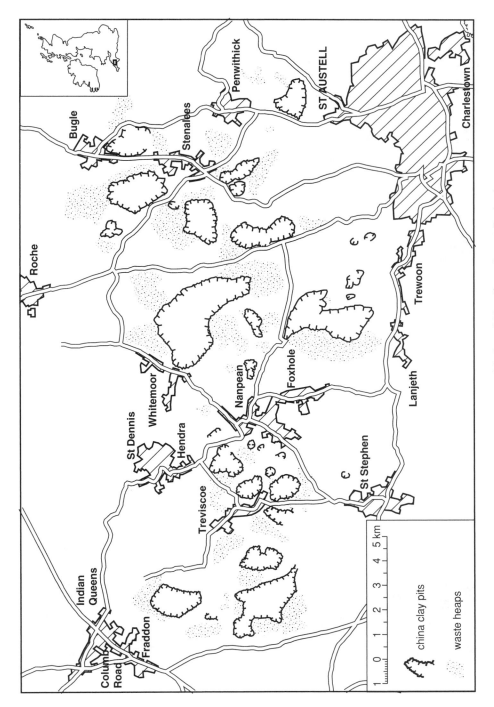

Fig. 10.2 Impact of china clay mining on the natural landscape of St Austell Massif, south-west England

Plate 10.9 Granite rock carvings in Mahabalipuram, southern India (Photo courtesy of Yanni Gunnell)

25 m long and 6 m high, of a mythical story of the River Ganges, It abounds in fine detail of animal beasts, semi-divine creatures, and various deities. Further granite outcrops have been sculpted into elaborate temples. In nearby Sri Lanka, a somewhat similar example is offered by the twelfth century AD Gal Vihara ('Stone Shrine') site, near the town of Polonnaruwa. The face of a low granite hill has been cut and transformed into a Buddhist shrine, containing four colossal Buddha statues. The largest of these, the reclining Buddha, is 14 m long, while the standing statue next to it is 7 m high. Both Mahabalipuram and Gal Vihara have been inscribed onto the UNESCO World Heritage List.

Plate 10.10 Large-scale rock carvings at Mount Rushmore, South Dakota, USA (Photo courtesy of Stawek Tutaczyk)

More recent works of this kind include the carving of General Robert Lee on the slope of the Stone Mountain bornhardt in Georgia, and Mount Rushmore in the Black Hills, South Dakota, USA, which is perhaps the most widely known example (Plate 10.10). The sculpture, called 'the shrine for democracy', consists of four colossal heads of US Presidents: George Washington, Thomas Jefferson, Abraham Lincoln, and Theodore Roosevelt. The work was done by a sculptor, John Gutzon de la Mothe Borglum, and lasted, with breaks, 14 years, from 1927 to 1941. The scale of engineering work involved was huge, since the completion of the project required the removal of 450,000 tonnes of granite. Interestingly, according to the original plan, the monument would have had to be carved out of the picturesque group of granite towers and spires called the Needles, in another part of the Black Hills. However, this location was dismissed because the granite there was too fractured and the tall spires did not offer enough rock surface for carving the human heads. Today, Mount Rushmore is a National Monument, visited by over 2 million people annually. Another large-scale work is in progress nearby, of the Indian Chief Crazy Horse. His head was completed in 1998 and measures 26 m high, but the whole sculpture is designed to be 195 m long and 171 m high. Such colossal carvings will undoubtedly rival and surpass many natural granite features of the Black Hills, formed by thousands and millions of years, illustrating the ability of humans to transform even the toughest rocks.

References

ADAMEK, H., and KUBIČEK, P. (1990). 'The development of granite weathering pits in the Žulovská pahorakitina (hilly land)', in *Fourth Pseudokarst Symposium—Proceedings, Podolanký 1990*, Podolanký 16–22.

ALDERTON, D. H. M., and RANKIN, A. H. (1983). 'The character and evolution of hydrothermal fluids associated with the kaolinised St Austell granite, SW England'. *Journal of the Geological Society, London*, 140: 297–309.

ALLISON, R. J., and GOUDIE, A. S. (1994). 'The effects of fire on rock weathering: an experimental study', in D. A. Robinson and R. B. G. Williams (eds.), *Rock Weathering and Landform Evolution*. Chichester: Wiley, 41–56.

AMARAL, C. (1997). 'Landslides disasters management in Rio de Janeiro', in *Second Pan-American Symposium on Landslides*. Rio de Janeiro, 209–12.

AMBROSE, J. W. (1964). 'Exhumed palaeoplains of the Precambrian Shield of North America'. *American Journal of Science*, 262: 817–57.

ANDRÉ, M.-F. (1995). 'Postglacial microweathering of granite roches moutonnées in Northern Scandinavia (Riksgränsen area, 68° N)', in O. Slaymaker (ed.), *Steepland Geomorphology*. Chichester: Wiley, 103–27.

—— (2002). 'Rates of postglacial rock weathering on glacially scoured outcrops (Abisko-Riksgränsen area, 68°N)'. *Geografiska Annaler*, 84A: 139–50.

—— (2003). 'Do periglacial landscapes evolve under periglacial conditions?' *Geomorphology*, 52: 149–64.

—— (2004). 'The geomorphic impact of glaciers as indicated by tors in North Sweden (Aurivaara, 68° N)'. *Geomorphology*, 57: 403–21.

AREL, E., and TUĞRUL, A. (2001). 'Weathering and its relation to geomechanical properties of Cavusbasi granitic rocks in northwestern Turkey'. *Bulletin of Engineering Geology and the Environment*, 60: 123–33.

ATHERDEN, M. (1992). *Upland Britain: A Natural History*. Manchester: Manchester University Press.

AU, S. W. C. (1996). 'The influence of joint-planes on the mass strength of Hong Kong saprolitic soils'. *Quarterly Journal of Engineering Geology*, 29: 199–204.

—— (1998). 'Rain-induced slope instability in Hong Kong'. *Engineering Geology*, 51: 1–36.

BAKKER, J. P. (1960). 'Some observations in connection with recent Dutch investigations about granite weathering in different climates'. *Zeitschrift für Geomorphologie N.F., Supplement-Band*, 1: 69–92.

—— (1967). 'Weathering of granites in different climates, particularly in Europe', in P. Macar (ed.), *L'Evolution des Versants*, Liège: Université de Liège et Académie Royale de Belgique, 51–68.

—— and LEVELT, T. W. M. (1970). 'An inquiry into probability of a polyclimatic development of peneplains and pediments (etchplains) in Europe during the Senonian and Tertiary period'. *Publications Service Géologique, Luxembourg*, 14: 27–75.

BALLANTYNE, C. K. (1986). 'Landslides and slope failures in Scotland: a review'. *Scottish Geographical Magazine*, 102: 134–50.

—— (1994). 'The tors of the Cairngorms'. *Scottish Geographical Magazine*, 110: 54–9.

—— and HARRIS, C. (1994). *The Periglaciation of Great Britain*. Cambridge: Cambridge University Press.

BARBARIN, B. (1999). 'A review of the relationships between granitoid types, their origins and their geodynamic environments'. *Lithos*, 46: 605–26.

BARTELS, G. (1975). 'Über Glockenberge und verwandte Formen'. *Catena*, 1: 57–70.

BARTON, N., and CHOUBEY, V. (1977). 'The shear strength of rock joints in theory and practice'. *Rock Mechanics*, 10: 1–54.

BATEMAN, P. C., and WAHRHAFTIG, C. (1966). 'Geology of the Sierra Nevada', in E. H. Bailey (ed.), *Geology of Northern California*, California Division of Mines and Geology, Bulletin, 190: 107–72.

BATES, R. L., and JACKSON, J. A. (eds.) (1987). *Glossary of Geology*, 3rd edn. Alexandria: American Geological Institute.

BEAVIS, S. G. (2000). 'Structural controls on the orientation of erosion gullies in mid-western New South Wales, Australia'. *Geomorphology*, 33: 59–72.

BELL, M., and WALKER, M. J. C. (1992). *Late Quaternary Environmental Change*. Harlow: Longman.

BENNETT, M., MATHER, A., and GLASSER, N. (1996). 'Earth hummocks and boulder runs at Merrivale, Dartmoor', in D. J. Charman, R. M. Newnham, and D. G. Croot (eds.), *Devon and East Cornwall, Field Guide*, London: Quarternary Research Association, 81–96.

BERG, G. (1927). 'Zur Morphologie des Riesengebirges'. *Zeitschrift für Geomorphologie*, II: 1–20.

BERRY, L., and RUXTON, B. P. (1960). 'The evolution of Hong Kong Harbour'. *Zeitschrift für Geomorphologie N. F.*, 4: 97–115.

BEST, M. (2003). *Igneous and Metamorphic Petrology*. Oxford: Blackwell.

BIERMAN, P. R. (1993). '*Cosmogenic isotopes and the evolution of granitic landforms*'. Ph.D. thesis, Seattle: University of Washington.

—— and CAFFEE, M. (2002). 'Cosmogenic exposure and erosion history of Australian bedrock landforms'. *Geological Society of America Bulletin*, 114: 787–803.

—— and TURNER, J. (1995). '^{10}Be and ^{26}Al evidence for exceptionally low rates of Australian bedrock erosion and the likely existence of pre-Pleistocene landscapes'. *Quaternary Research*, 44: 378–82.

BIGARELLA, J. J., and BECKER, R. D. (eds.) (1975). *International Symposium on the Quaternary*. Boletin Paranaense de Geociências, 33.

BIRD, E. C. F. (1968). *Coasts*. Canberra: Australian National University Press.

BIRD, E.C.F (1970). 'The steep coast of Macalister Range, North Queensland, Australia'. *Journal of Tropical Geography*, 31: 33–9.
—— (1998). *The Coast of Cornwall*. Fowey: Alexander.
—— (2000). *Coastal Geomorphology: An Introduction*. Chichester: Wiley.
—— and SCHWARTZ, M. L. (eds.) (1985). *The World's Coastline*. New York: Van Nostrand Reinhold.
BIROT, P. (1958). 'Les dômes crystallines'. *Mémoires et Documents, CNRS*, 6: 8–34.
—— GODARD, A., and PELLETIER, J. (1983). 'L'érosion différentielle dans les relief d l'Estrie et de la Nouvelle-Angleterre, entre Montréal et la piedmont sud-est des White Mountains'. *Géographie physique et Quaternaire*, 37: 3–25.
BJORNSON, J., LAURIOL, B. (2001). 'Météorisation des blocs de granite à la surface des pédiments dans le nord du Yukon, Canada'. *Permafrost and Periglacial Processes*, 12: 289–98.
BLACKWELDER, E. (1925). 'Exfoliation as a phase of rock weathering'. *Journal of Geology*, 33: 793–806.
—— (1926). 'Fire as an agent in rock weathering'. *Journal of Geology*, 35: 134–40.
—— (1929). 'Cavernous rock surfaces of the deserts'. *American Journal of Science*, 17: 393–99.
—— (1933). 'The insolation hypothesis of rock weathering'. *American Journal of Science*, 26: 97–113.
BLENKINSOP, T. G. (1993). 'Joint', in P. Kearey (ed.), *The Encyclopedia of the Solid Earth Science*, Oxford: Blackwell, 349–50.
BOAST, R. (1990). 'Dambos: a review'. *Progress in Physical Geography*, 14: 153–77.
BOUCHARD, M., and GODARD, A. (1984). 'Les altérites du bouclier canadien: Premier bilan d'une campagne de reconnaissance'. *Geographie Physique et Quaternaire*, 38: 149–63.
—— JOLICOEUR, S., and PIERRE, G. (1995). 'Characteristics and significance of two pre-late-Wisconsinan weathering profiles (Adirondacks, USA and Miramichi Highlands, Canada)'. *Geomorphology*, 12: 75–89.
—— —— (2000). 'Chemical weathering studies in relation to geomorphological research in southeastern Canada'. *Geomorphology*, 32: 213–38.
BOURMAN, R. P., and OLLIER, C. D. (2002). 'A critique of the Schellmann definition and classification of "laterite" '. *Catena*, 47: 117–31.
BOVIS, M. J. (1982). 'Uphill-facing (antislope) scarps in the Coast Mountains, southwest British Columbia'. *Geological Society of America Bulletin*, 93: 804–12.
—— and JAKOB, M. (1999). 'The role of debris supply conditions in predicting debris flow activity'. *Earth Surface Processes and Landforms*, 24: 1039–53.
BOYE, M., and FRITSCH, P. (1973). 'Dégagement artificial d'un dome crystalline au Sud-Cameroun'. *Travaux et Documents de Géographie Tropicale*, 8: 31–62.
BRADLEY, R. (1998). 'Ruined buildings, ruined stones: enclosures, tombs and natural places in the Neolithic of south-west England'. *World Archaeology*: 30: 13–22.
BRADLEY, W. C., HUTTON, J. T., and TWIDALE, C. R. (1978). 'Role of salts in development of granitic tafoni, South Australia'. *Journal of Geology*, 86: 647–54.
BRANNER, J. C. (1896). 'Decomposition of rocks in Brazil'. *Geological Society of America Bulletin*, 7: 255–314.

—— (1913). 'The fluting and pitting of granites in the tropics'. *Proceedings of the American Philosophical Society*, 52: 163–74.

BREMER, H. (1965). 'Ayers Rock, ein Beispiel für klimagenetische Morphologie'. *Zeitschrift für Geomorphologie N.F.*, 9: 249–84.

—— (1971). 'Flüsse, Flächen- und Stufenbildung in den feuchten Tropen'. *Würzburger Geographische Arbeiten*, 35: 1–194.

—— (1975). 'Intramontane Ebenen, Prozesse der Flächenbildung'. *Zeitschrift für Geomorphologie N.F., Supplement-Band*, 23: 26–48.

—— (1981a). 'Inselberge—Beispiele für eine ökologische Geomorphologie'. *Geographische Zeitschrift*, 69: 199–216.

—— (1981b). 'Reliefformen and reliefbildende Prozesse in Sri Lanka'. *Relief, Boden and Paläoklima*, 1: 7–183.

—— (1982). 'Verwitterungsformen als Stabilitätszeugen'. *Geographische Zeitschrift*, 70: 69–78.

—— (1989). *Allgemeine Geomorphologie*. Berlin and Stuttgart: Gebrüder Borntraeger.

—— (1993). 'Etchplanation, review and comments on Büdel's model'. *Zeitschrift für Geomorphologie N.F., Supplement-Band*, 92: 189–200.

—— (2004). 'Climato-genetic geomorphology', in A. S. Goudie (ed.), *Encyclopedia of Geomorphology*. London: Routledge, 164–5.

—— and SANDER, H. (2000). 'Inselbergs: geomorphology and geoecology', in S. Porembski and W. Barthlott (eds.), *Inselbergs*. Ecological Studies, 146, Berlin and Heidelberg: Springer, 7–35.

BRISTOW, C. R. (1969). 'Kaolin deposits of the United Kingdom of Great Britain and Northern Ireland', in *Twenty-Third International Geological Congress, Prague*, 15: 275–88.

—— (1998). 'China clay', in E. B. Selwood, E. M. Durrance, and C. M. Bristow (eds.), *The Geology of Cornwall*. Exeter: University of Exeter Press, 167–78.

—— and EXLEY, C. S. (1994). 'Historical and geological aspects of the China Clay industry of South-West England'. *Transactions of the Royal Geological Society of Cornwall*, 21: 247–314.

BROCKLEHURST, S. H., and WHIPPLE, K. X. (2002). 'Glacial erosion and relief production in the Eastern Sierra Nevada, California'. *Geomorphology*, 42: 1–14.

BROOK, G. A. (1978). 'A new approach to the study of inselberg landscapes'. *Zeitschrift für Geomorphologie N.F., Supplement-Band*, 31: 138–60.

BRUM FERREIRA, A. (1991). 'Neotectonics in Northern Portugal: a geomorphological approach'. *Zeitschrift für Geomorphologie N.F., Supplement-Band*, 82: 73–85.

BRUNNER, H. (1969). 'Verwitterungstypen auf den Granitgneisen (Peninsular Gneis) des östlichen Mysore-Plateaus (Südindien)'. *Petermanns Geographische Mitteilungen*, 113: 241–8.

BRUNSDEN, D. (1964). 'The origin of decomposed granite on Dartmoor', in I. G. Simmons (ed.), *Dartmoor Essays*, Exeter: Devonshire Association, 97–116.

BRYAN, K. (1925). 'Pedestal rocks in the arid Southwest'. *US Geological Survey, Bulletin*, 760: 1–11.

BUCKLE, C. (1978). *Landforms in Africa*. London: Longman.

BUDDINGTON, A. F. (1959), 'Granite emplacement with special reference to North America'. *Geological Society of America Bulletin*, 70: 671–747.

BÜDEL, J. (1937). 'Eiszeitliche and und rezente Verwitterung und Abtragung im ehemals nicht vereisten Teil Mitteleuropas'. *Petermanns Geographische Mitteilungen, Ergänzungsheft*, 229: 5–71.

—— (1957). 'Die "Doppelten Einebnungsflächen" in den feuchten Tropen'. *Zeitschrift für Geomorphologie N.F.*, 1: 201–28.

—— (1977). *Klima-Geomorphologie*. Stuttgart and Berlin: Gebrüder Borntraeger.

—— (1978). 'Das Inselberg-Rumpfflächenrelief der heutigen Tropen und das Schicksal seiner fossilen Altformen in anderen Klimazonen'. *Zeitschrift für Geomorphologie N.F., Supplement-Band*, 31: 79–110.

BURBANK, D. W., LELAND, J., FIELDING, E., ANDERSON, R. S., BRZOZOVIC, N., REID, M. R., and DUNCAN, C. (1996). 'Bedrock incision, rock uplift and threshold hillslopes in the northwestern Himalayas'. *Nature*, 379: 505–10.

BUSTIN, R. M., and MATHEWS, W. H. (1979). 'Selective weathering of granitic clasts'. *Canadian Journal of Earth Sciences*, 16: 215–23.

CAINE, N. (2004). 'Mechanical and chemical denudation in mountain systems', in P. N. Owens and O. Slaymaker (eds.), *Mountain Geomorphology*, London: Arnold, 132–52.

CALCATERRA, D., PARISE, M., and DATTOLA, L. (1987). 'Debris flows in deeply weathered granitoids (Serre Massif—Calabria, Southern Italy)', in K. Senneset (ed.), *Proceedings of the Seventh International Symposium on Landslides, Trondheim*, Rotterdam: Balkema, 171–6.

CALKIN, P., and CAILLEUX, A. (1962). 'A quantitative study of cavernous weathering (taffonis) and its application to glacial chronology in Victoria Valley, Antarctica'. *Zeitschrift für Geomorphologie N.F.*, 6: 317–24.

CAMPBELL, S., GERRARD, A. J., and GREEN, C. P. (1998). 'Granite landforms and weathering products', in S. Campbell, C. O. Hunt, J. D. Scourse, and D. H. Keen (eds.), *Quaternary of South-West England*, London: Chapman and Hall, 73–90.

CENTENO, J. D. (1989). 'Evolucíon cuaternaria del relieve en la vertiente sur del sistena central Español. Las Formas residuals como indicatoras morfológicas'. *Cuadernos Laboratorio Xeoloxico de Laxe*, 13: 79–88.

CHÁBERA, S., and HUBER, K. H. (1996). 'Polygonalstrukturen (polygonal cracking) auf Felsoberflächen aus Eisgarner Granit'. *Sborník Jihočeského muzea v Českých Budějovicích. Přírodní vědy*, 36: 5–22.

—— —— (1998). 'Pseudoschichung (pseudobedding) in Granitoiden des Südböhmischen Plutons'. *Sborník Jihočeského muzea v Českých Budějovicích. Přírodní vědy*, 38: 5–17.

CHAPMAN, C. A., and RIOUX R. L. (1958). 'Statistical study of topography, sheeting, and jointing in granite, Acadia National Park, Maine'. *American Journal of Science*, 256: 111–27.

CHAPMAN, R. W., and GREENFIELD, M. A. (1949). 'Spheroidal weathering of igneous rocks'. *American Journal of Science*, 247: 407–29.

CHAPPELL, B. W., and WHITE, A. J. R. (1974). 'Two contrasting granite types'. *Pacific Geology*, 8: 173–4.

CHAU, K. T., WONG, W. Y., FONG, E. L., CHAN, L. C. P., SZE, Y. L., and FUNG, M. K. (2004). 'Landslide hazard analysis for Hong Kong using landslide inventory and GIS'. *Computers and Geosciences*, 30: 429–43.

CHEN, H., and LEE, C. F. (2005). 'Geohazards of slope mass movement and its prevention in Hong Kong'. *Engineering Geology*, 76: 3–25.

CHIGIRA, M. (2001). 'Micro-sheeting of granite and its relationship with landsliding specifically after the heavy rainstorm in June 1999, Hiroshima Prefecture, Japan'. *Engineering Geology* 59: 219–31.

CHMAL, H., and TRACZYK, A. (1993). 'Plejstoceńskie lodowce gruzowe w Karkonoszach'. *Czasopismo Geograficzne*, 64: 253–62.

CLARKE, D. B. (1992). *Granitoid Rocks*. London: Chapman and Hall.

CLEAVES, E. T., FISHER, D. W., and BRICKER, O. P. (1970). 'Geochemical balance of a small watershed and its geomorphic implications'. *Geological Society of America Bulletin*, 85: 437–44.

CLOOS, H. (1925). *Einführung in die tektonische Behandlung magmatischer Erscheinungen (Granittektonik). Teil I. Das Riesengebirge in Schlesien*. Berlin: Gebrüder Borntraeger.

COBBING, J. (2000). *The Geology and Mapping of Granite Batholiths*. Berlin: Springer.

COCKBURN, H. A. P., SEIDL, M. A., and SUMMERFIELD, M. A. (1999). 'Quantifying denudation rates on inselbergs in the central Namib Desert using in situ-produced cosmogenic ^{10}Be and ^{26}Al'. *Geology*, 27: 399–402.

COE, M. (1986). 'The ecology of rock outcrops in the Kora National Reserve, Kenya', in M. Coe and N. M. Collins (eds.), *Kora: An Ecological Inventory of the Kora National Reserve, Kenya*, Norwich: Page Bros, 159–71.

COELHO NETTO, A. L. (1999). 'Catastrophic landscape evolution in a humid region (SE Brasil): inheritances from tectonic, climatic and land use induced changes'. *Supplementi di Geografia Fisica e Dinamica Quaternaria*, 3(3): 21–48.

—— FERNANDES, N. F., and DEUS, C. E. (1988). 'Gullying in the southeastern Brazilian Plateau, Bananal, S. P.' *IAHS Publications*, 174: 35–42.

CONCA, J. L., and ROSSMAN, G. R. (1985). 'Core softening in cavernous weathered tonalite'. *Journal of Geology*, 93: 59–73.

COOKE, R. U., and MASON, P. (1973). 'Desert Knolls pediment and associated landforms in the Mojave Desert, California'. *Revue Géomorphologie dynamique*, 20: 71–8.

—— WARREN, A., and GOUDIE, A. (1993). *Desert Geomorphology*. London: University College Press.

COQUE-DELHUILLE, B. (1978). *Les formations superficielles des plateaux de la Margeride occidentale: étude géomorphologique*. Thèse de doctorat de 3é cycle, Université Paris I.

—— (1979). 'Les formations superficielles et leur signification géomorphologique dans les régions de roches cristallines: l'exemple des plateaux de la Margeride occidentale'. *Revue de Géographie Physique et Géologie Dynamique*, 21: 127–46.

—— (1987). *Le massif du Sud-Ouest anglais et sa bordure sédimentaire*. Thèse d'Etat, Université du Paris I.

—— (1988). 'Altérations deutériques et géomorphologie: l'exemple du Sud-Ouest anglais'. *Zeitschrift für Geomorphologie N.F.*, 32: 195–216.

COSANDEY, C., BILLARD, A., and MUXART, T. (1987). 'Present-day evolution of gullies formed in historical times in the Montagne du Lingas (southern Cévennes, France)', in V. Gardiner (ed.), *International Geomorphology 1986*, II. Chichester: Wiley, 523–31.

COUDE, A. (1983). 'Géomorphologie structurale du batholite de Galway (Connemara, Irlande occidentale'. *Revue de Géographie Physique et Géologie Dynamique*, 24: 177–92.

CRICKMAY, G. W. (1935). 'Granite pedestal rocks in the southern Appalachian Piedmont'. *Journal of Geology*, 43: 745–58.

CROUCH, R. J., and BLONG, R. J. (1989). 'Gully sidewall classification: methods and applications'. *Zeitschrift für Geomorphologie N.F.*, 33: 291–305.

CUNNINGHAM, F. F. (1965). 'Tor theories in the light of South Pennine evidence'. *East Midlands Geographer*, 3: 424–33.

—— (1969). 'The Crow Tors, Laramie Mountains, Wyoming, USA'. *Zeitschrift für Geomorphologie N.F.*, 13: 56–74.

—— (1971). 'The Silent City of Rocks, a bornhardt landscape in the Cotterell Range, South Idaho'. *Zeitschrift für Geomorphologie N.F.*, 15: 404–29.

CZERWIŃSKI, J., and MIGOŃ, P. (1993). 'Mikroformy wietrzenia granitów w masywie karkonosko-izerskim'. *Czasopismo Geograficzne*, 64: 265–84.

CZUDEK, T. (1964). 'Periglacial slope development in the area of the Bohemian Massif in Northern Moravia'. *Biuletyn Peryglacjalny*, 14: 169–93.

—— (1989). 'Kryoplanationsterrasen im rezenten Dauerfrostboden'. *Přírodovědné práce ústavů Československé Akademie Věd v Brně*, 23(8): 1–41.

—— and DEMEK, J. (1970). 'Některé problémy interpretace povrchových tvarů České vysočiny'. *Zprávy Geografického Ústavu ČSAV*, 7(1): 9–28.

—— —— MARVAN, P., PANOŠ, V., and RAUŠER, J. (1964). 'Verwitterungs- und Abtragungsformen des Granits in der Böhmischen Masse'. *Petermanns Geographische Mitteilungen*, 108: 182–92.

DAHL, R. (1966). 'Block fields, weathering pits and tor-like forms in the Narvik Mountains, Nordland, Norway'. *Geografiska Annaler*, 48A: 55–85.

—— (1967). 'Post-glacial micro-weathering of bedrock surfaces in the Narvik District of Norway'. *Geografiska Annaler*, 49A: 155–66.

DAI, F. C., and LEE, C. F. (2001). 'Frequency-volume relation and prediction of rainfall-induced landslides'. *Engineering Geology*, 59: 253–66.

DAVEAU, S. (1969). 'Structure et relief de la Serra da Estrela'. *Finisterra*, 7: 31–63, 159–97.

—— (1971). 'La glaciation de la Serra da Estrela'. *Finisterra*, 11: 5–40.

DALE, T. N. (1923). 'The commercial granites of New England'. *US Geological Survey, Bulletin*, 738.

DE DAPPER, M. (1989). 'Pedisediments and stone-line complexes in Peninsular Malaysia'. *Geo-Eco-Trop*, 11: 37–59.

DE MEIS, M. R. M., and MONTEIRO, A. M. F. (1979). 'Upper Quaternary "rampas": Doce river valley, Southeastern Brazilian plateau'. *Zeitschrift für Geomorphologie N.F.*, 23: 132–51.

DE OLIVEIRA, M. A. T. (1990). 'Slope geometry and gully erosion development: Bananal, Sao Paulo, Brazil'. *Zeitschrift für Geomorphologie N.F.*, 34: 423–34.

DE PLOEY, J., and CRUZ, O. (1979). 'Landslides in the Serra do Mar, Brazil'. *Catena*, 6: 111–22.

DEARMAN, W. R., BAYNES, F. J., and IRFAN, T. Y. (1978). 'Engineering grading of weathered granite'. *Engineering Geology*, 12: 345–74.

DeGraff, J. V., Bryce, R., Jibson, R. W., Mora, S., and Rogers, C. T. (1989). 'Landslides: their extent and significance in the Caribbean', in E. E. Brabb and B. L. Harrod (eds.), *Landslides: Extent and Economic Significance*. Rotterdam: A. A. Balkema, 51–80.

Demek, J. (1964). 'Castle koppies and tors in the Bohemian Highland (Czechoslovakia)'. *Biuletyn Peryglacjalny*, 14: 195–216.

—— (1969a). 'Cryogene processes and the development of cryoplanation terraces'. *Biuletyn Peryglacjalny*, 18: 115–25.

—— (1969b). 'Cryoplanation terraces, their geographical distribution, genesis and development'. *Rozpravy Československé Akademie Věd, ř. MPV*, 79(4): 1–80.

Derbyshire, E. (1972). 'Tors, rock weathering and climate in southern Victoria Land'. *Institute of British Geographers, Special Publication*, 4: 93–105.

Didier, J., and Barbarin, B. (eds.) (1991). *Enclaves and Granite Petrology*. Amsterdam: Elsevier.

Dixon, J. C., and Young, R. W. (1981). 'Character and origin of deep arenaceous weathering mantles on the Bega batholith, southeastern Australia'. *Catena*, 8: 87–109.

Dohrenwend, J. C. (1987). 'Basin and Range', in W. L. Graf (ed.), *Geomorphic Systems of North America*. Geological Society of America, Centennial Special Volume, 2: 303–42.

—— (1994). 'Pediments in arid environments', in A. D. Abrahams and A. J. Parsons (eds.), *Geomorphology of Desert Environments*. London: Chapman and Hall, 321–53.

—— McFadden, L. D., Turrin, B. D., and Wells, S. G. (1984). 'K-Ar dating of the Cima volcanic field, eastern Mojave Desert, California: late volcanic history and landscape evolution'. *Geology*, 12: 163–7.

Doornkamp, J. C. (1968). 'The role of inselbergs in the geomorphology of southern Uganda'. *Institute of British Geographers, Transactions*, 44: 151–62.

—— (1974). 'Tropical weathering and the ultra-microscopic characteristics of regolith quartz on Dartmoor'. *Geografiska Annaler*, 56A: 73–82.

Dorn, R. I. (2004). 'Fire', in A. S. Goudie (ed.), *Encyclopedia of Geomorphology*. London: Routledge, 367–70.

Dragovich, D. (1967). 'Flaking, a weathering process operating on cavernous rock surfaces'. *Geological Society of America Bulletin*, 78: 801–04.

—— (1968). 'Granite lapies at Remarkable Rocks, South Australia'. *Revue de Géomorphologie dynamique*, 18: 1–16.

—— (1969). 'The origin of cavernous surfaces (tafoni) in granitic rocks of southern South Australia'. *Zeitschrift für Geomorphologie N.F.*, 13: 163–81.

—— (1993). 'Fire-accelerated boulder weathering in the Pilbara, Western Australia'. *Zeitschrift für Geomorphologie N.F.*, 37: 295–307.

Dredge, L. A. (2000). 'Age and origin of upland block fields on Melville Peninsula, Eastern Canadian Arctic'. *Geografiska Annaler*, 82A: 443–54.

Dresch, J. (1959). 'Notes sur la géomorphologie de l'Air'. *Association Géographie Français, Bulletin*, 280: 2–20.

Dumanowski, B. (1960). 'Comment on origin of depressions surrounding granite massifs in the Eastern Desert in Egypt'. *Bulletin de l'Académie Polonaise des Sciences*, 8(4): 305–12.

DUMANOWSKI, B. (1961). 'Cover deposits of the Karkonosze Mountains'. *Zeszyty Naukowe Uniwersytetu Wrocławskiego*, B8: 31–55.

—— (1963). 'Stosunek rzeźby do struktury w granicie Karkonoszy'. *Acta Universitatis Wratislaviensis*, 9, *Studia Geograficzne*, 1: 27–35.

—— (1964). 'Problem of the development of slopes in granitoids'. *Zeitschrift für Geomorphologie N.F.*, Supplement-Band, 5: 30–40.

—— (1968). 'Influence of petrographical differentiation of granitoids on land forms'. *Geographia Polonica*, 14: 93–8.

DURGIN, P. B. (1977). 'Landslides and the weathering of granitic rocks'. *Geological Society of America. Reviews in Engineering*, 3: 127–31.

DZULYNSKI, S., and KOTARBA, A. (1979). 'Solution pans and their bearing on the development of pediments and tors in granite'. *Zeitschrift für Geomorphologie N.F.*, 23: 172–91.

EDEN, M. J. (1971). 'Some aspects of weathering and landforms in Guyana (formerly British Guiana)'. *Zeitschrift für Geomorphologie N.F.*, 15: 181–98.

—— and GREEN, C. P. (1971). 'Some aspects of granite weathering and tor formation on Dartmoor, England'. *Geografiska Annaler*, 53A: 92–9.

EGGLER, D. H., LARSON, E. E., and BRADLEY, W. C. (1969). 'Granite, grusses, and the Sherman Erosion Surface, southern Laramie Range, Colorado–Wyoming'. *American Journal of Science*, 267: 510–22.

EHLEN, J. (1991). 'Significant geomorphic and petrographic relations with joint spacing in the Dartmoor granite'. *Zeitschrift für Geomorphologie N.F.*, 35: 425–38.

—— (1992). 'Analysis of spatial relationships among geomorphic, petrographic and structural characteristics of the Dartmoor tors'. *Earth Surface Processes and Landforms*, 17: 53–67.

—— (1994). 'Classification of Dartmoor tors', in D. A. Robinson and R. B. G. Williams (eds.), *Rock Weathering and Landform Evolution*. Chichester: Wiley, 393–412.

—— (1999). 'Fracture characteristics in weathered granites'. *Geomorphology*, 31: 29–45.

ELVHAGE, C., and LIDMAR-BERGSTRÖM, K. (1987). 'Some working hypotheses on the geomorphology of Sweden in the light of a new relief map'. *Geografiska Annaler*, 69A: 343–58.

EMERY, K. O., and KUHN, G. G. (1982). 'Sea cliffs: their processes, profiles and classification'. *Geological Society of America Bulletin*, 93: 644–54.

ENGELDER, T. (1987), 'Joints and shear fractures in rock', in B. K. Atkinson (ed.), *Fracture Mechanics of Rocks*, London: Academic Press, 27–69.

ERICSON, K., and OLVMO, M. (2004). 'A-tents in central Sierra Nevada, California: A geomorphological indicator of tectonic stress'. *Physical Geography*, 25: 291–312.

—— MIGOŃ, P., and OLVMO, M. (2005). 'Fractures and drainage in the granite mountainous area. A study from Sierra Nevada, USA'. *Geomorphology*, 64: 97–116.

ESTEOULE-CHOUX, J. (1983). 'Kaolinitc weathering profiles in Brittany: genesis and economic importance'. *Geological Society, Special Publication*, 11: 33–8.

ETLICHER, B. (1983). 'Structure du socle et morphogenèse dans les monts du Forez'. *Revue de Géographie Physique et Géologie Dynamique*, 24: 75–85.

EVANS, C. V., and BOTHNER, W. A. (1993). 'Genesis of altered Conway granite (grus) in New Hampshire, USA'. *Geoderma*, 58: 201–18.

EVANS, I. S. (1970). 'Salt crystallization and rock weathering: a review'. *Revue de Géomorphologie dynamique*, 19: 153–77.
FAHEY, B. D. (1986). 'Weathering pit development in the central Otago Mountains of southern New Zealand'. *Arctic and Alpine Research*, 18: 337–48.
FAIRBRIDGE, R. W. (1968). 'Solution pits and pans', in R. W. Fairbridge (ed.), *The Encyclopedia of Geomorphology*. New York: Reinhold, 1033–6.
—— and FINKL, C. W., Jr. (1980). 'Cratonic erosional unconformities and peneplains'. *Journal of Geology*, 88: 69–86.
FALCONER, J. D. (1911). *The Geology and Geography of Northern Nigeria*. London: Macmillan.
FANIRAN, A. (1974). 'Nearest-neighbour analysis of inter-inselberg distance: a case study of inselbergs of south-western Nigeria'. *Zeitschrift für Geomorphologie N.F., Supplement-Band*, 20: 150–67.
FARMIN, R. (1937). 'Hypogene exfoliation in rock masses'. *Journal of Geology*, 45: 625–35.
FEININGER, T. (1969). 'Pseudo-karst on quartz diorite, Colombia'. *Zeitschrift für Geomorphologie N.F.*, 13: 287–96.
FERNANDES, N. F., GUIMARÃES, R. F., GOMEZ, R. A. T., VIEIRA, B. C., MONTGOMERY, D. R., and GREENBERG, H. (2004). 'Topographic controls of landslides in Rio de Janeiro: field evidence and modeling'. *Catena*, 55: 163–81.
FERREIRA, N., and VIEIRA, G. T. (1999). *Guia Geológico e Geomorfológico do Parque Natural da Serra da Estrela*. Lisboa: Instituto da Conservação da Natureza and Instituto Geológico e Mineiro.
FINKL, C. W., Jr. (1979). 'Stripped (etched) landsurfaces in southern Western Australia'. *Australian Geographical Studies*, 17: 33–52.
—— and CHURCHWARD, H. M. (1973). 'The etched surfaces of southwestern Australia'. *Journal of the Geological Society of Australia*, 20: 295–307.
FLAGEOLLET, J. C. (1977). *Origine des reliefs, altérations et formations superficielles: contribution à l'étude géomorphologique des massifs anciens cristallins du Limousin et de la Vendée du Nord-Ouest*. Thèse Doctorate État, Université du Paris VII.
FLOHR, E. (1934). 'Alter, Entstehung und Bewegungserscheinungen der Blockmeere des Riesengebirges'. *Veröffentlichungen der schlesischen Gesellschaft für Länderkunde*, 21: 395–434.
FOLK, R. L., and PATTON, E. B. (1982). 'Buttressed expansion of granite and development of grus in Central Texas'. *Zeitschrift für Geomorphologie N.F.*, 26: 17–32.
FREISE, F. W. (1938). 'Inselberge und Inselberglandschaften in Granit und Gneisgebieten Brasiliens'. *Zeitschrift für Geomorphologie*, 10: 137–68.
FRENCH, H. M. (1996). *The Periglacial Environment*. Harlow: Longman.
—— and GUGLIELMIN, M. (2000). 'Cryogenic weathering of granite, Northern Victoria Land, Antarctica'. *Permafrost and Periglacial Processes*, 11: 305–14.
GANGLOFF, P. (1983). 'Les fondements géomorphologiques de la théorie des paléonunataks: le cas des Monts Torngats'. *Zeitschrift für Geomorphologie N.F., Supplement-Band*, 47: 109–36.
GARREAU, J. (1985). 'Alvéoles circulaires ou ellipsoïdales des roches granitiques et métamorphiques de Bretagne occidentale'. *Physio-Géo*, 13: 31–8.

GAVRILOVIĆ, D. (1968). 'Kamenice im magmatischen Gestein Jugoslawiens'. *Zeitschrift für Geomorphologie N.F.*, 12: 43–59.

GELLERT, J. F. (1970). Climatomorphology and palaeoclimates of the Central European Tertiary', in M. Pecsi (ed.), *Problems of Relief Planation*. Budapest: Akadémiai Kiado, 107–12.

GERRARD, A. J. (1974). 'The geomorphological importance of jointing in the Dartmoor granite'. *Institute of British Geographers, Special Publication*, 7: 39–51.

—— (1978). 'Tors and granite landforms of Dartmoor and eastern Bodmin Moor'. *Proceedings of the Ussher Society*, 4(2): 204–10.

—— (1982). 'Granite structures and landforms', in B. H. Adlam, C. R. Fenn, and L. Morris (eds.), *Papers in Earth Studies, Lovatt Lectures—Worcester*. Norwich: Geo Books, 69–105.

—— (1988a). 'Periglacial modification of the Cox Tor—Staple Tors area of western Dartmoor, England'. *Physical Geography*, 9: 280–300.

—— (1988b). *Rocks and Landforms*. London: Unwin Hyman.

—— (1989). 'The nature of slope materials on the Dartmoor granite, England'. *Zeitschrift für Geomorphologie N.F.*, 33: 179–88.

—— (1994a). 'Classics in physical geography revisited: Linton, D. L. 1955: The problem of tors'. *Progress in Physical Geography*, 18: 559–63.

—— (1994b). 'Weathering of granitic rocks: Environment and clay mineral formation', in D. A. Robinson and R. B. G. Williams (eds.), *Rock Weathering and Landform Evolution*. Chichester: Wiley, 3–20.

GIBBONS, C. L. M. H. (1981). 'Tors in Swaziland'. *Geographical Journal*, 147: 72–8.

GILBERT, G. K. (1904). 'Domes and dome structure of the High Sierra'. *Geological Society of America Bulletin*, 15: 29–36.

GILG, H. A., HULMEYER, S., MILLER, H., and SHEPPARD, S. M. F. (1999). 'Supergene origin of the Lastarria kaolin deposit, south-central Chile, and paleoclimatic implications'. *Clay and Clay Minerals*, 47: 201–11.

GLASSER, N. F. (1997). 'The origin and significance of sheet joints in the Cairngorm granite'. *Scottish Journal of Geology*, 33: 125–31.

GODARD, A. (1966). 'Les "tors" et le problème de leur origin'. *Revue Géographie de l'Est*, 6: 153–70.

—— (1969). 'L'ile d'Arran (Écosse): contribution a l'étude géomorphologique des racines de volcans'. *Revue de Géographie Physique et de Géologie dynamique*, 11: 3–30.

—— (1977). *Pays et paysages du granite*. Paris: Presses Universitaires de France.

—— and COQUE-DELHUILLE, B. (1982). 'L'ile de Lundy (Bristol Channel—G.B.): bilan d'une reconnaissance géomorphologique'. *Hommes et Terres du Nord*, 1982(3): 27–38.

—— LAGASQUIE, J. J., and LAGEAT, Y. (eds.) (1993). *Les régions de socle*. Faculté des Lettres et Sciences humaines de l'Université Blaise-Pascal, Nouvelle série, 43.

—— —— —— (eds.) (2001). *Basement Regions*. Berlin: Springer.

GOLDICH, S. S. (1938). 'A study in rock-weathering'. *Journal of Geology*, 46: 17–58.

GOODMAN, R. E. (1993). *Engineering Geology: Rock in Engineering Construction*. New York: Wiley.

GOUDIE, A. S. (1974). 'Further experimental investigation of rock weathering by salt and other mechanical processes'. *Zeitschrift für Geomorphologie N.F., Supplement-Band*, 21: 1–12.

—— (1984). 'Salt efflorescences and salt weathering in the Hunza Valley, Karakoram Mountains, Pakistan', in K. J. Miller (ed.), *The International Karakoram Project*, II. Cambridge: Cambridge University Press, 607–15.

—— (ed.) (1994). *The Encyclopedic Dictionary of Physical Geography*. Oxford: Blackwell.

—— (2004). 'Tafoni', in A. S. Goudie (ed.), *Encyclopedia of Geomorphology*. London: Routledge, 1034–5.

—— ALLISON, R. J., and MCLAREN, S. J. (1992). 'The relations between modulus of elasticity and temperature in the context of the experimental simulation of rock weathering by fire'. *Earth Surface Processes and Landforms*, 17: 605–15.

—— and BULL, P. A. (1984). 'Slope process change and colluvium deposition in Swaziland: an SEM analysis'. *Earth Surface Processes and Landforms*, 9: 289–99.

—— BRUNSDEN, D., COLLINS, D. N., DERBYSHIRE, E., FERGUSON, R. I., HASHMET, Z., JONES, D. K. C., PERROTT, F. A., SAID, M., WATERS, R. S., and WHALLEY, W. B. (1984). 'The geomorphology of the Hunza Valley, Karakoram mountains, Pakistan', in K. J. Miller (ed.), *The International Karakoram Project*, II, Cambridge: Cambridge University Press, 359–410.

—— and ECKARDT, F. (1999). 'The evolution of the morphological framework of the Central Namib Desert, Namibia, since the Early Cretaceous'. *Geografiska Annaler*, 81A: 443–58.

—— and MIGOŃ, P. (1997). 'Weathering pits in the Spitzkoppe area, Central Namib Desert'. *Zeitschrift für Geomorphologie N.F.*, 41: 417–44.

—— and VILES, H. (1997). *Salt Weathering Hazards*. Chichester: Wiley.

GREEN, C. P., and EDEN, M. J. (1973). 'Slope deposits on the weathered Dartmoor granite, England'. *Zeitschrift für Geomorphologie N.F., Supplement-Band*, 18: 26–37.

GRELOU-ORSINI, C., and PETIT, M. (eds.) (1985). 'Alvéoles et bassins dans les socles et leur couverture gréseuse'. *Physio-Géo*, 17.

GRIGG, P. V., and WONG, K. M. (1987). 'Stabilization of boulders at a hillslope site in Hong Kong'. *Quarterly Journal of Engineering Geology*, 20: 5–14.

GRIGGS, G. B., and JOHNSON, R. E. (1979). 'Coastline erosion, Santa Cruz County'. *California Geology*, 32: 67–76.

GUAN, P., NG, C. W. W., SUN, M., and TANG, W. (2001). 'Weathering indices for rhyolitic tuff and granite in Hong Kong'. *Engineering Geology*, 59: 147–59.

GUILCHER, A. (1958). *Coastal and Submarine Morphology*. London: Methuen.

—— (1985). 'France', in E. C. F. Bird and M. L. Schwartz (eds.), *The World's Coastline*. New York: Van Nostrand Reinhold, 385–96.

GUNNELL, Y. (2000). 'The characterization of steady state in Earth surface systems: findings from the gradient modelling of an Indian climosequence'. *Geomorphology*, 35: 11–20.

GUPTA, A. S., and RAO K. S. (2001). 'Weathering indices and their applicability for crystalline rocks'. *Bulletin of Engineering Geology and the Environment*, 60: 201–21.

GUZZETTI, F., REICHENBACH, P., and WIECZOREK, G. F. (2003). 'Rockfall hazard and risk assessment in the Yosemite Valley, California, USA'. *Natural Hazards and Earth System Sciences*, 3: 491–503.

HALL, A. (1996). *Igneous Petrology*. Harlow: Longman.

HALL, A. M. (1986). 'Deep weathering patterns in north-east Scotland and their geomorphological significance'. *Zeitschrift für Geomorphologie N.F.*, 30: 407–22.

—— (1987). 'Weathering and relief development in Buchan, Scotland', in V. Gardiner (ed.), *International Geomorphology 1986*, II. Chichester: Wiley, 991–1005.

—— (1991). 'Pre-Quaternary landscape evolution in the Scottish Highlands'. *Transactions of the Royal Society of Edinburgh, Earth Sciences*, 82: 1–26.

—— (1996). 'The paleic relief of the Cairngorm Mountains', in N. F. Glasser and M. R. Bennett (eds.), *The Quaternary of the Cairngorms: Field Guide*. London: Quaternary Research Association, 13–27.

HALL, K. (1995). 'Freeze-thaw weathering: the cold region "Panacea"'. *Polar Geography and Geology*, 19: 79–87.

—— (1999). 'The role of thermal stress fatigue in the breakdown of rock in cold regions'. *Geomorphology*, 31: 47–63.

—— and ANDRÉ, M.-F. (2001). 'New insights into rock weathering from high-frequency rock temperature data: an Antarctic study of weathering by thermal stress'. *Geomorphology*, 41: 23–35.

—— and OTTE, W. (1990). 'A note on biological weathering of nunataks of the Juneau Icefield, Alaska'. *Permafrost and Periglacial Processes*, 1: 189–96.

—— THORN, C. E., MATSUOKA, N., and PRICK, A. (2002). 'Weathering in cold regions: some thoughts and perspectives'. *Progress in Physical Geography*, 26: 577–603.

HALLET, B., WALDER, J. S., and STUBBS, C. W. (1991). 'Weathering by segregation ice growth in microcracks at sustained subzero temperatures: verification from an experimental study using acoustic emissions'. *Permafrost and Periglacial Processes*, 2: 283–300.

HAMDAN, J., and BURNHAM, C. P. (1996). 'The contribution of nutrients from parent material in three deeply weathered soils of Peninsular Malaysia'. *Geoderma*, 74: 219–33.

HAMILTON, W., and MEYERS, B. (1967). 'The nature of batholith'. *U.S. Geological Survey, Professional Papers*, 554–C.

HARBOR, J. (1995). 'Development of glacial-valley cross sections under conditions of spatially variable resistance to erosion'. *Geomorphology*, 14: 99–107.

HARRISON, S., ANDERSON, E., and WINCHESTER, V. (1996). 'Large boulder accumulations and evidence for permafrost creep, Great Mis Tor, Dartmoor', in D. J. Charman, R. M. Newnham, and D. G. Croot (eds.), *Devon and East Cornwall, Field Guide*, London: Quaternary Research Association, 97–100.

HASLETT, S. K., and CURR, R. H. F. (1998). 'Coastal rock platforms and Quaternary sea-levels in the Baie d'Audierne, Brittany, France'. *Zeitschrift für Geomorphologie N.F.*, 42: 507–16.

HEDGES, J. (1969). 'Opferkessel'. *Zeitschrift für Geomorphologie N.F.*, 13: 22–55.

HILL, I. D., and RACKHAM, I. J. (1978). 'Indications of mass movement on the Jos Plateau, Nigeria'. *Zeitschrift für Geomorphologie N.F.*, 22: 258–74.

HILLS, E. S. (1971). 'A study of cliffy coastal profiles based on examples in Victoria, Australia'. *Zeitschrift für Geomorphologie N.F.*, 15: 137–80.

HOEK, E., and BRAY, J. (1981). *Rock Slope Engineering*. London: Institute of Mining and Metallurgy.

Holzhausen, G. R. (1989). 'Origin of sheet structure, 1. Morphology and boundary conditions'. *Engineering Geology*, 27: 225–78.

Hoskin, C. M., and Sundeen, D. A. (1985). 'Grain size of granite and derived grus, Enchanted Rock Pluton, Texas'. *Sedimentary Geology*, 42: 25–40.

Hövermann, J. (1953). 'Die Periglazial-Erscheinungen im Harz'. *Göttinger Geographische Abhandlungen*, 14: 7–44.

—— (1978a). 'Formen und Formung in der Pränamib (Flächen-Namib)'. *Zeitschrift für Geomorphologie N.F., Supplement-Band*, 30: 55–73.

—— (1978b). 'Untersuchungen und Darlegungen zum Inselbergproblem in der deutchen Literatur der 1. Hälfte des 20. Jahrhunderts'. *Zeitschrift für Geomorphologie N.F., Supplement-Band*, 31: 64–78.

Huber, N. K. (1987). 'The geologic story of Yosemite National Park'. *US Geological Survey, Bulletin*, 1595.

Hurault, J. (1963). 'Recherches sur les inselbergs granitiques nus en Guyane Française'. *Revue de Géomorphologie dynamique*, 14: 49–61.

Ikeda, H. (1998). *The World of Granite Landforms*. Tokyo: Kokon-Shoin.

Innes, J. L. (1983), 'Lichenometric dating of debris flow deposits in the Scottish Highlands'. *Earth Surface Processes and Landforms*, 8: 579–88.

Irfan, T. Y. (1996). 'Mineralogy, fabric properties and classification of weathered granites in Hong Kong'. *Quarterly Journal of Engineering Geology*, 29: 5–35.

—— and Dearman, W. R. (1978). 'Engineering classification and index properties of a weathered granite'. *Bulletin of the International Association of Engineering Geology*, 17: 79–90.

Isherwood, D., and Street, A. (1976). 'Biotite-induced grussification of the Boulder Creek, granodiorite, Boulder County, Colorado'. *Geological Society of America Bulletin*, 87: 366–70.

Ivan, A. (1983). 'Geomorfologické poměry Žulovské pahorkatiny'. *Zprávy Geografického Ústavu ČSAV*, 20(4): 49–69.

—— and Kirchner, K. (1994). 'Geomorphology of the Podyjí National Park in the southeastern part of the Bohemian Massif (South Moravia)'. *Moravian Geographical Reports*, 2: 2–25.

Iverson, R. M., Reid, M. E., and LaHusen, R. G. (1997). 'Debris-flow mobilization from landslides'. *Annual Reviews in Earth and Planetary Sciences*, 25: 85–138.

Jackson, M. D., and Pollard, D. D. (1988). 'The laccolith stock controversy: new results from the southern Henry Mountains, Utah'. *Geological Society of America Bulletin*, 100: 117–39.

Jahn, A. (1962). 'Geneza skałek granitowych'. *Czasopismo Geograficzne*, 33: 19–44.

—— (1968). 'Peryglacjalne pokrywy stokowe Karkonoszy i Gór Izerskich'. *Opera Corcontica*, 5: 9–25.

—— (1974). 'Granite tors in the Sudeten Mountains'. *Institute of British Geographers, Special Publication*, 7: 53–61.

—— (1975). *Problems of the Periglacial Zone*. Warszawa: PWN.

Jahns, R.H. (1943). 'Sheet structure in granites: its origin and use as a measure of glacial erosion in New England'. *Journal of Geology*, 51: 71–98.

JEJE, L. K. (1973). 'Inselbergs' evolution in a humid tropical environment: the example of south western Nigeria'. *Zeitschrift für Geomorphologie N.F.*, 17: 194–225.

—— (1974). 'Effects of rock composition and structure on landform development: The example of the Idanre Hills of western Nigeria'. *Singapore Journal of Tropical Geography*, 39: 43–53.

JIBSON, R. W. (1989). 'Debris flows in southern Puerto Rico', in A. P. Schultz and R. W. Jibson (eds.), *Landslide Processes of the Eastern United States and Puerto Rico*, Geological Society of America, Special Paper, 236: 29–55.

JOHANSSON, M., MIGOŃ, P., and OLVMO, M. (2001a). 'Development of joint-controlled rock basins in Bohus granite, SW Sweden'. *Geomorphology*, 40: 145–61.

—— OLVMO, M., and LIDMAR-BERGSTRÖM, K. (2001b). 'Inherited landforms and glacial impact of different palaeosurfaces in southwest Sweden'. *Geografiska Annaler*, 83A: 67–89.

JOHNSON, A. M. (1970). *Physical Processes in Geology*. San Francisco: Freeman, Cooper and Company.

JONES, F. O. (1973). 'Landslides of Rio de Janeiro and the Serra das Araras escarpment, Brazil'. *US Geological Survey, Professional Paper*, 697.

JUTSON, J. T. (1914). 'An outline of the physiographic geology (physiography) of Western Australia'. *Geological Survey of Western Australia, Bulletin*, 61.

KAITANEN, V. (1985). 'Problems concerning the origin of inselbergs in Finnish Lapland'. *Fennia*, 163: 359–64.

KALVODA, J. (1994). 'Rock slopes of the High Tatras Mountains'. *Acta Universitatis Carolinae*, 29(2): 13–33.

Kaolin Deposits of the World (1969). Twenty-Third International Geological Congress, vol. 15, Prague.

KEJONEN, A. (1985). 'Weathering in the Wyborg rapakivi area, southeastern Finland'. *Fennia*, 163: 309–13.

—— KIELOSTO, S., and LAHTI, S. I. (1988). 'Cavernous weathering forms in Finland'. *Geografiska Annaler*, 70A: 315–22.

KESEL, R. H. (1973). 'Inselberg landform elements: definition and synthesis'. *Revue Géomorphologie Dynamique*, 22: 97–108.

—— (1977). 'Some aspects of the geomorphology of inselbergs in central Arizona, USA'. *Zeitschrift für Geomorphologie N.F.*, 21: 119–46.

KIM, S., and PARK, H.-D. (2003). 'The relationship between physical and chemical weathering indices of granites around Seoul, Korea'. *Bulletin of Engineering Geology and the Environment*, 62: 207–12.

KING, L. C. (1949). 'A theory of bornhardts'. *Geographical Journal*, 112: 83–7.

—— (1953). 'Canons of landscape evolution'. *Geological Society of America, Bulletin*, 64: 721–52.

—— (1958). 'Correspondence: the problem of tors'. *Geographical Journal*, 124: 289–91.

—— (1962). *The Morphology of the Earth*. Edinburgh: Oliver and Boyd.

—— (1966). 'The origin of bornhardts'. *Zeitschrift für Geomorphologie N.F.*, 10: 97–8.

—— (1975). 'Bornhardt landforms and what they teach'. *Zeitschrift für Geomorphologie N.F.*, 19: 299–318.

KLAER, W. (1956). 'Verwitterungsformen in Granit auf Korsika'. *Petermanns Geographische Mitteilungen, Ergänzungsheft*, 261: 1–146.

KLEMAN, J. (1994). 'Preservation of landforms under ice sheets and ice caps'. *Geomorphology*, 9: 19–32.

—— and STROEVEN, A. (1997). 'Preglacial surface remnants and Quaternary glacial regimes in northwestern Sweden'. *Geomorphology*, 19: 35–54.

KLIMASZEWSKI, M. (1964). 'On the effect of preglacial relief on the course and the magnitude of glacial erosion in the Tatra Mountains'. *Geographia Polonica*, 2: 11–21.

KÖRBER, E., and ZECH, W. (1984). 'Zur Kenntnis tertiärer Verwitterungsreste und Sedimente in der Oberpfalz und ihrer Ungebung'. *Relief, Boden and Paläoklima*, 3: 67–150.

KOTARBA, A. (1986). 'Granite hillslope morphology and present-day processes in semiarid zone of Mongolia'. *Geographia Polonica*, 52: 125–33.

—— (1987). 'Glacial cirques transformation under differentiated maritime climate'. *Studia Geomorphologica Carpatho-Balcanica*, 21: 77–92.

—— KRZEMIEŃ, K., and KASZOWSKI L. (1987). 'High-mountain denudational system of the Polish Tatra Mountains'. *Polish Academy of Sciences, Geographical Studies*, Special Issue 3: 1–106.

KRANZ, R. L. (1983). 'Microcracks in rocks: a review'. *Tectonophysics*, 100: 449–80.

KROONENBERG, S. B., and MELITZ, P. J. (1983). 'Summit levels, bedrock control and the etchplain concept in the basement of Suriname'. *Geologie en Mijnbouw*, 62: 389–99.

KRYGOWSKI, B., and KOSTRZEWSKI, A. (1971). 'Kilka geograficznych problemów podsofijskiej Witoszy'. *Czasopismo Geograficzne*, 42: 125–36.

KRÝSTKOVÁ, L. (1971). 'Kaolinová ložiska na Znojemsku'. *Časopis pro Mineralogie a Geologie*, 16: 159–72.

KUBINIOK, J. (1988). 'Kristallinvergrusung an Beispielen aus Südostaustralien und deutschen Mittelgebirgen'. *Kölner Geographische Arbeiten*, 48: 1–178.

KUMP, L. R., BRANTLEY, S. L., and ARTHUR, M. A. (2000). 'Chemical weathering, atmospheric CO_2, and climate'. *Annual Review of Earth and Planetary Science*, 28: 611–67.

KURAL, S. (1979). 'Geologiczne warunki występowania kaolinów w zachodniej części masywu strzegomskiego'. *Biuletyn Instytutu Geologicznego*, 313: 9–68.

KUŽVART, M. (1969). 'Kaolin deposits of Czechoslovakia', in *Kaolin Deposits of the World*, Twenty-Third International Geological Congress, Prague, 15: 47–73.

LAGASQUIE, J. J. (1984). *Géomorphologie des granites. Les massifs granitiques de la moitié orientale des Pyrénées françaises*. Thèse Doctorate État, Université du Paris I.

LAGEAT, Y. (1978). 'L'érosion différentielle dans les roches cristallines: l'exemple de la region de Barberton dans le Transvaal oriental, République d'Afrique du Sud'. *Géographie physique et Quaternaire*, 32: 105–17.

—— (1994). 'Le desert du Namib central'. *Annales de Géographie*, 103: 339–60.

—— SELLIER, D., and TWIDALE, C. R. (1994). 'Mégalithes et météorisation des granites en Bretagne littorale, France'. *Géographie Physique et Quaternaire*, 48: 107–13.

LAHTI, S. (1985). 'Porphyritic pyroxene-bearing granitoids—a strongly weathered rock group in central Finland'. *Fennia*, 163: 315–21.

LAN, H. X., HU, R. L., YUE, Z. Q., LEE, C. F., and WANG, S. J. (2003). 'Engineering and geological characteristics of granite weathering profiles in South China'. *Journal of Asian Earth Sciences*, 21: 353–64.

LASALLE, P., and DE KIMPE, C. (1989). 'Saprolites and related materials in Québec'. *Zeitschrift für Geomorphologie N.F., Supplement-Band*, 72: 139–47.

LAUTENSACH, H. (1950). 'Granitische Abtragungsformen auf der Iberischen Halbinsel und in Korea, ein Vergleich'. *Petermanns Geographische Mitteilungen*, 94: 187–96.

LAUTRIDOU, J. P., and SEPPÄLÄ, M. (1986). 'Experimental frost shattering of some Precambrian rocks, Finland'. *Geografiska Annaler*, 68A: 89–100.

LEAKE, B. E. (1990). 'Granite magmas: their sources, initiation and consequences of emplacement'. *Journal of the Geological Society in London*, 147: 579–89.

LE COEUR, C. (1989). 'La question des altérites profondes dans la region des Hébrides internes (Ecosse occidentale)'. *Zeitschrift für Geomorphologie N.F., Supplement-Band*, 72: 109–24.

LE PERA, E., and SORRISO-VALVO, M. (2000). 'Weathering and morphogenesis in a Mediterranean climate, Calabria, Italy'. *Geomorphology*, 34: 251–70.

LEITE MAGALHÃES, S., and SEQUEIRA BRAGA, M. A. (2000). 'Biological colonization features on a granite monument from Braga (NW, Portugal)', in V. Fassina (ed.), *Ninth International Congress on Deterioration and Conservation of Stone, Venice 19–24 June 2000*. Amsterdam: Elsevier, 521–9.

LEONARD, R. J. (1927). 'Pedestal rocks resulting from disintegration'. *Journal of Geology*, 35: 469–74.

—— (1929). 'Polygonal cracking in granite'. *American Journal of Science*, 18: 487–92.

LIDMAR-BERGSTRÖM, K. (1986). 'Flint and Pre-Quaternary geomorphology in south Sweden and south-west England', in G. De Sieveking and M. B. Hart (eds.), *The Scientific Study of Flint and Chert, Proceedings of the Fourth International Flint Symposium*. Cambridge: Cambridge University Press, 191–9.

—— (1989). 'Exhumed Cretaceous landforms in south Sweden'. *Zeitschrift für Geomorphologie N.F., Supplement-Band*, 89: 21–40.

—— (1995). 'Relief and saprolites through time on the Baltic Shield'. *Geomorphology*, 12: 45–61.

—— (1996). 'Long-term morphotectonic evolution in Sweden'. *Geomorphology*, 16: 33–59.

—— (1997). 'A long-term perspective on glacial erosion'. *Earth Surface Processes and Landforms*, 22: 297–306.

—— OLSSON, S., and OLVMO, M. (1997). 'Palaeosurfaces and associated saprolites in southern Sweden', in M. Widdowson (ed.), *Palaeosurfaces: Recognition, Reconstruction and Palaeoenvironmental Interpretation*. Geological Society Special Publication, 120: 95–124.

—— —— and ROALDSET, E. (1999). 'Relief features and palaeoweathering remnants in formerly glaciated Scandinavian basement areas'. *International Association of Sedimentologists, Special Publication*, 27: 275–301.

LINTON, D. L. (1955). 'The problem of tors'. *Geographical Journal*, 121: 470–487.

—— (1964). 'The origin of the Pennine tors—An essay in analysis'. *Zeitschrift für Geomorphologie N.F.*, 8, Sonderheft, 5–24.

Lippert, H. J., Lob, F., Meisl, S., Rée, C., Salger, M., Stadler, G., and Teuscher, E. O. (1969). 'Die Kaolinlagerstätten der Bundesrepublik Deutschland', in *Kaolin Deposits of the World*, Twenty-Third International Geological Congress, Prague, 15: 85–105.

Lugo, J., Vázquez Conde, M. T., Zamorano Orozco, J. J., Matías Ramírez, G., and Gómez Arizmendi, A. (2002). 'El huracán Pauline en Acapulco, octubre de 1997', in J. Lugo and M. Inbar (eds.), *Desastres naturales en América Latina*. México: Fondo de Cultura Económica, 267–88.

Luk, S. H., Yao, Q. Y., Gao, J. Q., Zhang, J. Q., He, Y. G., and Huang, S. M. (1997). 'Environmental analysis of soil erosion in Guangdong Province: a Deqing case study'. *Catena*, 29: 97–113.

Lukniš, M. (1973). *Reliéf Vysokých Tatier a ich predpolia*. Bratislava: Vydavatel'stvo Slovenskej akadémie vied.

Lumb, P. (1962). 'The properties of decomposed granite'. *Géotechnique*, 12: 226–43.

—— (1975). 'Slope failures in Hong Kong'. *Quarterly Journal of Engineering Geology*, 8: 31–65.

Lundqvist, J. (1985). 'Deep-weathering in Sweden'. *Fennia*, 163: 287–92.

Mabbutt, J. A. (1952). 'A study of granite relief from South West Africa'. *Geological Magazine*, 89: 87–96.

—— (1955). 'Pediment landforms in Little Namaqualand'. *Geographical Journal*, 121: 77–83.

—— (1961a). 'A stripped landsurface in Western Australia'. *Institute of British Geographers, Transactions*, 29: 101–14.

—— (1961b). ' "Basal Surface" or "Weathering Front" '. *Proceedings of the Geologists' Association*, 72: 357–59.

—— (1965). 'The weathered landsurface of central Australia'. *Zeitschrift für Geomorphologie N.F.*, 9: 82–114.

—— (1966). 'Mantle-controlled planation of pediments'. *American Journal of Science*, 264: 78–91.

McConnell, R. B. (1968). 'Planation surfaces in Guyana'. *Geographical Journal*, 134: 506–20.

Mäckel, R. (1985). 'Dambos and related landforms in Africa—an example for the ecological approach to tropical geomorphology'. *Zeitschrift für Geomorphologie N. F., Supplement-Band*, 52: 1–23.

Maniar, P. D., and Piccoli, P. M. (1989). 'Tectonic discrimination of granitoids'. *Geological Society of America Bulletin*, 101: 635–43.

Marmo, V. (1956). 'On the porphyroblastic granite of central Sierra Leone'. *Acta Geographica* (Helsinki), 15: 1–26.

—— (1971). *Granite Petrology and the Granite Problem*. Amsterdam: Elsevier.

Martel, S. J., Pollard, D. D., and Segall, P. (1988). 'Development of simple strike-slip-fault zones, Mount Abbott quadrangle, Sierra Nevada, California'. *Geological Society of America Bulletin*, 100: 1451–65.

Martínez Lope, M. J., García González, M. T., and Molina, E. (1995). 'Relationships between geomorphology and palaeoweatherings on the hercynian basement in central Spain. A mineralogical and geochemical approach'. *Revista de la Sociedad Geológica de España*, 8(1–2): 127–36.

MARTINI, A. (1967). 'Preliminary experimental studies on frost weathering of certain rock types from the West Sudetes'. *Biuletyn Peryglacjalny*, 16: 147–94.

—— (1969). 'Sudetic tors formed under periglacial conditions'. *Biuletyn Peryglacjalny*, 19: 351–69.

MARUTANI, T., KASAI, M., EBISU, N., and TRUSTRUM, N. A. (2000). 'Sediment generation from numerous shallow landslides related with clear cutting at granite mountain, Mt. Ichifusa, Japan'. *Proceedings INTERPRAEVENT*, 1: 271–79.

MASSELINK, G., and HUGHES, M. G. (2003). *Introduction to Coastal Processes and Geomorphology*. London: Arnold.

MATIAS, J. M. S., and ALVES, C. A. S. (2002), 'The influence of petrographic, architectural and environmental factors in decay patterns and durability of granite stones in Braga monuments (NW Portugal)', in S. Siegesmund, Weiss, T. and Vollbrecht, A. (eds.), *Natural Stone, Weathering Phenomena, Conservation Strategies and Case Studies*. Geological Society Special Publication, 205: 273–81.

MATSUKURA, Y., and TANAKA, Y. (2000). 'Effect of rock hardness and moisture content on tafoni weathering in the granite of Mount Doeg-sung, Korea'. *Geografiska Annaler*, 82A: 59–67.

MATSUOKA, N. (2001a). 'Microgelivation versus macrogelivation: towards bridging the gap between laboratory and field frost weathering'. *Permafrost and Periglacial Processes*, 12: 299–313.

—— (2001b). 'Solifluction rates, processes and landforms: a global review'. *Earth Science Reviews*, 55: 107–34.

MATTHES, F. E. (1950). *The Incomparable Valley. A geologic interpretation of the Yosemite*. Berkeley and Los Angeles: University of California Press.

—— (1956). *Sequoia National Park. A Geological Album*. Berkeley and Los Angeles: University of California Press.

MAYER, E. (1992). 'Inventaire géomorphologique: Dynamique externe et evolution des formes au Cameroun septentrional'. *Zeitschrift für Geomorphologie N.F.*, 36: 385–99.

MEADOWS, M. E. (1985). 'Dambos and environmental change in Malawi, central Africa'. *Zeitschrift für Geomorphologie N.F., Supplement-Band*, 52: 147–69.

MEINECKE, F. (1957). 'Granitverwitterung, Entstehung und Alter der Granitklippen'. *Zeitschrift der deutchen geologischen Gesellschaft*, 109: 483–98.

MELTON, M. A. (1965). 'Debris-covered hillslopes of the southern Arizona desert—consideration of their stability and sediment contributions'. *Journal of Geology*, 73: 715–29.

MENSCHING, H. (1978). 'Inselberge, Pedimente und Rumpfflächen im Sudan (Republik): Ein Beitrag zur morphogenetischen Sequenz in den ariden Subtropen and Tropen Afrikas'. *Zeitschrift für Geomorphologie N.F., Supplement-Band*, 30: 1–19.

MEYER, R. (1967). 'Studien über Inselberge und Rumpfflächen in Nordtransvaal'. *Münchner Geographische Hefte*, 31.

MICHELL, J. (1982). *Megalithomania*, Ithaca, NY: Cornell University Press.

MIGOŃ, P. (1993). 'Kopułowe wzgórza granitowe w Kotlinie Jeleniogórskiej'. *Czasopismo Geograficzne*, 64: 3–23.

—— (1996). 'Granite landscapes of the Sudetes Mountains: some problems of interpretation: a review'. *Proceedings of the Geologists' Association*, 107: 25–38.

—— (1997a). 'Palaeoenvironmental significance of grus weathering profiles: a review with special reference to northern and central Europe', in M. Widdowson (ed.), Palaeosurfaces: Recognition, Reconstruction and Palaeoenvironmentel Interpretation. *Proceedings of the Geologists' Association*, 108: 57–70.

—— (1997b). 'Tertiary etchsurfaces in the Sudetes Mountains, SW Poland: a contribution to the pre-Quaternary morphology of Central Europe', in M. Widdowson (ed.), Palaeosurfaces: Recognition, Reconstruction and Palaeoenvironmental Interpretation. *Geological Society, Special Publication*, 120: 187–202.

—— (1997c). 'The geologic control, origin and significance of inselbergs in the Sudetes, NE Bohemian Massif, Central Europe'. *Zeitschrift für Geomorphologie N.F.*, 41: 45–66.

—— (1999). 'The role of "preglacial" relief in the development of mountain glaciation in the Sudetes, with the special reference to the Karkonosze mountains'. *Zeitschrift für Geomorphologie, N.F., Supplement-Band*, 113: 33–44.

—— (2000). 'Geneza jaskiń granitowych na Witoszy w Kotlinie Jeleniogórskiej'. *Kras i Speleologia*, 10(19): 143–54.

—— (2004a). 'Bornhardt', in A. S. Goudie (ed.), *Encyclopedia of Geomorphology*. London: Routledge, 92–3.

—— (2004b). 'Grus', in A. S. Goudie (ed.), *Encyclopedia of Geomorphology*. London: Routledge, 501–3.

—— and DACH, W. (1995). 'Rillenkarren on granite outcrops, SW Poland, age and significance'. *Geografiska Annaler*, 77A: 1–9.

—— and GOUDIE, A. S. (2003). 'Granite landforms of the Central Namib'. *Acta Universitatis Carolinae, Geographica 35, Supplement*, 17–38.

—— and JOHANSSON, M. (2004). 'Lithological and structural influence on the development of basin-and-hill landscape within a basement complex in SW Sweden'. *Zeitschrift für Geomorphologie, N.F.*, 48: 305–22.

—— and LIDMAR-BERGSTRÖM, K. (2001). 'Weathering mantles and their significance for geomorphological evolution of central and northern Europe since the Mesozoic'. *Earth Science Reviews*, 56: 285–324.

—— and ROŠTINSKÝ, P. (2003). 'Granite landscape of Krumlovský les, South Moravia: an example of a variety of structural controls'. *Moravian Geographical Reports*, 11(1): 36–44.

—— and THOMAS, M. F. (2002). 'Grus weathering mantles—problems of interpretation'. *Catena*, 49: 5–24.

—— and VIEIRA, G. T. (In preparation). 'Rock–landform relationships in a granite terrain, Serra da Estrela, Portugal'.

MÖBUS, G. (1956). *Einführung in die geologische Geschichte der Oberlausitz*. Berlin: Deutscher Verlag der Wissenschaften.

MODENESI, M. C. (1988). 'Quaternary mass movements in a tropical plateau (Campos do Jordão, São Paulo, Brazil)'. *Zeitschrift für Geomorphologie, N.F.*, 32: 425–40.

MONROE, W. H. (1979). 'Some tropical landforms of Puerto Rico'. *US Geological Survey, Professional Paper*, 1159: 1–39.

MONTGOMERY, D. R., and DIETRICH, W. E. (1994). 'A physically based model for the topographic control on shallow landsliding'. *Water Resources Research*, 30: 1153–71.

MORGAN, R. P. C., and MNGOMEZULU, D. (2003). 'Threshold conditions for initiation of valley-side gullies in the Middle Veld of Swaziland'. *Catena*, 50: 401–14.

MORGAN, R. P. C, and RICKSON, R. J., MCINTYRE, K., BREWER, T. R., and ALTSHUL, H. J. (1997). 'Soil erosion survey of the central part of the Swaziland Middleveld'. *Soil Technology*, 11: 263–89.

MÖRNER, N. A. (2003). *Paleoseismicity of Sweden: a novel paradigm*. Stockholm: Paleogeophysics and Geodynamics, Stockholm University.

MOSS, J. H. (1977). 'The formation of pediments: scarp backwearing or surface downwearing', in D. O. Doehring (ed.), *Geomorphology in Arid Regions*. Boston: Allen and Unwin, 51–75.

MOTTERSHEAD, D., and LUCAS, G. (2004). 'The role of mechanical and biotic processes in solution flute development', in B. J. Smith and A. V. Turkington (eds.), *Stone Decay: Its Causes and Controls*. Shaftesbury: Donhead, 273–91.

MOYERSONS, J. (1977). 'Joint patterns and their influence on the form of granitic residuals in NE Nigeria'. *Zeitschrift für Geomorphologie N.F.*, 21: 14–25.

MUSTOE, G. E. (1982). 'The origin of honeycomb weathering'. *Geological Society of America, Bulletin*, 93: 108–15.

MUXART, T., COSANDEY, C., and BILLARD, A. (1990). *L'érosion sur les hautes terres du Lingas*. Mémoires et documents de géographie. Nouvelle collection, Paris: Editions du CRNS.

NESBITT, H. W., and YOUNG, G. M. (1989). 'Formation and diagenesis of weathering profiles'. *Journal of Geology*, 97: 129–47.

NORRIS, R. M. (1996). 'Eastern Mojave desert pediments'. *California Geology*, 49: 3–10.

O'BEIRNE-RYAN, A.-M., and ZENTILLI, M. (2003). 'Paleoweathered surfaces on granitoids of southern Nova Scotia: paleoenvironmental implications of saprolites'. *Canadian Journal of Earth Sciences*, 40: 805–17.

OBERLANDER, T. (1972). 'Morphogenesis of granitic boulder slopes in the Mojave Desert, California'. *Journal of Geology*, 80: 1–20.

—— (1974). 'Landscape inheritance and the pediment problem in the Mojave Desert of southern California'. *American Journal of Science*, 274: 849–75.

—— (1989). 'Slope and pediment system', in D. S. G. Thomas (ed.), *Arid Zone Geomorphology*. London: Belhaven, 56–84.

OEN, I. S. (1965). 'Sheeting and exfoliation in the granites of Sermersoq, South Greenland'. *Meddelelser om Grønland*, 176(6): 1–40.

OH, K.-S., and KEE, K.-D. (2001). 'Genesis of granitic regolith and its deformation by morphogenetic processes in actual periglacial environment of middle latitude (Daegwallyong area in Taebaek Range, Korea)'. *Transactions, Japanese Geomorphological Union*, 22: 337–49.

OHMORI, H. (2000). 'Morphotectonic evolution of Japan', in M. A. Summerfield (ed.), *Geomorphology and Global Tectonics*. Chichester: Wiley, 147–66.

OLIVA, P., VIERS, J., and DUPRÉ, B. (2003). 'Chemical weathering in granitic environments'. *Chemical Geology*, 202: 225–56.

OLLIER, C. D. (1960). 'The inselbergs of Uganda'. *Zeitschrift für Geomorphologie N.F.*, 4: 43–52.

—— (1963). 'Insolation weathering: examples from Central Australia'. *American Journal of Science*, 261: 376–87.

—— (1965). 'Some features of granite weathering'. *Zeitschrift für Geomorphologie N.F.*, 9: 265–84.

—— (1967). 'Spheroidal weathering, exfoliation and constant volume alteration'. *Zeitschrift für Geomorphologie N.F.*, 11: 103–8.

—— (1976). 'Catenas in different climates', in E. Derbyshire (ed.), *Geomorphology and Climate*. London: Wiley, 137–69.

—— (1977). 'Applications of weathering studies', in J. R. Hails (ed.), *Applied Geomorphology*. Amsterdam: Elsevier, 9–50.

—— (1978a). 'Induced fracture and granite landforms'. *Zeitschrift für Geomorphologie N.F.*, 22: 249–57.

—— (1978b). 'Inselbergs of the Namib Desert: process and history'. *Zeitschrift für Geomorphologie N.F., Supplement-Band*, 31: 161–76.

—— (1983). 'Weathering or hydrothermal alteration'. *Catena*, 10: 57–9.

—— (2004). 'The evolution of mountains on passive continental margins', in P. N. Owens and O. Slaymaker (eds.), *Mountain Geomorphology*, London: Arnold, 59–88.

—— and ASH, J. E. (1983). 'Fire and rock breakdown'. *Zeitschrift für Geomorphologie N.F.*, 27: 363–74.

OLVMO, M., and JOHANSSON, M. (2002). 'The significance of rock structure, lithology and preglacial deep weathering for the shape of intermediate-scale glacial erosional landforms'. *Earth Surface Processes and Landforms*, 27: 251–68.

—— LIDMAR-BERGSTRÖM, K., and LINDBERG, G. (1999). 'The glacial impact on an exhumed sub-Mesozoic etch surface in southwestern Sweden'. *Annals of Glaciology*, 28: 153–60.

ONDA, Y. (2004). 'Hillslope hydrology and mass movement in the Japanese Alps', in P. N. Owens and O. Slaymaker (eds.), *Mountain Geomorphology*, London: Arnold, 153–64.

OYAGI, N. (1989). 'Geological and economic extent of landlides in Japan and Korea', in E. E. Brabb and B. L. Harrod (eds.), *Landslides. Extent and Economic Significance*. Rotterdam: A. A. Balkema, 289–302.

PAIN, C. F., and OLLIER, C. D. (1981). 'Geomorphology of a Pliocene granite in Papua New Guinea'. *Zeitschrift für Geomorphologie N.F.*, 25: 249–58.

PALMER, J., and NEILSON, R. A. (1962). 'The origin of granite tors on Dartmoor, Devonshire'. *Proceedings of the Yorkshire Geological Society*, 33: 315–40.

—— and RADLEY, J. (1961). 'Gritstone tors of the English Pennines'. *Zeitschrift für Geomorphologie N.F.*, 5: 37–52.

PANZER, W. (1954). 'Verwitterungs- und Abtragungsformen im Granit von Hongkong', in *Ergebnisse und Probleme moderner geographischer Forschung. Hans Mortensen zu seinem 60. Geburtstag*. Bremen: Walter Dorn, 41–60.

PARADISE, T. R., and ZHI-YONG YIN (1993). 'Weathering pit characteristics and topography on Stone Mountain, Georgia'. *Physical Geography*, 14: 68–80.

PARTSCH, J. (1894). 'Die Vergletscherung des Riesengebirges zur Eiszeit'. *Forschungen zur deutschen Landes- und Volkskunde*, 8(2): 103–94.

PAVICH, M. J. (1985). 'Appalachian piedmont morphogenesis: weathering, erosion and Cenozoic uplift', in M. Morisawa and J. T. Hack (eds.), *Tectonic Geomorphology*. London: George Allen and Unwin, 27–51.

PAVICH, M.J. (1986). 'Processes and rates of saprolite production and erosion on a foliated granitic rock of the Virginia Piedmont', in S. M. Colman and D. P. Dethier (eds.), *Rates of Chemical Weathering of Rocks and Minerals*. Orlando: Academic Press, 552–90.

—— (1989). 'Regolith residence time and the concept of surface age of the Piedmont "peneplain" '. *Geomorphology*, 2: 181–96.

—— and OBERMEIER, S. F. (1985). 'Saprolite formation beneath Coastal Plain sediments near Washington, D.C.'. *Geological Society of America, Bulletin*, 96: 886–900.

PEDRAZA, J., ANGEL SANZ, M., and MARTÍN, A. (1989). *Formas graniticas de la Pedriza*. Madrid: Agencia de Medio Ambiente.

PĘKALA, K., and ZIĘTARA, T. (1980). 'Present-day slope modelling in the southern Khangai Mountains'. *Polish Academy of Sciences, Geographical Studies*, 136: 52–64.

PENCK, A. (1894). *Morphologie der Erdoberfläche*. Stuttgart: Engelhorn.

PENCK, W. (1924). *Die Morphologische Analyse*. Stuttgart: Engelhorn.

PETFORD, N., and CLEMENS, J. D. (2000). 'Granites are not diapiric!'. *Geology Today*, 16: 180–4.

PEULVAST, J. P. (1989). 'Les altérites et l'identification des reliefs préglaciaires dans une montagne de haute latitude: l'exemple des Scandes'. *Zeitschrift für Geomorphologie N.F., Supplement-Band*, 89: 55–78.

PIERRE, G. (1990). 'Générations d'altérites dans le Massif central français (Auvergne, Aubrac, Velay) du Miocene au Quaternaire: implications paléoclimatologiques et géomorphologiques'. *Physio-Géo*, 20: 31–50.

PILOUS, V. (1993). 'Pseudokrasové jeskyní v Labském dole v Krkonoších'. *Opera Corcontica*, 30: 117–31.

PITCHER, W. S. (1978). 'The anatomy of a batholith'. *Journal of the Geological Society in London*, 135: 157–82.

—— (1982). 'Granite type and tectonic environment', in K. J. Hsü (ed.), *Mountain Building Processes*. London: Academic Press, 19–40.

—— (1997). *The Nature and Origin of Granite*. London: Chapman and Hall.

PITTS, J. (1984). 'A review of geology and engineering geology in Singapore'. *Quarterly Journal of Engineering Geology*, 17: 93–101.

PLAFKER, G., and ERICKSEN, G. E. (1978). 'Nevados Huascarán avalanches, Peru', in B. Voight (ed.), *Rockslides and Avalanches, 1*. Amsterdam: Elsevier, 277–314.

PLAYFAIR, J. (1802). *Illustrations of the Huttonian Theory of the Earth*. Edinburgh: Creech.

POWER, E. T., and SMITH, B. J. (1994). 'A comparative study of deep weathering and weathering products: Case studies from Ireland, Corsica and Southeast Brazil', in D. A. Robinson and R. B. G. Williams (eds.), *Rock Weathering and Landform Evolution*. Chichester: Wiley, 21–40.

—— —— and WHALLEY, W. B. (1990). 'Fracture patterns and grain release in physically weathered granitic rocks', in L. A. Douglas (ed.), *Soil Micromorphology: A Basic and Applied Science. Developments in Soil Science*, Amsterdam: Elsevier, 545–50.

PRICE, N. J., and COSGROVE, J. (1990). *Analysis of Geological Structures*. Cambridge: Cambridge University Press.

PRICE WILLIAMS, D., WATSON, A., and GOUDIE, A. S. (1982). 'Quaternary colluvial stratigraphy, archaeological sequences and palaeoenvironment in Swaziland'. *Geographical Journal*, 148: 50–67.

PSYRILLOS, A., MANNING, D. A. C., and BURLEY, S. D. (1998). 'Geochemical constraints on kaolinisation in the St. Austell granite, Cornwall, England'. *Journal of Geological Society, London*, 155: 829–40.

PUGH, J. C. (1956). 'Fringing pediments and marginal depressions in the inselberg landscape of Nigeria'. Institute of British Geographers, Transactions 22: 15–31.

PULLAN, R. A. (1959). 'Tors'. *Scottish Geographical Magazine*, 75: 51–55.

PYE, K. (1986). 'Mineralogical and textural controls on the weathering of granitoid rocks'. *Catena*, 13: 47–57.

—— GOUDIE, A. S., and THOMAS, D. S. G. (1984). 'A test of petrological control in the development of bornhardts and koppies on the Matopos Batholith, Zimbabwe'. *Earth Surface Processes and Landforms*, 9: 455–67.

—— —— and WATSON, A. (1986). 'Petrological influence on differential weathering and inselberg development in the Kora area of Central Kenya'. *Earth Surface Processes and Landforms*, 11: 41–52.

RAHN, P. H. (1967). 'Sheetfloods, streamfloods and the formation of pediments'. *Annals of the Association of American Geographers*, 57: 593–604.

RAUKAS, A. (1992). 'Estland—das Land der großen erratischen Blöcke'. *Der Geschiebesammler*, 25: 81–90.

REA, B. R., WHALLEY, W. B., RAINEY, M. M., and GORDON, J. E. (1996). 'Blockfields, old or new? Evidence and implications from some plateaus in northern Norway'. *Geomorphology*, 15: 109–21.

READ, H. H. (1957). *The Granite Controversy*. London: Murby.

REICHE, P. (1943). 'Graphic representation of chemical weathering'. *Journal of Sedimentary Petrology*, 13: 58–68.

RICHTER, H. (1963). 'Das Vorland des Erzgebirges. Die Landformung während des Tertiärs'. *Wissenschaftliche Veröffentlichungen des deutschen Instituts für Länderkunde*, N.F., 19–20: 5–231.

RINGROSE, P. S., and MIGOŃ, P. (1997). 'Analysis of digital elevation data for the Scottish Highlands and recognition of Pre-Quaternary elevated surfaces', in M. Widdowson (ed.), Palaeosurfaces: Recognition, Reconstruction and Palaeoenvironmental Interpretation. *Geological Society, Special Publication*, 120: 25–35.

RISER, J. (1975). 'Les modelés des granites du Jebel Sarhro oriental (Anti-Atlas oriental)'. *Revue de Géographie Physique et Géologie Dynamique*, 17: 61–72.

RIVAS, T., PRIETO, B., SILVA, B., and BIRGINIE, J. M. (2003). 'Weathering of granitic rocks by chlorides: effect of the nature of the solution on weathering morphology'. *Earth Surface Processes and Landforms*, 28: 425–36.

ROBERTSON, I. D. M., and BUTT, C. R. M. (1997). *Atlas of Weathered Rocks*. CRC LEME Open File Report 1, CSIRO, Wembley, W. Australia.

ROMÁN-BERDIEL, T., GAPAIS, D., and BRUN, J.-P. (1997). 'Granite intrusion along strike-slip zones in experiment and nature'. *American Journal of Science*, 297: 651–78.

Romão, P. M. S., and Rattazzi, A. (1996). 'Biodeterioration on megalithic monuments. Study of lichens' colonization on Tapadão and Zambujeiro dolmens (Southern Portugal)'. *International Biodeterioration and Biodegradation*, 32: 23–35.

Roštinský, P. (2004). 'Geomorphology of the Diendorf Fault area on the SE margin of the Bohemian Massif in SW Moravia and NE Austria'. *Studia Geomorphologca Carpatho-Balcanica*, 38: 67–81.

Ruxton, B. P., and Berry, L. (1957). 'Weathering of granite and associated erosional features in Hong Kong'. *Geological Society of America, Bulletin*, 68: 1263–82.

—— —— (1961). 'Weathering profiles and geomorphic position on granite in two tropical regions'. *Revue de Géomorphologie dynamique*, 12: 16–31.

Sarapää, M. (1996). *Proterozoic primary kaolin deposits at Virtasalmi, south-eastern Finland*. Espoo: Geological Survey of Finland.

Schaefer, C., and Dalrymple, J. D. (1995). 'Landscape evolution in Roraima, North Amazonia: planation, paleosols and paleoclimates'. *Zeitschrift für Geomorphologie N.F.*, 39: 1–28.

Schaffer, J. P. (1997). *The Geomorphic Evolution of the Yosemite Valley and Sierra Nevada Landscapes*. Berkeley: Wilderness Press.

Schattner, I. (1961). 'Weathering phenomena in the crystalline of the Sinai in the light of current notions'. *Bulletin of the Research Council of Israel*, 10: 247–66.

Schiavon, N. (2002). 'Biodeterioration of calcareous and granitic building stones in urban environments', in S. Siegesmund, Weiss, T. and Vollbrecht, A. (eds.), *Natural Stone, Weathering Phenomena, Conservation Strategies and Case Studies*. Geological Society Special Publication, 205: 195–205.

Schnütgen, A. (1991). 'Spheroidal weathering, granular disintegration and loamification of compact rock under different climatic conditions'. *Zeitschrift für Geomorphologie N.F., Supplement-Band*, 91: 79–94.

Schrepfer, H. (1933). 'Inselberge in Lappland und Neufundland'. *Geologische Rundschau*, 24: 137–43.

Schülke, H. (1973). 'Schildkrötenmuster und andere Polygonalstrukturen auf Felsoberflächen'. *Zeitschrift für Geomorphologie N.F.*, 17: 474–88.

Schuster, R. L., Salcedo, D. A., and Valenzuela, L. (2002). 'Overview of catastrophic landslides of South America in the twentieth century'. *Geological Society of America, Reviews in Engineering Geology*, 15: 1–34.

Scourse, J. D. (1987). 'Periglacial sediments and landforms in the Isles of Scilly and West Cornwall', in J. Boardman (ed.), *Periglacial Processes and Landforms in Britain and Ireland*. Cambridge: Cambridge University Press, 225–36.

Segall, P., McKee, E. H., Martel, S. J., and Turrin, B. D. (1990). 'Late Cretaceous age of fractures in the Sierra Nevada batholith'. *Geology*, 18: 1248–51.

—— and Pollard, D. D. (1983). 'Joint formation in granitic rock of the Sierra Nevada'. *Geological Society of America Bulletin*, 94: 563–75.

Sekyra, J. (1964). 'Kvartérně geologické a geomorfologické problémy krkonošského krystalinika'. *Opera Corcontica*, 1: 7–24.

SELBY, M. J. (1971). 'Slopes and their development in an ice-free, arid area of Antarctica'. *Geografiska Annaler*, 53A: 235–45.

—— (1972). 'Antarctic tors'. *Zeitschrift für Geomorphologie N.F., Supplement-Band*, 13: 73–86.

—— (1977). 'Bornhardts of the Namib Desert'. *Zeitschrift für Geomorphologie N.F.*, 21: 1–13.

—— (1980). 'A rock-mass strength classification for geomorphic purposes: with tests from Antarctica and New Zealand'. *Zeitschrift für Geomorphologie N.F.*, 24: 31–51.

—— (1982*a*). 'Controls on the stability and inclinations of hillslopes formed on hard rock'. *Earth Surface Processes and Landforms*, 7: 449–67.

—— (1982*b*). 'Form and origin of some bornhardts of the Namib Desert'. *Zeitschrift für Geomorphologie N.F.*, 26: 1–15.

SELLIER, D. (1997). 'Utilisation des megaliths comme marqueurs de la vitesse de l'érosion des granites en milieu tempéré: enseignements apportés par les alignements de Carnac (Morbihan)'. *Zeitschrift für Geomorphologie N.F.*, 41: 319–56.

—— (1998). 'Signification de quelques marqueurs des rythmes de la météorisation des granites en milieu tempéré océanique'. *Cahiers Nantais*, 49: 87–110.

SEMMEL, A. (1985). *Periglazialgeomorphologie*. Darmstadt: Erträge der Forschung, 231.

SEN, D. (1983). 'Geomorphology of the Aravalli Range, Rajasthan and a reinterpretation of residual erosion surfaces', in K. R. Dikshit (ed.), *Contributions to Indian Geography II: Geomorphology*, New Delhi: Heritage Publishers, 53–71.

SEO, Y. S., JEONG, G. C., KIM, J. S., and ICHIKAWA, Y. (2002). 'Microscopic observation and contact stress analysis of granite under compression'. *Engineering Geology*, 63: 259–75.

SEQUEIRA-BRAGA, M. A., NUNES, J. E., PAQUET, H., and MILLOT, G. (1990). 'Climatic zonality of coarse granitic saprolites ("arenes") in Atlantic Europe from Scandinavia to Portugal', in V. C. Farmer and Y. Tardy (eds.), *Proceedings of the Ninth International Clay Conference, Strasbourg 1989*. Sciences Géologiques, Memoir, 85: 99–108.

SHAW, R. (1997). 'Variations in sub-tropical deep weathering profiles over the Kowloon Granite, Hong Kong'. *Journal of the Geological Society, London*, 154: 1077–85.

SHENG, J., and LIAO, A. (1997). 'Erosion control in South China'. *Catena*, 29: 211–21.

SHEPARD, F. P., and WANLESS, H. R. (1971). *Our Changing Coastlines*. New York: McGraw-Hill.

SHEPPARD, S. M. F. (1977). 'The Cornubian batholith SW England: D/H and $^{18}O/^{16}O$ studies of kaolinite and other alteration minerals'. *Journal of the Geological Society, London*, 133: 573–91.

SHRODER, J. F., Jr. (1976). 'Mass movements on the Nyika Plateau, Malawi'. *Zeitschrift für Geomorphologie N.F.*, 20: 56–77.

SIMMONS, G., RICHTER, D. (1976), 'Microcracks in rocks', in R. G. J. Strens (ed.), *The Physics and Chemistry of Minerals and Rocks*, New York: Wiley, 105–37.

SIMON-COINÇON, R., THIRY. M., and SCHMITT, J.-M. (1997). 'Variety and relationships of weathering features along the early Tertiary palaeosurface in the southwestern French Massif Central and the nearby Aquitaine Basin'. *Palaeogeography, Palaeoclimatology, Palaeoecology*, 129: 51–79.

SJÖBERG, R. (1981). 'Tunnel caves in Swedish Archean rocks'. *Transactions of the British Cave Research Association*, 8(3): 159–67.

—— (1987). 'Caves indicating neotectonic activity in Sweden'. *Geografiska Annaler*, 68A: 393–8.

SLAYMAKER, O. (1988). 'The distinctive attributes of debris torrents'. *Hydrological Sciences Journal*, 33: 567–73.

SMITH, B. J., and MCALISTER, J. J. (1987). 'Tertiary weathering environments and products in northeast Ireland', in V. Gardiner (ed.), *International Geomorphology 1986*, II. Chichester: Wiley, 1007–31.

SO, C. L. (1971). 'Mass movements associated with the rainstorm of June 1966 in Hong Kong'. *Institute of British Geographers, Transactions*, 53: 55–66.

—— (1987). 'Coastal forms in granite, Hong Kong', in V. Gardiner (ed.), *International Geomorphology 1986*, I. Chichester: Wiley, 1213–29.

SÖDERMAN, G. (1985). 'Planation and weathering in eastern Fennoscandia'. *Fennia*, 163: 347–52.

—— KEJONEN, A., and KUJANSUU, R. (1983). 'The riddle of the tors at Lauhavuori, western Finland'. *Fennia*, 161: 91–144.

STEERS, J. A. (1948). *The Coastline of England and Wales*. Cambridge: Cambridge University Press.

STETTNER, G. (1958). *Erläuterungen zur Geologischen Karte von Bayern, 1: 25 000, Blatt Fichtelberg*. Munich: Bayerisches Geologisches Landesamt.

STODDART, D. R. (1969). 'Climatic geomorphology', in R. J. Chorley (ed.), *Water, Earth and Man*. London: Methuen, 473–85.

STÖRR, M. (1983). 'Die Kaolinlagerstätten der Deutschen Demokratischen Republik'. *Schriftenreihe der Geologischen Wissenschaften*, 18.

—— KÖSTER, H. M., KUŽVART, M., SZPILA, K., and WIEDEN, P. (1977). 'Kaolin deposits of central Europe', in *Proceedings of the Eighth International Kaolin Symposium and Meeting on Alunite, K-20*. Madrid and Rome, 1–21.

STRECKEISEN, A. (1976). 'To each plutonic rock its proper name'. *Earth Science Reviews*, 12: 1–33.

STRIEBEL, T. (1999). 'Working meeting "Caves in Sandstone and in Granite" '. *Mitteilungsheft der Höhlenforschungsgruppe Blaustein*, 15(1): 45–53.

STROEVEN, A. P., FABEL, D., HÄTTESTRAND, C., and HARBOR, J. (2002). 'A relict landscape in the centre of Fennoscandian glaciation: cosmogenic radionuclide evidence of tors preserved through multiple glacial cycles'. *Geomorphology*, 44: 145–54.

STROM, A. L. (1996). 'Some morphological types of long-runout rockslides: effect of the relief on their mechanism and the rockslide deposits distribution', in K. Senneset (ed.), *Landslides*, Rotterdam: Balkema, 1977–82.

SUEOKA, T. (1988). 'Identification and classification of granitic residual soils using chemical weathering index', in *Geomechanics in Tropical Soils. Proceedings of the Second International Conference on Geomechanics in Tropical Soils, Singapore*, vol. 1. Rotterdam : A. A. Balkema, 55–61.

SUGDEN, D. E. (1968). 'The selectivity of glacial erosion in the Cairngorm mountains'. *Institute of British Geographers, Transactions*, 45: 79–92.

—— (1974). 'Landscapes of glacial erosion in Greenland and their relationship to ice, topographic and bedrock conditions'. *Institute of British Geographers, Special Publication*, 7: 177–95.

—— (1978). 'Glacial erosion by the Laurentide ice sheet'. *Journal of Glaciology*, 20: 367–91.

—— GLASSER, N., and CLAPPERTON, C. M. (1992). 'Evolution of large roches moutonnées'. *Geografiska Annaler*, 74A: 253–64.

—— and WATTS, S. (1977). 'Tors, felsenmeer, and glaciation in northern Cumberland peninsula, Baffin Island'. *Canadian Journal of Earth Sciences*, 14: 2817–23.

SUMMERFIELD, M. A. (1991). *Global Geomorphology*. London: Longman.

SWAN, S. B. (1970). 'Analysis of residual terrain, Johor, Malaya'. *Annals of the Association of American Geographers*, 60: 124–33.

—— (1971). 'Coastal geomorphology in a humid tropical low energy environment: the islands of Singapore'. *Journal of Tropical Geography*, 33: 43–61.

SWANTESSON, J. O. H. (1989). *Weathering Phenomena in a Cool Temperate Climate*. Göteborg: Department of Physical Geography, GUNI Rapport 28.

—— (1992). 'Recent microweathering phenomena in southern and central Sweden', *Permafrost and Periglacial Processes*, 3: 369–78.

TANAKA, Y., and MATSUKURA, Y. (2001). 'Some characteristics of Korean granite and gneiss landforms'. *Transactions of the Japanese Geomorphological Union*, 22: 361–79.

TARDY, Y. (1971). 'Characterisation of the principal weathering types by the geochemistry of waters from some European and African crystalline massifs'. *Chemical Geology*, 7: 253–71.

—— BOCQUIER, G., PAQUET, H., and MILLOT, G. (1973). 'Formation of clay from granite and its distribution in relation to climate and topography'. *Geoderma*, 10: 271–84.

TAYLOR, G., and EGGLETON, R. A. (2001). *Regolith Geology and Geomorphology*. Chichester: Wiley.

TAYLOR, R. G., and HOWARD, K. W. F. (1999). 'Lithological evidence for the evolution of weathered mantles in Uganda by tectonically controlled cycles of deep weathering and stripping'. *Catena*, 35: 65–94.

TEEUW, R. M. (1991). 'Comparative studies of adjacent drainage basins in Sierra Leone: some insights into tropical landscape evolution'. *Zeitschrift für Geomorphologie N.F.*, 35: 257–68.

TERNAN, J. L., and WILLIAMS, A. G. (1979). 'Hydrological pathways and granite weathering on Dartmoor', in A. F. Pitty (ed.), *Geographical Approaches to Fluvial Processes*. Norwich: Geo Books, 530.

THOMAS, M. F. (1965). 'Some aspects of the geomorphology of domes and tors in Nigeria'. *Zeitschrift für Geomorphologie N.F.*, 9: 63–81.

—— (1966). 'Some geomorphological implications of deep weathering patterns in crystalline rocks in Nigeria'. *Institute of British Geographers, Transactions*, 40: 173–93.

—— (1967). 'A bornhardt dome in the plains near Oyo, Western Nigeria'. *Zeitschrift für Geomorphologie N.F.*, 11: 239–61.

—— (1968). 'Tor', in R. W. Fairbridge (ed.), *The Encyclopedia of Geomorphology*. New York: Reinhold, 1157–9.

THOMAS, M.F. (1974). 'Granite landforms: a review of some recurrent problems of interpretation'. *Institute of British Geographers, Special Publication*, 7: 13–37.
—— (1976). 'Criteria for the recognition of climatically induced variations in granite landforms', in E. Derbyshire (ed.), *Geomorphology and Climate*. London: Wiley, 411–45.
—— (1978). 'The study of inselbergs'. *Zeitschrift für Geomorphologie N.F., Supplement-Band*, 31: 1–41.
—— (1980). 'Timescales of landform development on tropical shields: a study from Sierra Leone', in R. A. Cullingford, D. A. Davidson, and J. Lewin (eds.), *Timescales in Geomorphology*. Chichester: Wiley, 333–54.
—— (1989a). 'The role of etch processes in landform development. I. Etching concepts and their applications'. *Zeitschrift für Geomorphologie N.F.*, 33: 129–42.
—— (1989b). 'The role of etch processes in landform development. II. Etching and the formation of relief'. *Zeitschrift für Geomorphologie N.F.*, 33: 257–74.
—— (1994a). 'Ages and geomorphic relationships of saprolite mantles', in D. A. Robinson and R. B. G. Williams (eds.), *Rock Weathering and Landform Evolution*. Chichester: Wiley, 287–301.
—— (1994b). *Geomorphology in the Tropics: A Study of Weathering and Denudation in Low Latitudes*. Chichester: Wiley.
—— (1995). 'Models for landform development on passive margins: some implications for relief development in glaciated areas'. *Geomorphology*, 12: 3–15.
—— (1997). 'Weathering and landslides in the humid tropics: a geomorphological perspective'. *Journal of the Geological Society of China*, 40: 1–16.
—— and THORP, M. B. (1980). 'Some aspects of the geomorphological interpretation of Quaternary alluvial sediments in Sierra Leone'. *Zeitschrift für Geomorphologie N.F., Supplement-Band*, 36: 140–61.
—— —— (1985). 'Environmental change and episodic etchplanation in the humid tropics of Sierra Leone', in I. Douglas and T. Spencer (eds.), *Environmental Change and Tropical Geomorphology*. London: Allen and Unwin, 239–67.
THORN, C. E. (1992). 'Periglacial geomorphology: what, where, when?', in J. C. Dixon and A. D. Abrahams (eds.), *Periglacial Geomorphology*. Chichester: Wiley, 1–30.
—— and HALL, K. (2002). 'Nivation and cryoplanation: the case for scrutiny and integration'. *Progress in Physical Geography*, 26: 533–50.
THORP, M. B. (1967a). 'Closed basins in Younger Granite Massifs, northern Nigeria'. *Zeitschrift für Geomorphologie N.F.*, 11: 459–80.
—— (1967b). 'The geomorphology of the Younger Granite Kuduru Hills'. *Journal of the Geographical Association of Nigeria*, 10: 77–90.
—— (1969). 'Some aspects of the geomorphology of the Aïr Mountains, southern Sahara'. *Institute of British Geographers, Transactions*, 47: 25–46.
—— (1975). 'Geomorphic evolution in the Liruei Younger Granite hills, Nigeria'. *Savanna*, 4: 139–54.
TILLEY, C. (1996). 'The powers of rocks: topography and monument construction on Bodmin Moor'. *World Archaeology*, 28: 161–76.

TJIA, H. D. (1973). 'Geomorphology', in D. J. Gobbett and C. S. Hutchinson (eds.), *Geology of the Malay Peninsula*. New York: Wiley, 13–24.

TRACZYK, A. (1996). 'Geneza i znaczenie stratygraficzne rytmicznie warstwowanych utworów stokowych w Sudetach'. *Acta Universitatis Wratislaviensis*, 1808, *Prace Instytutu Geograficznego*, A8: 93–104.

TRENHAILE, A. S. (1987). *The Geomorphology of Rock Coasts*. Oxford: Clarendon Press.

—— (1997). *Coastal Dynamics and Landforms*. Oxford: Clarendon Press.

TRUDGILL, S. (2004). 'Kaolinization', in A. S. Goudie (ed.), *Encyclopedia of Geomorphology*. London: Routledge, 582–3.

TSCHANG HSI-LIN (1961). 'The pseudokarren and exfoliation forms of granite on Pulau Ubin, Singapore'. *Zeitschrift für Geomorphologie N.F.*, 5: 302–12.

—— (1962). 'Some geomorphological observations on the region of Tampin, Southern Malaya'. *Zeitschrift für Geomorphologie N.F.*, 6: 253–9.

TUĞRUL, A., and ZARIF, I. H. (1999). 'Correlation of mineralogical and textural characteristics with engineering properties of selected granitic rocks from Turkey'. *Engineering Geology*, 51: 303–17.

TURKINGTON, A. (2004). 'Cavernous weathering', in A. S. Goudie (ed.), *Encyclopedia of Geomorphology*. London: Routledge, 128–30.

TUTTLE, O. F., and BOWEN, N. L. (1958). 'Origin of granite in the light of experimental studies in the system $NaAlSi_3O_8$-$KAlSi_3O_8$-SiO_2H_2O'. *Geological Society of America, Memoir*, 74.

TWIDALE, C. R. (1962). 'Steepened margins of inselbergs from north-western Eyre Peninsula, South Australia'. *Zeitschrift für Geomorphologie N.F.*, 6: 51–69.

—— (1964). 'A contribution to the general theory of domed inselbergs'. *Institute of British Geographers, Transactions*, 34: 91–113.

—— (1968). 'Inselberg', in R. W. Fairbridge (ed.), *The Encyclopedia of Geomorphology*. New York: Reinhold, 556–9.

—— (1973). 'On the origin of sheet jointing'. *Rock Mechanics*, 5: 163–87.

—— (1978a). 'Early explanations of granite boulders'. *Revue de Géomorphologie dynamique*, 27: 133–42.

—— (1978b). 'On the origin of pediments in different structural settings'. *American Journal of Science*, 278: 1138–76.

—— (1981a). 'Early explanations of inselbergs'. *Revue de Géomorphologie dynamique*, 30: 49–61.

—— (1981b). 'Granite inselbergs: domed, block-strewn and castellated'. *Geographical Journal*, 147: 54–71.

—— (1982). *Granite Landforms*. Amsterdam: Elsevier.

—— (1983). 'Pediments, peneplains and ultiplains'. *Revue de Géomorphologie dynamique*, 32: 1–35.

—— (1986). 'Granite platforms and low domes: newly exposed compartments or degraded remnants?'. *Geografiska Annaler*, 68A: 399–411.

—— (1994). 'Gondwanan (Late Jurassic and Cretaceous) palaeosurfaces of the Australian craton'. *Palaeogeography, Palaeoclimatology, Palaeoecology*, 112: 157–86.

TWIDALE, C.R. (1997a). 'Comment on ^{10}Be and ^{26}Al evidence for exceptionally low rates of Australian bedrock erosion and the likely existence of pre-Pleistocene landscapes'. *Quaternary Research*, 48: 386–7.

—— (1997b). 'The great age of some Australian landforms: examples of, and possible explanation for, landscape longevity', in M. Widdowson (ed.), Palaeosurfaces: Recognition, Reconstruction and Palaeoenvironmental Interpretation. *Geological Society, Special Publication*, 120: 13–23.

—— (2002). 'The two-stage concept of landform and landscape development involving etching: origin, development and implications of an idea'. *Earth Science Reviews*, 57: 37–74.

—— and BOURNE, J. A. (1975). 'Episodic exposure of inselbergs'. *Geological Society of America Bulletin*, 86: 1473–81.

—— —— (1976). 'Origin and significance of pitting on granitic rocks'. *Zeitschrift für Geomorphologie N.F.*, 20: 405–16.

—— —— (1978). 'Bornhardts'. *Zeitschrift für Geomorphologie N.F., Supplement-Band*, 31: 111–37.

—— —— (1993), 'Fractures: a double edged sword. A note on fracture density and its importance'. *Zeitschrift für Geomorphologie N.F.*, 37: 459–75.

—— —— (1998). 'Multistage landform development, with particular reference to a cratonic bornhardt'. *Geografiska Annaler*, 80A: 79–94.

—— and CAMPBELL, E. M. (1992). 'On the origin of pedestal rocks'. *Zeitschrift für Geomorphologie N.F.*, 36: 1–13.

—— and CORBIN, E. M. (1963). 'Gnammas'. *Revue de Géomorphologie dynamique*, 14: 1–20.

—— and LAGEAT, Y. (1994). 'Climatic geomorphology: a critique'. Progress in Physical Geography, 18: 319–34.

—— SCHUBERT, C., and CAMPBELL, E. M. (1991). 'Dislodged blocks'. *Revue de Géomorphologie dynamique*, 40: 119–29.

—— and VIDAL-ROMANI, J. R. (1998). *Formas y Paisajes Graníticos*. A Coruña: Universidade da Coruña Servicio de Publicacións.

—— —— CAMPBELL, E. M., and CENTENO, J. D. (1996). 'Sheet fractures: response to erosional offloading or to tectonic stress?'. *Zeitschrift für Geomorphologie N.F., Supplement-Band*, 106: 1–24.

VALADAS, B. (1987). 'Morphodynamiques récentes dans le Massif Central français: étude comparée des massifs granitiques du Limousin, de Margeride et du Mont Lozère'. *Zeitschrift für Geomorphologie N.F., Supplement-Band*, 65: 85–99.

—— and VEYRET, Y. (1974). 'Quelques aspects des modelés d'origine glaciaire, périglaciaire et nivale sur les confines méridionaux de la Margeride'. *Revue de Géomorphologie dynamique*, 23: 163–77.

VAN DER WATEREN, F. M, and DUNAI, T. J. (2001). 'Late Neogene passive margin denudation history: cosmogenic isotope measurements from the central Namib desert'. *Global and Planetary Change*, 30: 271–307.

VAN VLIET-LANOË, B. (1988). *Le rôle de la glace de ségrégation dans les formations superficielles de l'Europe de l'Ouest. Processus et héritages'*. Thèse Doctorate État, Université du Paris I.

—— Coque-Delhuille, B., and Valadas, B. (1981). 'Les structures derivés de la formation de glace de ségrégation dans les arenas déplacées. Analyse et application à la Margeride occidentale'. *Physio-Géo*, 2: 17–38.

Varnes, D. J., Radbruch-Hall, D. H., and Savage, W. Z. (1989). 'Topographic and structural conditions in areas of gravitational spreading of ridges in the Western United States'. *US Geological Survey, Professional Paper*, 1496.

Velbel, M. A. (1985). 'Geochemical mass balances and weathering rates in forested watersheds of the southern Blue Ridge'. *American Journal of Science*, 285: 904–30.

Veldkamp, A., and Oosterom A. P. (1994). 'The role of episodic plain formation and continuous etching and stripping processes in the End-Tertiary landscape development of SE Kenya'. *Zeitschrift für Geomorphologie, N.F.* 38: 75–90.

Veyret, Y. (1978). *Les modelés et formations d'origine glaciaire dans le Massif Central français*. Thèse d'Etat, Université de Paris I.

Vidal Romani, J. R. (1989). 'Granite geomorphology in Galicia (NW Spain)'. *Cuadernos Laboratorio Xeoloxico de Laxe*, 13: 89–163.

—— and Twidale, C. R. (1999), 'Sheet fractures, other stress forms and some engineering implications', *Geomorphology* 31: 13–27.

Vieira, B. C., and Fernandes, N. F. (2004). 'Landslides in Rio de Janeiro: the role played by variations in soil hydraulic conductivity'. *Hydrological Processes*, 18: 791–805.

Vieira, G. T. (1998). 'Periglacial research in the Serra da Estrela: an overview', in G. T. Vieira (ed.), *Glacial and Periglacial Geomorphology of the Serra da Estrela, Portugal: Guidebook for the field-trip: International Geographical Union*. Lisbon: CEG and Department of Geography, 49–65.

—— (1999). 'Coarse sand accumulations in granite mountains: the case-studies of the Serra do Gerêsand Serra sa esrela (Portugal)'. *Zeitschrift für Geomorphologie N.F., Supplement-Band*, 119: 105–18.

Viles, H. (2001). 'Scale issues in weathering studies'. *Geomorphology*, 41: 63–72.

Völkel, J. (1995). 'Periglaziale Deckschichten und Böden im Bayerischen Wald und seinen Randgebieten'. *Zeitschrift für Geomorphologie N.F., Supplement-Band*, 96: 1–301.

—— and Leopold, M. (2001). 'Age determination of the youngest phase of periglacial morphodynamics on slopes in Central European Highlands'. *Zeitschrift für Geomorphologie N.F.*, 45: 273–94.

Votýpka, J. (1964). 'Tvary zvětrávání a odnosu žuly v severní části Novobystřické vrchoviny'. *Sborník Československé Společnosti Zeměpisné*, 69: 243–58.

—— (1971). 'Ukázky zvětrávání žul Českého masivu'. *Acta Universitatis Carolinae, Geographica*, 2: 75–91.

—— (1974). 'Vznik a vývoj mezoreliéfu a mikroreliéfu Sedmihoří'. *Acta Universitatis Carolinae, Geographica*, 2: 17–34.

—— (1979). 'Geomorfologie granitové oblasti masívu Plechého'. *Acta Universitatis Carolinae, Geographica*, 16(2): 55–83.

Wahrhaftig, C. (1965). 'Stepped topography of the southern Sierra Nevada, California'. *Geological Society of America, Bulletin*, 76: 1165–90.

WAKABAYASHI, J., and SAWYER, T. L. (2001). 'Stream incision, tectonics, uplift, and evolution of topography of the Sierra Nevada, California'. *Journal of Geology*, 109: 539–62.

WANG, C., and ROSS, G. J. (1989). 'Granitic saprolites: their characteristics, identification and influence on soil properties in the Appalachian region of Canada'. *Zeitschrift für Geomorphologie N.F., Supplement-Band*, 72: 149–61.

WARAGAI, T. (1999). 'Weathering processes on rock surfaces in the Hunza Valley, Karakoram, North Pakistan'. *Zeitschrift für Geomorphologie N.F., Supplement-Band*, 119: 119–36.

WARD, J. D., SEELY, M. K., and LANCASTER, N. (1983). 'On the antiquity of the Namib'. *South African Journal of Science*, 79: 175–83.

WARKE, P. A., and SMITH, B. J. (1994). 'Short-term rock temperature fluctuations under simulated hot desert conditions: some preliminary data', in D. A. Robinson and R. B. G. Williams (eds.), *Rock Weathering and Landform Evolution*. Chichester: Wiley, 57–70.

WASHBURN, A. L. (1979). *Geocryology*. London: Arnold.

WATERS, R. S. (1954). 'Pseudo-bedding in the Dartmoor granite'. *Royal Society of Cornwall, Transactions*, 18: 456–62.

—— (1964). 'The Pleistocene legacy to the geomorphology of Dartmoor', in I. G. Simmons (ed.), *Dartmoor Essays*, Exeter: Devonshire Association, 73–96.

WATSON, A., and PYE, K. (1985). 'Pseudokarstic micro-relief and other weathering features on the Mswati Granite (Swaziland)'. *Zeitschrift für Geomorphologie N.F.*, 29: 285–300.

WATTS, S. H. (1986). 'Intensity versus duration of bedrock weathering under periglacial conditions in High Arctic Canada'. *Biuletyn Peryglacjalny*, 30: 141–52.

WAYLAND, E. J. (1933). 'Peneplains and some erosional landforms'. *Annual Report Bulletin, Protectorate of Uganda, Geological Survey, Department of Mines*, 1: 77–79; repr. in G. F. Adams (ed.) (1975), *Planation Surfaces*, Benchmark Papers in Geology 22, Stroudsburg: Hutchinson and Ross, 355–7.

WEISCHET, W. (1969). 'Zur Geomorphologie des Glatthangreliefs in der ariden Subtropenzone des Kleinen Nordens von Chile'. *Zeitschrift für Geomorphologie N.F.*, 13: 1–21.

WELLS, N. A., ANDRIAMIHAJA, B., and SOLO RAKOTOVOLOLONA, H. F. (1991). 'Patterns of development of lavaka, Madagascar's unusual gullies'. *Earth Surface Processes and Landforms*, 16: 189–206.

WESSMAN, L. (1996). 'Studies on salt-frost attack on natural stone', in J. Riederer (ed.), *Proceedings of the Eighth International Congress on Deterioration and Conservation of Stone*. Berlin: Ernst und Sohn, 563–71.

WHALLEY, W. B., DOUGLAS, G. R., and MCGREEVY, J. P. (1982). 'Crack propagation and associated weathering in igneous rocks'. *Zeitschrift für Geomorphologie N.F.*, 26: 33–54.

—— REA, B. R., RAINEY, M. M., and MCALISTER, J. J. (1997). 'Rock weathering in blockfields: some preliminary data from mountain plateaus in North Norway', in M. Widdowson (ed.), *Palaeosurfaces: Recognition, Reconstruction and Palaeoenvironmental Interpretation. Geological Society, Special Publication*, 120: 133–45.

WHITAKER, C. R. (1979). 'The use of the term "pediment" and related terminology'. *Zeitschrift für Geomorphologie N.F.*, 23: 427–39.

WHITE, A. F. (2002). 'Determining mineral weathering rates based on solid and solute weathering gradients and velocities: application to biotite weathering in saprolites'. *Chemical Geology*, 190: 69–89.

—— and BLUM, A. E. (1995). 'Effects of climate on chemical weathering in watersheds'. *Geochimica et Cosmochimica Acta*, 59: 1729–47.

—— —— BULLEN, T. D., VIVIT, D. V., SCHULZ, M., and FITZPATRICK J. (1999). 'The effect of temperature on experimental and natural chemical weathering rates of granitoid rocks'. *Geochimica et Cosmochimica Acta*, 63: 3277–91.

—— —— SCHULZ, M. S., VIVIT, D. V., LARSEN, M., and MURPHY, S. F. (1998). 'Chemical weathering in a tropical watershed, Luquillo Mountains, Puerto Rico: I. Long-term versus short-term chemical fluxes'. *Geochimica et Cosmochimica Acta*, 62: 209–26.

WHITE, W. A. (1945). 'Origin of granite domes in the south-eastern Piedmont'. *Journal of Geology*, 53: 276–82.

WHITLOW, J. R. (1979). 'Bornhardt terrain on granitic rocks in Zimbabwe: a preliminary assessment'. *Zambian Geographical Journal*, 33–4: 75–93.

—— and SHAKESBY, R. A. (1988). 'Bornhardt micro-geomorphology: form and origin of micro-valleys and rimmed gutters, Domboshava, Zimbabwe'. *Zeitschrift für Geomorphologie N.F.*, 32: 179–94.

WHITTOW, J. B. (1986). *Landscapes of Stone*. London: Whittet Books.

WIDDOWSON, M. (2004). 'Ferricrete', in A. S. Goudie (ed.), *Encyclopedia of Geomorphology*. London: Routledge, 365–7.

WIECZOREK, G. F. (2002). 'Catastrophic rockfalls and rockslides in the Sierra Nevada, USA'. *Geological Society of America, Reviews in Engineering Geology*, 15, 165–90.

—— and JÄGER, S. (1996). 'Triggering mechanisms and depositional rates of postglacial slope-movement processes in the Yosemite Valley, California'. *Geomorphology*, 15: 17–31.

—— SNYDER, J., and WAITT, R. (2000). 'Unusual July 10, 1996, rock fall at Happy Isles, Yosemite National Park, California'. *Geological Society of America Bulletin*, 112: 75–85.

WILHELMY, H. (1958). *Klimamorphologie der Massengesteine*. Braunschweig: Westermann.

—— (1977). 'Verwitterungskleinformen als Anzeichen stabiler Grossformung'. *Würzburger Geographische Arbeiten*, 35: 177–98.

WILLETT, S. D., and BRANDON, M. T. (2002). 'On steady states in mountain belts'. *Geology*, 30: 175–8.

WILLIAMS, A. G., TERNAN, L., and KENT, M. (1986). 'Some observations on the chemical weathering of the Dartmoor granite'. *Earth Surface Processes and Landforms*, 11: 557–74.

WILLIAMS, R. B. G., and ROBINSON, D. A. (1989). 'Origin and distribution of polygonal cracking of rock surfaces'. *Geografiska Annaler*, 71A: 145–59.

—— —— (1994). 'Weathering flutes on siliceous rocks in Britain and Europe', in D. A. Robinson and R. B. G. Williams (eds.), *Rock Weathering and Landform Evolution*. Chichester: Wiley, 413–32.

WILLIS, B. (1936). *East African Plateaus and Rift Valleys*. Washington: Carnegie Institute Publications, 470.

WINKLER, E. M. (1973). *Stone: Properties, Durability in Man's Environment*. New York: Springer.

WISSMAN, H. (1954). 'Karrenähnliche Rillen im Granit im Gipfelbereich des Hoaschan an der Grenze von Schansi und Honan', in *Ergebnisse und Probleme moderner geographischer Forschung. Hans Mortensen zu seinem 60. Geburtstag*. Bremen: Walter Dorn, 61.

WONG, P. P. (2005), 'The Coastal Environment in Southeast Asia', in A. Gupta (ed.), *The Physical Geography of Southeast Asia*. Oxford: Oxford University Press, 177–92.

WRIGHT, J. (2002). 'Particle size characteristics and quartz microfracture patterns in I- and S-type granitoid weathering profiles: some preliminary observations from Eastern Australia'. *Transactions, Japanese Geomorphological Union*, 23: 309–33.

XU, J. (1996). 'Benggang erosion: the influencing factors'. *Catena*, 27: 249–63.

—— and ZENG, G. (1992). 'Benggang erosion in subtropical granite crust geoecosystems: an example from Guangdong Province'. *IAHS Publication*, 209: 455–63.

YOUNG, A. (1972). *Slopes*. Edinburgh: Oliver and Boyd.

ZHOU, C. H., XU, Z. W., LEE, C. F., and LI, J. (2002). 'On the spatial relationship between landslides and causative factors on Lantau Island, Hong Kong'. *Geomorphology*, 43: 197–207.

ZONNENVELD, J. I. S. (1993). 'Planation and summit levels in Suriname (S. America)'. *Zeitschrift für Geomorphologie N.F., Supplement-Band*, 93: 29–46.

ŻURAWEK, R. (1999). 'Relict rock glaciers in the Central European Mid-mountains: state-of-the-art'. *Biuletyn Peryglacjalny*, 38: 163–92.

Index

Africa, eastern 84, 118, 309, 310
all-slopes topography 260–1, 317–18
alveoles 139, 145
Angola 309
Antarctica 47, 97, 109, 141, 159, 166
aplite 22–3, 147, 277
Aruba 140
Ascension 16
Australia
    Australia, south 123, 135, 143, 152
    Bega, NSW 66, 70, 97, 263
    Eyre Peninsula 27, 130, 154, 216
    Hyden Rock 154
    Kangaroo Island 230
    Kulgera-Everard Range 130
    New South Wales 200, 201
    Pilbara 48, 130
    Queensland 227
    Victoria 89, 225
    Yilgarn 70, 307, 308

basins, topographic 66–7, 258, 263, 273, 280, 297, 311–13
biotite, role of 38, 207, 268, 273
block field 29, 44, 208–12, 241, 244, 324
block stream 208–12, 244
bornhardt 85, 112, 166–7, 175
Bougainville 16
boulders 33–4, 51, 84, 88–96, 166, 183, 188–9, 198, 207, 244, 264, 269, 287, 329
    split 33–4, 47–8, 243
Bowen's reaction series 9, 37

Brazil
    Brazil, SE 195, 197, 200–1, 214, 215, 310
    Rio de Janeiro 164, 176, 188, 194, 198, 225, 263, 271
    Serra de Mantiqueira 197, 262
    Serra do Mar 52, 70, 74, 78, 185, 193–4, 198, 200, 202, 262, 310, 325
building stone 46, 333
Bulgaria
    Vitosha 96

Cameroon 121, 213, 310
Canada
    British Columbia 16, 42, 197
    Canadian Arctic 109, 110, 212, 307
    Canadian Cordillera 178, 256
    Canadian Shield 237, 265
    Labrador Peninsula 110
    Yukon 29, 237
caves 169, 179–82
    marine 226–7
Chile
    Andes 166, 225, 256, 317
China
    China, SE 69, 185, 195, 196–7, 200, 201, 202, 258, 260, 310, 325
    Great Wall 337
    Hong Kong 52, 53, 69, 74–5, 77, 78, 184, 188, 190–3, 198, 200, 202, 214, 215, 235, 310, 324
    Tibetan Plateau, 237
clefts 179

cliffs 167, 220–6, 231
    retreat of 222–3
colluvium 194, 203, 212–15
Colombia 312
contact metamorphism 10, 51, 278
corestones 57–60, 90–2, 188, 283
crusting 143, 146, 149
cryoplanation 240–1, 243–4
Czech Republic
    Bohemian Massif 32, 69, 76, 91, 104, 106, 119, 139, 149, 157, 178, 211, 243, 257, 271, 309, 325
    Karlovy Vary 340
    Krumlovský les 285–7
    Skuteč 338
    Šumava 258, 299
    Žulova Highland 299

debris flow 189, 197–9, 263
debris torrent 197
deep weathering 52–82, 92–4, 196, 223–4, 227, 232–5, 260, 263, 310, 324–5
domes 112–13, 117, 120, 126, 160, 163–4, 177–9, 194, 221, 225, 251, 253, 260, 264, 269, 271, 275, 283

earth flow 190, 193, 196
earthquake effects 35, 178, 182, 188, 190
Egypt
    Aswan 30, 340
    Eastern Desert 276
    Sinai 31–2, 265, 318
enclave 8, 277
erratic 89
Estonia 89
etching 280, 294–306
etchplains and etchsurfaces 266, 295–306
    Koidu etchplain 302
exfoliation 20, 29–33
exhumation 130, 260, 265–6, 325

Finland 69, 85, 304–5
    Aland Islands 231

fire, effects of 47–8
flaking 30, 47–8, 137
flared slopes 153–6
fractures 16–22, 126–7, 164, 256–7, 278–83
    influence on slope form 164–6
    master fractures 19, 278–80, 315–16
France
    Brittany 69, 89, 90, 96, 135, 151, 159, 211, 221, 223–4, 227, 258, 328
    Carnac, Brittany 328–30
    Cévennes 208
    Corsica 29, 140, 141, 157, 159, 170, 233, 260, 323
    Massif Central 29, 49, 69, 88, 90, 97, 110, 149, 200, 204, 205, 207, 209, 211, 212, 237, 257, 273, 299, 312, 313, 314, 334, 335–6
    Pink Granite Coast, Brittany 232
    Pyrenees 273
    Sidobre 96, 106, 211
    Vosges 247

geochemical budget 38–9
Germany
    Bavarian Forest 204, 205
    Fichtelgebirge 75, 106, 181, 258, 299
    Harz 91, 108, 110, 204, 205, 237, 258, 299, 314
    Hirschau 340
    Lusatia 6, 17, 273, 338
    Schwarzwald 248, 257, 258
glacial erosion 231–2, 246, 261, 314, 318
glaciation 110–11, 163, 245–53, 258, 260–1, 308, 318
granitization 10–11
granular disintegration 27–8, 42–3, 50
Great Britain
    Bodmin Moor 66, 85, 97, 159, 205, 258, 313, 328, 331
    Carnmenellis 328, 331
    Dartmoor 2, 29, 32, 66–7, 69, 77, 85–6, 91, 96–7, 99, 100, 101, 102, 103, 104, 106, 108, 158, 204, 205, 207–8,

210–11, 239–41, 257, 258, 260, 269, 299, 313, 314, 328, 331–2, 339–40
Land's End 96, 204, 220–2, 226, 235, 328, 340
Lundy Island 167, 221
Penwith Peninsula 167
St Austell 66–7, 340–1
Scilly Islands 204, 232
Scotland 16, 44, 69, 75, 210, 251, 257, 312
   Aberdeen 338–9
   Arran, Isle of 91, 166, 246, 269, 318
   Buchan 130, 227, 260
   Cairngorms 21, 32, 97, 99, 110, 166, 170, 207, 212, 243, 251, 258, 314, 322
   Lochnagar 314
   Rannoch Moor 251–2
   Rhum, Isle of 170
   Skye, Isle of 318
Great Escarpments 262–4
Greenland 32, 308, 312
grus 50, 55, 61, 63–5, 67, 71, 75–6, 110, 183, 187, 206, 256, 275, 283, 320
   bedded 206–7
   geographical distribution 64–5
gully erosion 185, 199–202, 310
Guyana Shield 70, 266, 302–3, 308, 310

Himalaya 16, 170, 256, 317
hot spot 267–8
human impact, on granite landscapes 185, 188–9, 194, 197, 201, 331–2, 336
hydrothermal alteration 65–8, 263

India
   Deccan 266, 303, 308, 309
   India, southern 46
   Kerala 113,
   Mahabalipuram 340
inheritance 231–5, 247–8, 251, 318
inlet 226–7
inselberg 84–5, 111–31, 164, 167, 175, 177–8, 266–7, 271, 275, 299, 309–10, 323
   boulder inselberg 117–19

castellated inselberg 117–19
conical 120
geographical distribution 113–14
origin of 121–5
intrusion 11–13, 127, 258
Ireland 16, 328
   Donegal 318
   Galway Bay 232
   Mourne Mountains 69, 170–1, 253, 318
Italy
   Sardinia 233
   Serre Massif, Calabria 198
   Sila Massif 69

Japan
   Japanese Alps 69, 71, 184, 186–7, 198, 256–7, 318
   Rokko Mountains 35, 190
Jordan 265, 318

kaolin 61–2, 82, 339–41
karren 149–53, 229–30, 324
   geographical distribution 149–51
Kazakhstan 309
Kenya 2, 129, 267
knock-and-lochan 251, 312

landslide 184–7, 189–97, 213, 263, 310, 324–5
   hydrological conditions 194, 196
   turning into debris flow 198
lateral spreading 178–9
laterite 62
lightning 34–5

Macedonia 135
Madagascar 199, 200, 201, 266
Malawi
   Nyika Plateau 195
Malay Peninsula 69, 91, 135, 230, 233, 258, 260
mantle-controlled planation 183
megaliths 49, 90, 151, 328–31

meltwater effects 253
Mexico
   Acapulco 70, 188–9, 198, 215, 221, 318–19,
   Sierra Madre del Sur 91, 318
microcracks (microfractures) 22, 28, 283
migmatite 10
mineral-stability series 37–8
Mongolia 32, 44, 107, 135, 244, 309
multi-concave relief 291, 311–13, 316
multi-convex relief 195, 258, 260, 291, 310–11, 316

Namibia
   Brandberg 267
   Erongo 34, 89, 138, 147, 160, 167, 169, 176, 267, 278
   Gobabeb 94–5
   Mirabib 125, 127, 167
   Namib Desert 49, 90, 94, 107, 109, 117, 125, 127, 129, 135, 138, 149, 157–8, 159, 164, 183, 217, 266, 307, 308, 309, 323
   Spitzkoppe 94, 111, 113, 120, 125, 127, 140, 160, 167, 175, 178,
   Vogelfederberg 111–12
New Zealand
   New Zealand Alps 71, 78
   South Island 225, 322
Niger
   Air 268
Nigeria 16, 70, 112, 121, 124, 280
   Idanre Hills 316
   Jos Plateau 74, 213–14
   Kudaru Hills 268, 278
   Liruei Hills 268
   Oyo 85, 116–17
Norway
   Narvik 27, 135
   Norway, NW 96
   Oslo rift 16

Pakistan
   Karakoram 166, 170, 174, 256, 317
Papua New Guinea 16, 78, 318
pedestal rocks 107
pediment 182–4, 214, 216–17, 324
pegmatite 23, 340
perched blocks 34, 106
periglacial 103–4, 203–12, 237–45, 291, 314
Peru
   coastal batholith 4, 11, 13
   Nevados Huascaran 174, 178
piedmont angle 113–14
pitting 26–7
plain 265, 307–8, 323
planation surface 265–6
plateau 241, 247, 251, 269, 287, 313–15
plateau uplift 257, 313–14
Poland
   Jelenia Góra Basin 129, 177, 273–4, 299
   Karkonosze (Riesengebirge) 18, 26, 29, 31, 42, 52, 59, 78–9, 97, 99–100, 102, 105, 106, 110, 146, 159, 181, 204, 205, 210, 211, 212, 241–3, 247, 258, 269, 278, 299, 340
   Strzegom Hills 339
   Tychowo 89
   Witosza 177, 181
polygonal cracking 145–9
Portugal
   Alentejo 49, 328
   Braga 46
   Monsanto 333, 337
   Serra da Estrela 42–3, 48, 69, 89, 97, 99, 100, 105, 110, 133, 154, 170, 201, 204, 210, 211, 212, 247, 248–9, 258, 269, 286–9, 313, 322
   Serra do Geres 314
pseudobedding 30, 32
Puerto Rico 40, 70, 73, 189, 194–5, 198, 318, 324

quarrying 338–40

rampa complex 195, 197
Reunion 16
Richter denudation slope 166
ring dyke 278
roche moutonnée 181, 232, 253
rock fall 173
   in the Yosemite Valley 173–4
rock slide 171–5, 249
rock slope 161–5, 249
Russia
   Aldan Upland 243
   Kola Peninsula 225
   Oymiakon Upland 243
   Stanove Mountains 243

saprolite 55, 67–73, 184, 186–7, 299
Scandinavia 72, 93, 135, 212, 251, 265, 277,
scarp-foot depression 115
schlieren 8
scree 165, 174
sea arch 227
Serbia 135
Seychelles 149, 159, 230
sheet structures 111, 127–8, 181, 260
sheet wash 213
sheeting 20–2, 31–3, 129, 164, 176, 177
shore platforms 220, 225–6
Sierra Leone 70, 129, 301–2, 304, 310
Singapore 69, 96, 152
skerry 231–2
slope deposits 203–8, 239
Slovakia
   Tatra Mountains 165, 170, 248
solifluction 108, 207, 211, 239, 241, 243
Solomon Islands 16
South Africa 70, 129, 199, 280, 309, 315
   Barberton 2, 276–7
   Namaqualand 307
   Transvaal 120
   Valley of the Thousand Hills 57, 264

South Korea 141, 144, 159, 318
Spain
   Galicia 260
   Iberian Range 69, 135, 149, 257
   Sierra de Guadarrama 97, 170, 175, 260
spheroidal weathering 30–1
Sri Lanka 124, 315
   Gal Vihara 342
   Sigiriya 338
stack 222–4
steady-state 306
stepped topography 301, 318–22
strength-equilibrium 166–8, 173, 178
Sudan 70
Surinam 134
Swaziland 152, 160, 185, 199, 201, 213, 214–15
Sweden
   Blekinge 315–16
   Bohuslän 179, 232, 253, 280, 282, 297–9, 315
   Ivön 74, 96
   Stockholm Archipelago 231
   Sweden, N 109, 111, 181
   Sweden, S and SW 26, 69, 110, 130, 266, 294, 297, 304, 309

tafoni 139–45
   geographical distribution 141–3
talus 169–71, 175, 181
tectonic setting 14–16, 183–4, 254–68
Tian Shan 174
toppling 172, 221
tor 85–7, 95–111, 183, 239–40, 244, 258, 264, 269, 287–9, 331
   geographical distribution 98–9
trough, glacial 248–49
Turkey 69

Uganda 70
United States of America
   Alabama Hills, California 99
   Aleutian Arc 14

United States of America (*Contd.*)
    Appalachian Piedmont 70, 299, 304
    Appalachians 70, 275
    Arizona 85, 120, 125, 164, 183
    Basin and Range province 16
    Colorado, mountains of 179
    Enchanted Rock, Texas 106, 176, 181
    Grand Teton 170
    Joshua Tree NP, California 275
    Juneau Icefield, Alaska 50
    Laramie Range 70, 260, 275, 314
    Maine 225, 231, 251
    Mojave Desert 23, 28, 70, 93–4, 109–10, 117–19, 127, 130, 140, 182, 183, 216–17, 301
    Mono Lake, California 35,
    Monterey Bay, California 226
    Mount Rushmore, South Dakota 342–3
    Rocky Mountains 44, 237, 260, 300, 313, 314–15
    Sequoia NP, California 33
    Sierra Nevada, California 1, 13, 16, 19, 21, 33, 52, 54, 60–1, 70, 71, 76, 78, 85, 97, 127, 135, 246, 248–9, 253, 258, 261, 273, 277, 278, 280, 304, 317, 319–22
    Stone Mountain, Georgia 130, 343
    Yosemite, California 1, 20, 33, 163, 165, 168, 170–1, 173–4, 178, 246, 273, 317

unloading 20–1, 127

vegetation, role of 49–50, 152–3, 191–2, 194
Venezuela 93

weathering
    biological 48–50, 227
    chemical 35–40
    frost 44–5, 209
    grades 55–6
    in coastal setting 227
    indices 40–2
    rates of 39
    resistance to 269
    salt 45–6, 143, 227, 229
    thermal 46–7
weathering front 24, 40, 56–7, 74–5, 155, 300
weathering pits 107, 132–9
    geographical distribution 134–7
    rates of growth 139
weathering profile 53–6, 60–1
    evolutionary pathways 80–1
    thickness of 68–70

Zimbabwe 112, 280, 309
    Great Zimbabwe 335
    Matopos 160, 275, 315